Catalytic Materials: Emerging Trends

Catalytic Materials: Emerging Trends

Editor: Steve Garter

MURPHY & MOORE
www.murphy-moorepublishing.com

Published by Murphy & Moore Publishing
1 Rockefeller Plaza,
New York City, NY 10020, USA
www.murphy-moorepublishing.com

Catalytic Materials: Emerging Trends
Edited by Steve Garter

International Standard Book Number: 978-1-63987-095-0 (Hardback)

Cataloging-in-Publication Data

Catalytic materials : emerging trends / edited by Steve Garter.
p. cm.
Includes bibliographical references and index.
ISBN 978-1-63987-095-0
1. Catalysts. 2. Materials science. I. Garter, Steve.
TP159.C3 C38 2022
660.299 5--dc23

Contents

Preface

This book has been an outcome of determined endeavour from a group of educationists in the field. The primary objective was to involve a broad spectrum of professionals from diverse cultural background involved in the field for developing new researches. The book not only targets students but also scholars pursuing higher research for further enhancement of the theoretical and practical applications of the subject.

The solids which permit the chemical reaction to occur efficiently and cost-effectively are known as catalytic materials. Catalysis is defined as the process in which the rate of chemical reaction is increased by adding a substance called a catalyst. There are two kinds of catalytic reactions, namely, heterogeneous catalysis and homogeneous catalysis. Electrocatalyst is a form of heterogeneous catalysis whereas homogeneous catalysis is further divided into organocatalysis, photocatalysts, and enzymes and biocatalysts. Some of the industry applications of catalysis are energy processing, bulk chemicals, fine chemicals and food processing. The objective of this book is to give a general view of the different areas of catalytic materials, and its applications. It is a compilation of chapters that discuss the most vital concepts and emerging trends within this field. This book includes contributions of experts and scientists which will provide innovative insights on this area of study.

It was an honour to edit such a profound book and also a challenging task to compile and examine all the relevant data for accuracy and originality. I wish to acknowledge the efforts of the contributors for submitting such brilliant and diverse chapters in the field and for endlessly working for the completion of the book. Last, but not the least; I thank my family for being a constant source of support in all my research endeavours.

Editor

1

First-Transition Metal Oxocomplex–Surface-Modified Titanium(IV) Oxide for Solar Environmental Purification

Hiroaki Tada and Qiliang Jin

Abstract

The ongoing global energy and environmental issues warrant the development of environmental catalysts for decomposing pollutants in water and air by utilizing solar energy named as "solar environmental catalysts." This chapter describes the recent studies on a novel class of solar environmental catalysts consisting of TiO_2 and molecular-scale first-row transition metal oxide clusters (or metal oxocomplexes) on the surface (MOs/TiO_2). The TiO_2 surface modification with the oxocomplexes by the chemisorption–calcination cycle (CCC) technique presents a novel band engineering for fine-tuning the band energy. The unique physicochemical and electronic properties of MOs/TiO_2 give rise to the outstanding photocatalytic activity for the decomposition of organic pollutants. The combination with the rapidly growing technique for preparation of TiO_2 nanostructures allows us to expect further improvement of the performances and the wide application to the solar chemical transformation for producing useful substances.

Keywords: Visible-light photocatalyst, Titanium(IV) oxide, Surface modification, Metal oxide cluster, Metal oxocomplex, Band energy turning, Solar environmental purification

1. Introduction

Environmental pollution is beyond limits, and the development of environmental catalysts is a critical subject for scientists and engineers all over the world. Here, the catalysts for purifying polluted water and air by utilizing the solar energy are termed as "solar environmental catalyst." The TiO_2 photocatalyst possesses great potential as the "solar environmental catalyst" owing to its strong oxidation ability, high physicochemical stability, abundance in nature, and nontoxicity [1,2]. Scheme 1 shows the fundamental reaction mechanism on the TiO_2-photocatalyzed decomposition of organic pollutants with the characteristic time for each process [3]. UV-light absorption by TiO_2 causes the excitation of electrons in the valence band (VB) to the conduction band (CB) in the order of femtoseconds. Most of the photogenerated charge carriers are lost by the recombination within ~100 ns. The charge carriers surviving the recombination are trapped at the TiO_2 surface to induce the redox reactions. In general, the CB electrons reduce oxygen (O_2), whereas the VB holes oxidize organic pollutants. The VB holes have a highly positive potential (+2.67 V versus standard hydrogen electrode (SHE) at pH 7) to oxidize most organic compounds. Conversely, the driving force for one-electron O_2 reduction (standard potential, $E^0(O_2/O_2^-) = -0.284$ V versus SHE) by the CB electrons (−0.53 V versus SHE at pH 7) is small. Consequently, the O_2 reduction reaction (ORR) is much slower (~ms) than the oxidation process (~100 ns). TiO_2 usually takes crystal forms of rutile and anatase. The flatband potential of anatase is ~0.2 V, which is more negative than that of rutile, and anatase has a higher UV-light activity for the oxidation of organic compounds as compared with rutile [4]. This fact also points to the importance of the ORR in TiO_2 photocatalytic reactions. Also, the coupling of anatase and rutile TiO_2 can further increase the UV-light activity because of the enhancement of charge separation due to the interfacial electron transfer [5].

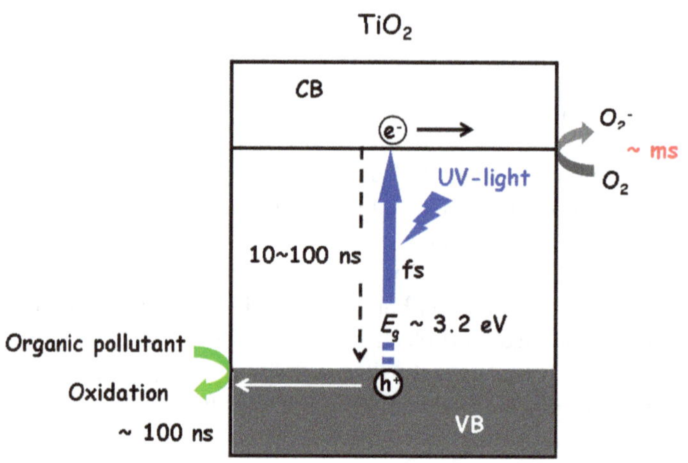

Sheme 1. The basic mechanism on the TiO_2 photocatalytic reaction (the surface trapping processes for the CB electrons and VB holes are abbreviated) with the characteristic time for each step shown.

Recently, the visible-light activation of TiO_2 by its surface modification with metal oxide nanoparticles (NPs) or oxocomplexes has been developed [6,7]. This approach has a major advantage over the conventional doping [8–14], in that visible-light activation can be achieved by a simple procedure without the introduction of the impurity/vacancy levels into the bulk TiO_2. To date, the impregnation method has been mainly used for the surface modification with metal oxide NPs, including chromium oxides [15], iron oxides [16–18], and copper oxides [19]. Unfortunately, the surface modification by the impregnation method is effective for rutile but less effective for anatase.

This chapter deals with our recent studies on the surface modification of anatase TiO_2 with the first (3d) transition metal oxocomplexes (MOCs) by the chemisorption–calcination cycle technique (MOCs/TiO_2) [20] and the characterization and photocatalytic activities for the degradation of organic pollutants. We show that some MOCs/TiO_2 fulfill the requirements for the "solar environmental catalyst."

2. Design for solar environmental catalysts

The requirements for the highly active TiO_2-based "solar environmental catalyst" are described below. As shown in Figure 1, the excitation from the VB electrons to the CB needs UV-light irradiation, occupying only 3% of the incident sunlight. From a viewpoint of the effective use of the sunlight, the response to the visible-light occupying 45% of the solar energy (Figure 1) should be particularly imparted to anatase TiO_2 and anatase–rutile-mixed TiO_2 (Requirement 1). However, even if TiO_2 can be endowed with the visible-light activity, it is usually much smaller than the UV-light activity. Therefore, the inherent excellent UV-light activity of TiO_2 should be compatible with the visible-light activity (Requirement 2). For high levels of visible and UV-light activities to be achieved, the enhancement of the ORR is crucial because it is usually the rate-determining step in the TiO_2-photocatalyzed reactions [21] (Requirement 3).

3. Catalyst preparation

3.1. Chemisorption–calcination cycle technique

The adsorption mechanism of metal acetylacetonates (acac) on TiO_2 depends on the kind of complexes. As an example, Fe oxocomplex formation by the CCC technique is represented in Scheme 2. In the first step, $Fe(acac)_3$ is chemisorbed on the TiO_2 surface via the ligand exchange between the acac ligand and the surface Ti-OH group from the nonaqueous solution (Eq. 1) [22]:

$$Fe\left(acac\right)_3 + l\left(Ti_s\text{-}OH\right) \rightarrow Fe\left(acac\right)_{3-l}\left(O\text{-}Ti_s\right)_l + lAcacH \ \left(l \approx 1\right) \quad (1)$$

where the subscript s denotes the surface atom.

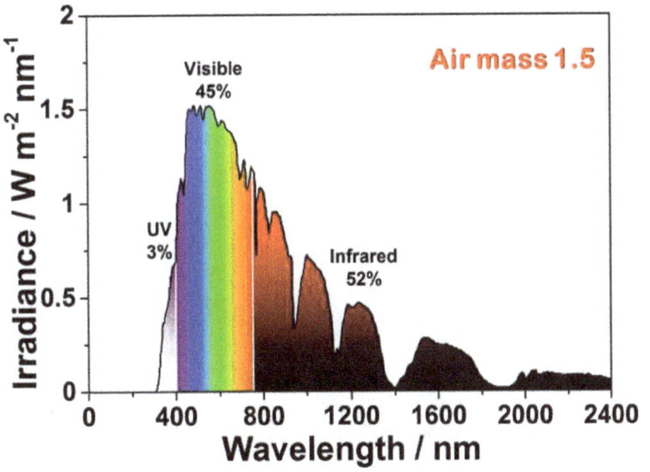

Figure 1. Solar spectrum (AM 1.5).

Conversely, $[Sn(acac)_2]Cl_2$ is adsorbed on TiO_2 via the ion exchange between H^+ and $[Sn(acac)_2]^{2+}$ ion (Eq. 2) [23]. In each case, the adsorption apparently obeys the Langmuir model. The saturated adsorption amount and the adsorption constant for the adsorption of various metal acetylacetonates on TiO_2 at 298 K are summarized in Table 1. The adsorption constants range from 10^2 to 10^4, indicating that they are strongly adsorbed on the TiO_2 surface by chemical bonds. Exceptionally, $Cr(acac)_3$ is not adsorbed because of its large ligand-field stabilization energy ($1.2\Delta_0$).

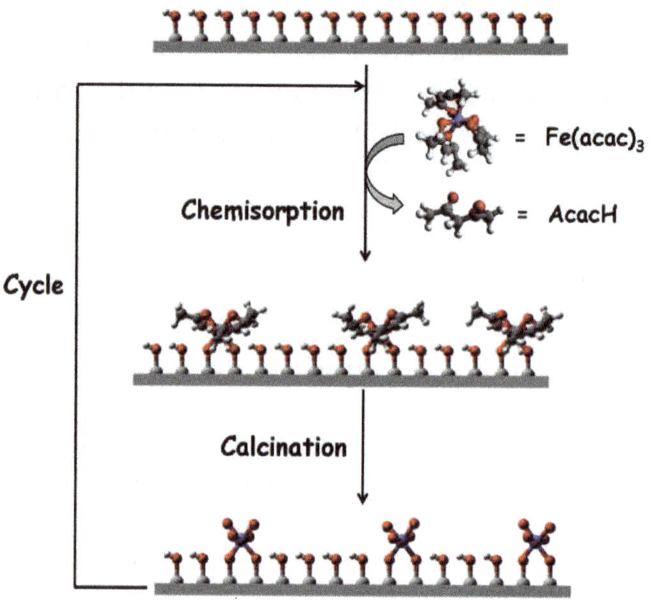

Sheme 2. Fe oxocomplex (Fe_2O_3) formation on the TiO_2 surface by the CCC technique.

TiO_2 (crystal form, specific surface area)	Precursor	Solvent	Saturated adsorption amount / ions nm^{-2}	Adsorption equilibrium constant / 10^3 mol^{-1} dm^3	Mechanism	Oxidation number of metal ion in oxide cluster [a]
ST-01 (A, 308 m^2 g^{-1})	$VO(acac)_2$	ethanol	0.18	1.67	ligand exchange	V^{5+} (V_2O_5)
P-25 (A-R, 50 m^2 g^{-1})	$Cr(acac)_3$	ethanol	0	0	–	–
P-25 (A-R, 50 m^2 g^{-1})	$Mn(acac)_3$	ethanol	1.65	0.17	ligand exchange	Mn^{3+} (Mn_2O_3)
P-25 (A-R, 50 m^2 g^{-1})	$Fe(acac)_3$	ethanol : n-hexane 3: 17 v/v	0.46	2.06	ligand exchange	Fe^{3+} (Fe_2O_3)
ST-01 (A, 308 m^2 g^{-1})			0.35	1.52		Fe^{3+} (Fe_2O_3)
ST-01 (A, 308 m^2 g^{-1})	$Co(acac)_2(H_2O)_2$	methanol	0.18	1.23	surface Ti-OH coordination	Co^{3+} (Co_2O_3)
P-25 (A-R, 50 m^2 g^{-1})	$Ni(acac)_2(H_2O)_2$	ethanol : n-hexane 3: 17 v/v	0.56	0.74	ligand exchange	Ni^{2+} (NiO)
P-25 (A-R, 50 m^2 g^{-1})	$Cu(acac)_2$	acetonitrile	0.54	1.34	surface Ti-OH coordination	Cu^{2+} (CuO)
A-100 (A, 8.1 m^2 g^{-1})	$[Sn(acac)_2]Cl_2$	ethanol	0.29	17.0	ion-exchange	Sn^{4+} (SnO_2)
MT-700B (R, 12.5 m^2 g^{-1})			0.10	46.0		Sn^{4+} (SnO_2)

Table 1. Adsorption properties of 3d metal acetylacetonates on TiO_2 at 298 K.

$$\left[Sn(acac)_2\right]Cl_2 + 2(Ti_s\text{-}OH) \rightarrow Sn(acac)_2(O\text{-}Ti_s)_2 + 2HCl \quad (2)$$

In the second step, the oxidation of the acac ligands by heating the samples in air at 773 K yields iron oxides on the TiO_2 surface. Further, these procedures are repeated to control the Fe-loading amount. Chemical analysis confirmed that all the Fe was confirmed to be present only on the TiO_2 surface. The Fe-loading amount is expressed by the number of Fe ions per unit TiO_2 surface area (Γ/ions nm^{-2}).

3.2. Control of loading amount

A feature of the CCC technique is precise control of the loading amount of metal oxides. As an example, the manner in which the Fe-loading amount is controlled in the iron oxide/TiO_2 system is described. Figure 2A shows the relation between Γ and initial complex concentration: black, $Fe(acac)_3$ and blue, $Mn(acac)_3$. In each case, the Γ gradually increases with an increase in the initial concentration. Figure 2B shows plots of Γ versus CCC cycle number (N): black, $Fe(acac)_3]_0$ = 0.65 mmol dm^{-3} (black) and blue, $Mn(acac)_3]_0$ = 8 mmol dm^{-3}. The Γ further increases in proportional to N in both the systems. The linear Γ–N relation is also observed in the other metal oxide/TiO_2 systems. In this manner, the loading amount of metal oxides can be controlled in a wide range using the precursor complex concentration and the cycle number.

4. Structure of surface metal oxocomplexes

Another feature of the CCC technique is the formation of molecular-scale metal oxide species on TiO_2. Figure 3 shows transmission electron micrographs (TEMs) of iron oxide/TiO_2 with varying Γ. No particulate deposits are observed on the TiO_2 surface at $\Gamma < 1$ ions nm^{-2}. This fact suggests that iron oxides exist as molecular-scale iron oxide species on the TiO_2 surface.

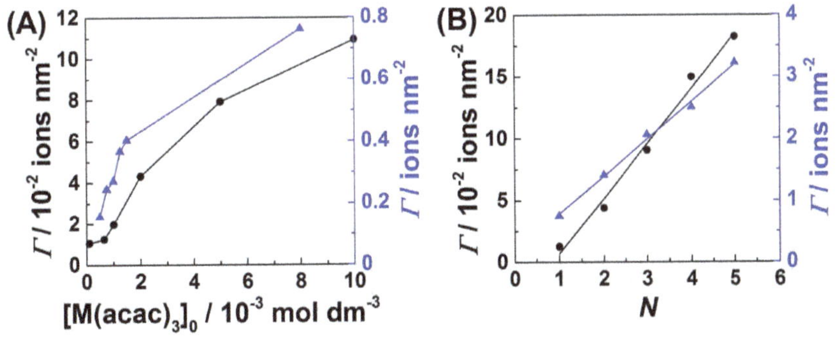

Figure 2. (A) Plots of Fe-loading amount (Γ/ions nm^{-2}) versus initial concentration of the complex ($[M(acac)_3]_0$): M = Fe (black) and M = Mn (blue). (B) Plots of Γ versus cycle number (N) at $Fe(acac)_3]_0$ = 0.65 mmol dm^{-3} (black) and $Mn(acac)_3]_0$ = 8 mmol dm^{-3} (blue).

To obtain the detailed structural information, Fe K-edge X-ray absorption fine-structure spectra were measured for the iron oxide/TiO_2 samples with varying Γ [24]. Figure 4A shows

X-ray absorption near-edge structure (XANES) spectra for the iron oxide/TiO_2 samples, and authentic Fe, Fe_3O_4, and α-Fe_2O_3 for comparison. The absorption edge of the iron oxide/TiO_2 sample is in agreement with that of α-Fe_2O_3, indicating the oxidation number of the iron to be +3 (Table 1). Figure 4B shows the Fourier transforms of the k^3-weighted X-ray absorption fine structure (EXAFS) for the iron oxide/TiO_2 samples. The peaks around 1.6 Å and 2.8 Å arise from the scattering from the nearest neighbor O and Fe, respectively. It is worth noting that the latter peak becomes very weak at $\Gamma \leq 0.5$. Evidently, the iron oxides exist as a mononuclear Fe oxocomplex on TiO_2 designated as Fe_2O_3/TiO_2 below.

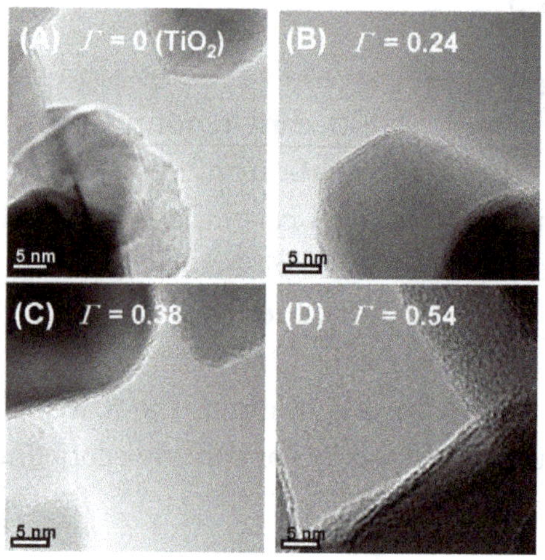

Figure 3. TEM images of Fe_2O_3/TiO_2 with varying Γ: (A) $\Gamma = 0$ (P-25); (B) $\Gamma = 0.23$, (C) $\Gamma = 0.38$, (D) $\Gamma = 0.54$.

In this manner, MOCs are formed on the TiO_2 surface in a highly dispersed state by the CCC technique, whereas the conventional impregnation method usually yields metal oxide NPs. As illustrated in Scheme 2, the molecular size of $Fe(acac)_3$ (~0.5 nm^2 $complex^{-1}$) is much larger than the reciprocal number density of the Ti-OH groups on the TiO_2 surface (~0.1 nm^2 group $^{-1}$) [25]. In the first step of the CCC process, $Fe(acac)_3$ complexes are chemisorbed isolatedly each other owing to the bulky acac-ligands. Also, the strong bond between the complexes and the TiO_2 surface suppresses the aggregation of the oxocomplexes during the second step. Consequently, mononuclear MOCs can be formed on the TiO_2 surface.

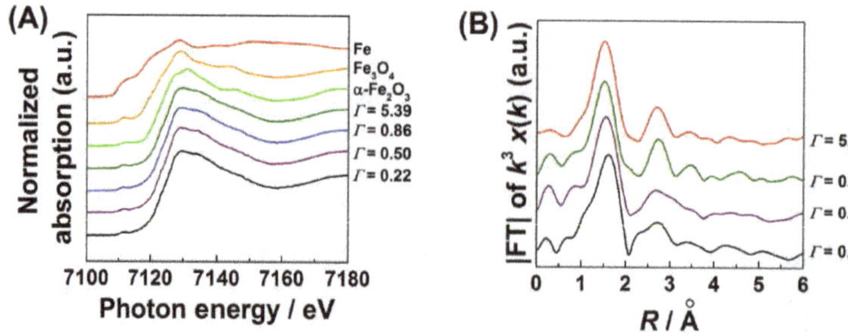

Figure 4. XANES and EXAFS spectra. (A) XANES spectra for Fe, Fe_3O_4, α-Fe_2O_3, and Fe_2O_3/TiO_2 with varying Γ. (B) Fourier transforms of the k^3-weighted EXAFS spectra for Fe_2O_3/TiO_2. The figures were taken from Ref. 24.

5. Characteristics of metal oxocomplex–surface-modified TiO_2

5.1. Optical property

The optical property is a fundamental factor affecting the photocatalytic activity. Figure 5 shows UV–visible absorption spectra for Fe oxocomplex–surface-modified mesoporous TiO_2 nanocrystalline films (Fe_2O_3/mp-TiO_2) with varying Γ. Impregnation samples usually have absorption approximately 470 nm in addition to absorption at 410 nm [16,17,26,27]. The former and latter absorptions can be attributed to the d–d transition and electronic transition from Fe^{3+} levels to the CB of TiO_2, respectively [28]. A strong d–d absorption is also observed for a heavy-loading CCC sample (Fe_2O_3(Γ = 5.5)/mp-TiO_2). In contrast, the absorption spectra of Fe_2O_3($\Gamma \leq 2.1$)/mp-TiO_2 apparently show a marked bandgap narrowing from 3.3 to 2.85 eV with an increase in Γ, whereas the d–d transition absorption is very weak [24]. Similar spectra were previously observed for TiO_2 doped with Cr [12] and N [13] prepared by ion implantation and magnetron sputtering. The weak d–d transition absorption is a common feature for the CCC samples including Co_2O_3/TiO_2 [29], NiO/TiO_2 [30,31], and CuO/TiO_2 [32]. Clearly, the unique optical properties of the CCC samples originate from the highly dispersed MOCs on the TiO_2 surface.

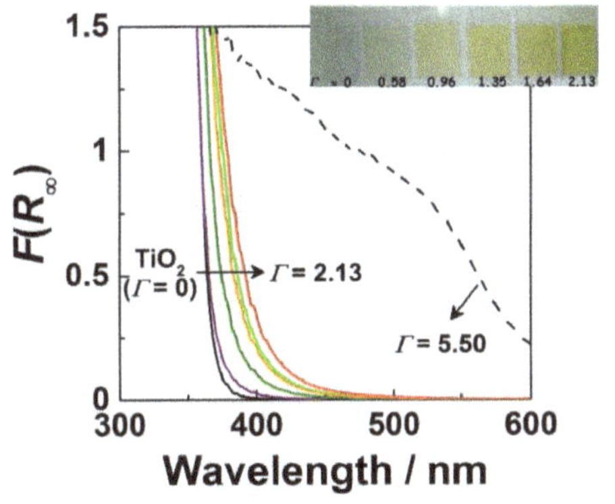

Figure 5. UV–Vis absorption spectra of Fe_2O_3/mp-TiO_2 prepared by the CCC technique.

5.2. Fine-tuning of band energy

The VB maximum determines the oxidation ability of the holes, and thus is a key factor for the decomposition of organic pollutants by semiconductor photocatalysts. The VB maximum level can be estimated from the VB-X-ray photoelectron spectroscopy (XPS) [24]. Since the VB maximum of TiO_2 is almost independent of its crystal form and size, the VB maximum of MOC-modified TiO_2 can be compared with respect to that of unmodified TiO_2. Figure 6 shows the VB-XPS spectra for Fe_2O_3/TiO_2 with varying Γ. The emission from the O 2*p*–VB extends from 2 to 9 eV. As a result of the surface modification, the top of VB rises, whereas the bottom remains invariant. Figure 6B compares the energy shift in the VB maximum level with respect to that of unmodified TiO_2 (ΔE_{VBM}) as a function of Γ for the Fe_2O_3/TiO_2, Co_2O_3/TiO_2, NiO/TiO_2, and SnO_2/TiO_2 systems. Interestingly, the ΔE_{VBM} for the MOC-modified TiO_2, except SnO_2/TiO_2, goes up with an increase in Γ, which means that the oxidation ability of the VB holes can be tuned by Γ. This is the most unique and important feature of the TiO_2 modification with MOCs using the CCC technique.

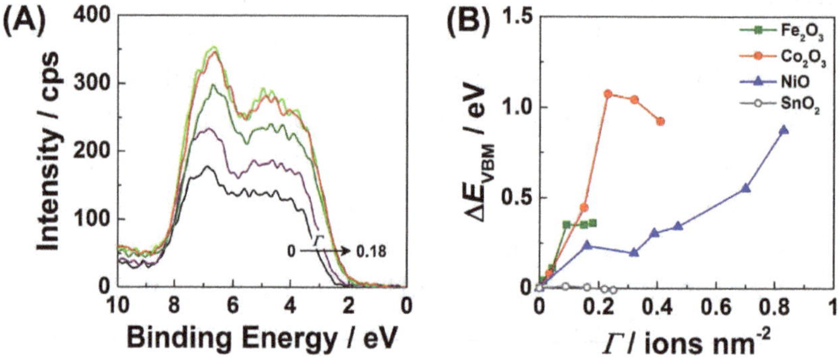

Figure 6. (A) VB-XPS spectra for Fe_2O_3/TiO_2 with varying Γ. (B) Energy shift in the VB maximum level (ΔE_{VBM}) as a function Γ for various MOCs/TiO_2. The figure (A) was taken from ref. 35.

These results were further simulated by the density functional theory (DFT) calculations [7]. In the DFT-optimized structure for a model Fe_2O_3 cluster-adsorbed TiO_2 system, plural Fe–O–Ti interfacial bonds were observed. The PEDOS (projected electronic density of states) plots showed that states from the adsorbed Fe_2O_3 clusters lie above the VB of TiO_2, that is, the iron oxide-derived states make it to the top of the VB. This changes the nature of the VB edge that moves the top of the VB to higher energy. The offsets between the TiO_2 VB edge and the iron oxide states around the VB are ~0.3 eV for Fe_2O_3-modified TiO_2, which is comparable with the experimental value. The effective mixing between the surface Fe_2O_3 levels and O 2*p* through the Ti_s–O–Fe interfacial bond is considered as yielding a d-band overlapping the VB of TiO_2. Thus, the excitation of Fe_2O_3/TiO_2 by the visible light with wavelength below 500 nm can induce the interfacial electron transfer from the surface d-band to the CB of TiO_2.

Conversely, the information about empty levels can be obtained by photoluminescence spectroscopy. TiO_2(ST-01) has a broad emission band centered at 538 nm (E_1) [22]. The E_1 signal intensity remarkably weakens with heating ST-01 at 773 K for 1 h in air. This PL band is assignable to the emission from the surface oxygen vacancy levels of anatase TiO_2 [33]. On modifying ST-01 with the Fe oxocomplexes, the intensity further decreases to disappear at $\Gamma > 0.044$ ions nm^{-2}, while a new emission appears at 423 nm (E_2). The E_2 signal can be attributed to the emissions from extrinsic levels. These findings strongly suggest that the excited electrons in the CB of TiO_2 are transferred to the empty surface Fe oxocomplex levels with the energy distributed around 0.27 V below the CB of TiO_2.

5.3. Electrocatalytic activity for oxygen reduction reaction

As stated above, the ORR is frequently the key process in the TiO_2-photocatalyzed reactions as well as low-temperature polymer electrolyte membrane fuel cells (PEMFCs) [34,35]. Figure 7 shows dark current (I)–potential (E) curves of the mp-TiO_2 film-coated F-doped tin oxide (FTO) electrodes (mp-TiO_2/FTO) with and without the surface modification by Fe oxocomplexes. In the absence of O_2, a small current due to water reduction is observed regardless of the surface modification. In the presence of O_2, the current for ORR markedly increases with the surface modification (Fe_2O_3/mp-TiO_2/FTO), whereas it remains weak without O_2. In this manner, the surface Fe oxocomplex has an electrocatalytic activity for the ORR, and a similar ORR-promoting effect is also observed for the NiO/TiO_2 [30] and Co_2O_3/TiO_2 [29] systems. This is also the unique feature of the MOC/TiO_2 systems.

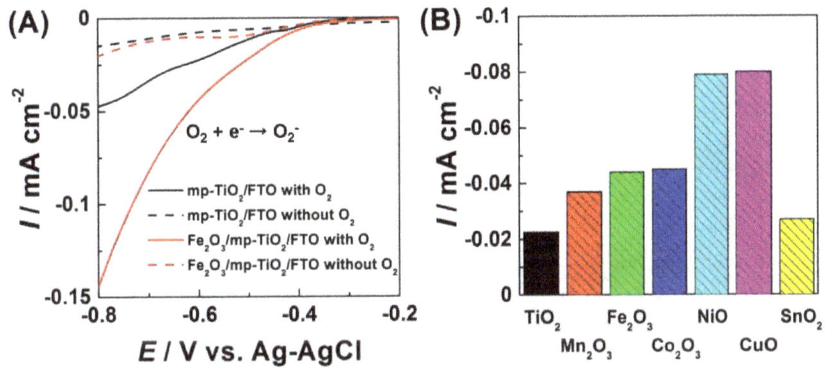

Figure 7. (A) Dark current (I)–potential (E) curves for the Fe_2O_3/mp-TiO_2/FTO electrodes. (B) Comparison of the electrocatalytic activity of the MOC/mp-TiO_2/FTO electrodes for the ORR.

6. Photocatalytic activity

Acetaldehyde is a toxic volatile organic compound (VOC), while 2-naphthol is widely used as the starting material of azo dyes. Both of them are optically transparent in the visible region, and then, acetaldehyde and 2-naphthol were used as model air and water pollutants, respectively. The relative photocatalytic activities of various MOCs/TiO_2 were evaluated with respect to that of highly active commercial TiO_2 particles with a crystal form of anatase (ST-01, Ishihara Sangyo Co.). The photocatalytic degradation of 2-naphthol and acetaldehyde apparently follows the first-order rate law, and the rate constants for irradiation of UV light ($330 < \lambda < 400$ nm, k_{UV}) and visible light ($\lambda > 400$ nm, k_{vis}) were used as the indicators for the photocatalytic activities. Figure 8 shows the k_{vis} and k_{UV} of Fe_2O_3/TiO_2 for the degradations of 2-naphthol (A) and acetaldehyde (B) as a function of Γ. Surprisingly, the surface modification of TiO_2 by the Fe-oxocomplex gives rise to a high level of visible-light activity and a concomitant increase in the UV-light activity of anatase TiO_2 [24]. Each plot exhibits a volcano-shaped curve, which is a general feature in the activity-Γ plots for the MOC/TiO_2 systems. By using an atomic layer deposition technique, Libera et al. have recently prepared Fe(III) oxospecies–surface-modified TiO_2 showing a reactivity for the decoloration of methylene blue under visible-light irradiation [18].

Particulate systems have high photocatalytic activity due to the large surface area, but needs the troublesome separation of the particles from purified water. Oppositely, in supported films, the photocatalytic activity is generally much lower due to the smaller surface area,

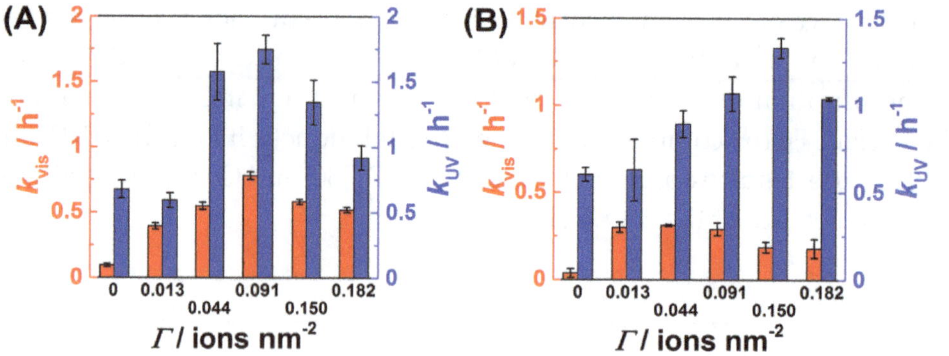

Figure 8. (A) UV-light activity (k_{UV}, blue) and visible-light activity (k_{vis}, red) of Fe_2O_3/TiO_2(ST-01) for the liquid-phase decomposition of 2-NAP as a function of Γ. (B) UV-light activity (k_{UV}, blue) and visible-light activity (k_{vis}, red) of Fe_2O_3/TiO_2(ST-01) for the gas-phase decomposition of CH_3CHO as a function of Γ.

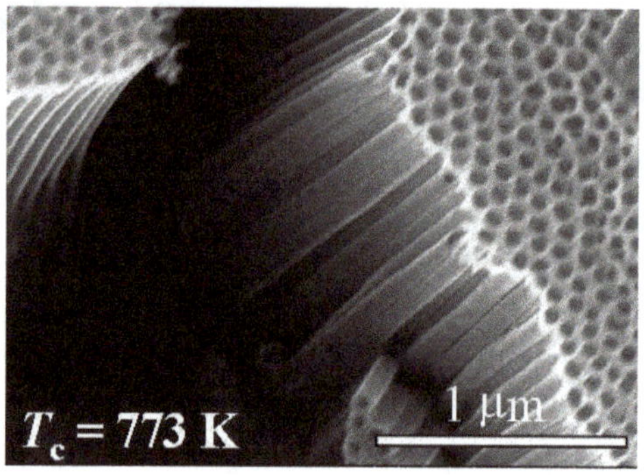

Figure 9. TiO_2 NTA parepared by a two-step anodization of Ti plate (first anodization 40 V, 0.5 h/second anodization 40 V-1 h/heating temperature, T_c = 773 K). The figure was taken from Ref. 35.

while the separation process is unnecessary. TiO_2 nanotube array (NTA) has the advantages of the particulate and film systems is promising. Figure 9 shows TEM image for TiO_2 NTA prepared by two-step anodization. The application of the CCC technique to the TiO_2 NTA led to a high visible-light activity for 2-naphthol degradation comparable with that of the particulate system [36].

Figure 10 compares the relative visible-light activity (k_{vis}) and UV-light activity (k_{UV}) of 3d MOCs/TiO_2(ST-01) with respect to the activities of unmodified TiO_2(ST-01) ($k_{vis}{}^0$ and $k_{UV}{}^0$) for the 2-naphthol degradation under the same conditions. Each Γ shows the optimum value for visible-light activity in each MOC/TiO_2 system. Among MOCs, the surface modification by Fe_2O_3 [24], Co_2O_3 [25], and NiO [30] is effective in the visible-light activation. Particularly, the Co_2O_3/TiO_2 system exhibits a very high level of visible-light activity [29]. The activity is on the order of $Co_2O_3 > Fe_2O_3 > NiO > CuO > V_2O_5 \approx Mn_2O_3 > SnO_2 \approx$ unmodified TiO_2. However, the surface modification with Fe_2O_3, NiO, and Co_2O_3 by the CCC technique can endow anatase TiO_2 with high levels of visible-light activity, with the high UV-light activity further increased (Fe_2O_3) or maintained (Co_2O_3, NiO). Although the effect of the surface modification by SnO_2 was small for anatase, a significant increase in the UV-light activity was induced for rutile [37]. Interestingly, Boppana and Lobo have recently reported that loading of SnO_x clusters on $ZnGa_2O_4$ by the impregnation method causes visible-light activity for the decomposition of p-cresol [38]. Besides metal oxides, the surface modification of TiO_2 with halogeno complexes of rhodium(III) and platinum(IV) on the TiO_2 surface is known to induce visible-light activity [39,40].

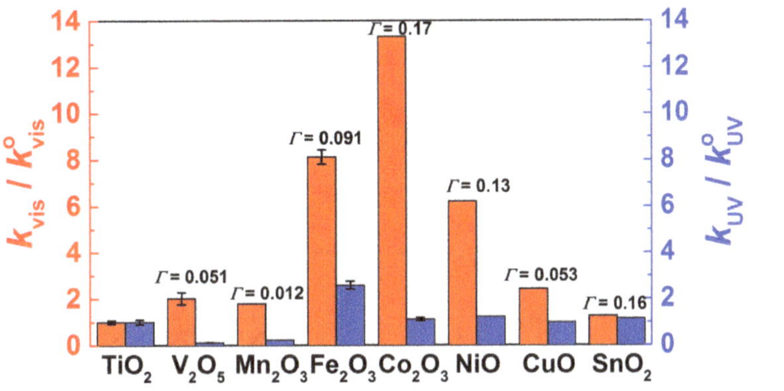

Figure 10. Comparison of the visible-light activities (k_{vis}) and UV-light activities (k_{UV}) of MOCs/TiO_2(ST-01) with respect to those of unmodified TiO_2 for the 2-naphthol degradation under the same conditions.

The degradation of formic acid was further carried out in the aqueous phase with Co_2O_3/TiO_2 at 298 K under visible-light irradiation. The Co_2O_3 surface modification greatly enhanced the decomposition of formic acid to CO_2. The visible-light activity reached a maximum at $\Gamma = 0.17$ with the conversion to CO_2 reaching ~100% within 5 h [29] (Eq. 3).

$$HCOOH+1/2O_2 \xrightarrow[h\nu(\lambda>400nm)]{CO_2O_3/TiO_2} CO_2+H_2O \quad (3)$$

Also, prolonging irradiation decomposed 2-naphthol to CO_2, but the conversion was only ~6% at 96 h. The decomposition of 2-naphthol to CO_2 would proceed stepwise via oxidative cleavage of the naphthalene ring.

On the basis of the energy band diagram, the action mechanism of MOCs in the TiO_2 photocatalysis can be explained. In the nanoscale Fe_2O_3–TiO_2 coupling system, Fe_2O_3 NP with a band gap of 2.2 eV is excited by the visible-light irradiation. However, the potential of the CB electrons is more positive than the TiO_2 CB minimum of TiO_2 (−0.48 V) and the standard redox potential of O_2 ($E^0(O_2/O_2^-) = -0.284$ V). Thus, the electron transfer from the CB electrons of Fe_2O_3 to neither TiO_2 nor O_2 can occur. Consequently, nano-coupling does not show visible-light activity [41].

Scheme 3 illustrates the surface modification effects of the Fe oxocomplex on the TiO_2 photocatalytic decomposition of organic pollutants. In this case, the surface modification raises the VB maximum with the CB minimum unchanged, due to the effective electronic coupling through the Fe–O–Ti interfacial bonds (Effect 1). The resulting decrease in the band gap shifts the light absorption to the visible region (Effect 2). The visible-light absorption triggers electronic excitation from the highest-energy oxocomplex-derived VB states to the empty CB

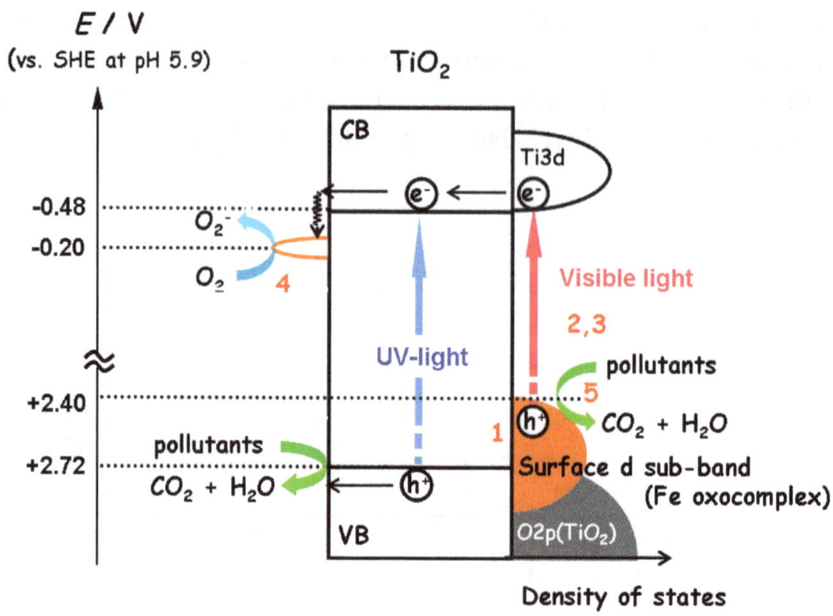

Sheme 3. Surface modification effects of the Fe oxocomplex on the TiO_2-photocatalyzed decomposition of organic pollutants. The levels around −0.2 V show the vacant Fe oxocomplex.

of TiO_2 in order to generate charge carriers. This surface-to-bulk interfacial electron transfer enhances charge separation (Effect 3). The surface modification permits the electron transfer from the CB of TiO_2 to shallow vacant surface oxocomplex levels, which distribute around ca. −0.2 V [22]. The formation of O_2^- radicals was confirmed by chemiluminescence photometry in the Cu^{2+}-grafted TiO_2 system under visible-light irradiation [42]. In this cathodic process, the electrons efficiently reduce adsorbed O_2 with the aid of the electrocatalytic activity of the surface-adsorbed oxocomplex (Effect 4). This effect should also contribute to the increase in the UV-light activity. In the anodic process, the holes generated in the VB could take part in the oxidation process without diffusion (Effect 5) [15]. Consequently, Fe_2O_3/TiO_2 as well as NiO/TiO_2 and Co_2O_3/TiO_2 satisfy the three requirements of the "solar environmental catalyst."

7. Conclusions and future prospect

The surface of TiO_2 can be modified by oxocomplexes of the first transition metals (MOCs/TiO_2) with the loading amount precisely controlled by using the CCC technique. Among the MOCs/TiO_2, Fe_2O_3-, Co_2O_3- and NiO-surface-modified TiO_2 possess unique physicochemical properties such as strong visible-light absorption and the excellent reduction ability of O_2. Spectroscopic experiments and first-principles DFT simulation have revealed that the surface modification with the MOCs raises the VB maximum of TiO_2 due to the formation of plural metal–O–Ti interfacial bonds. Surface-to-bulk and/or bulk-to-surface interfacial electron transfer induced by visible-light absorption enhances charge separation. This novel coupling system consisting of MOCs and TiO_2 would be promising as the "solar environmental catalyst."

The standard potentials of multiple-electron ORRs ($E^0(O_2/H_2O_2)$ = +0.695 V and $E^0(O_2/H_2O)$ = +1.229 V versus SHE) are much more positive than that of one-electron ORR. Therefore, the hybridization of appropriate electrocatalysts for the multiple-electron ORR can impart visible-light activity to many metal oxide semiconductors with E_g < 3 eV. The effectiveness of this approach has recently been verified in the Pt NP-WO_3 (E_g = 2.7 eV) [43] and Cu(acac)$_2$-$BiVO_4$ (E_g = 2.4 eV) hybrid systems [44], where Pt NP and O_2-bridged Cu complex work as excellent electrocatalysts for multiple ORRs, respectively.

As a future subject, we further suggest the importance of the effective use of the infrared ray occupying 52% of the solar energy for the catalytic reactions (Figure 1). For example, Co_2O_3/TiO_2 exhibits high levels of photocatalytic and thermocatalytic activities [29], whereas Mn_2O_3/TiO_2 exhibits a high thermocatalytic activity for the oxidation of organic compounds [45]. Further, MOCs/TiO_2 with the VB maximum level (or the oxidizing ability of the VB holes) fine-tuned by the loading amount may open up the application of MOCs/TiO_2 to "green" and selective chemical transformations [46–48].

Acknowledgements

A series of studies on metal oxocomplex–surface-modified TiO_2 have been performed in collaboration with Dr. Michael Nolan and Dr. Anna Iwaszuk (Tyndall National Institute, University College Cork). The authors are sincerely grateful for their very useful DFT simulations. Also, the authors acknowledge Dr. M. Fujishima (Kinki University) for a helpful discussion, and T. Hattori, S. Okuoka, and Y. Sumida (Nippon Shokubai Co.) for EXAFS measurements and a valuable discussion. Ishihara Sangyo Co. gifted us with ST-01, and K. Fujiwara aided us in the collection and arrangement of the materials. H.T. acknowledges the support from the Ministry of Education, Science, Sport, and Culture, Japan, through a Grant-in-Aid for Scientific Research (C) No. 24550239, No. 15K05654, MEXT-Supported Program for the Strategic Research Foundation at Private Universities, and Nippon Sheet Glass Foundation for Materials Science and Engineering, and by Sumitomo Foundation.

Author details

Hiroaki Tada[1*] and Qiliang Jin[2]

*Address all correspondence to: h-tada@apch.kindai.ac.jp

1 Department of Applied Chemistry, School of Science and Engineering, Kinki University, Kowakae, Higashi-Osaka, Osaka, Japan

2 R&D Placement Business Division, WORLD INTEC CO., Sakae, Naka-ku, Nagoya, Aichi, Japan

References

[1] Fujishima A, Zhang X, Tryk D A: TiO_2 photocatalysis and related surface phenomena. Surf. Sci. Rep. 2008; 63: 515–582. DOI: j.surfrep.2008.10.001

[2] Hashimoto K, Irie H, Fujishima A: TiO_2 photocatalysis: A historical overview and future prospects. Jpn. J. Appl. Phys. 2005; 44: 8269–8285.

[3] Hoffmann M R, Martin S T, Choi W, Bahnemann D W: Environmental applications of semiconductor photocatalysis. Chem. Rev. 1995; 95: 69–96. DOI: 10.1021/cr00033a004

[4] A. Y. Ahmed, T. A. Kandiel, T. Oekermann, D. Bahnemann: Photocatalytic activities of different well-defined single crystal TiO_2 surfaces: Anatase versus rutile. J. Phys. Chem. Lett. 2011; 2: 2461–2465. DOI: 10.1021/jz201156b

[5] Kawahara T, Konishi Y, Tada H, Tohge N, Nishii J, Ito S: A patterned TiO_2(anatase)/TiO_2(rutile) bilayer-type photocatalyst: Effect of the anatase/rutile junction on the photocatalytic activity. Angew. Chem. Int. Ed. 2002; 41: 2811–2813. DOI: 10.1002/1521-3757(20020802)114:15<2935:: AID-ANGE2935>3.0.CO;2-6

[6] Kisch H: Semiconductor photocatalysis–mechanistic and synthetic aspects. Angew. Chem. Int. Ed. 2013; 52: 812–847. DOI: 10.1002/anie.201201200

[7] Nolan M, Iwaszuk A, Tada H: Molecular metal oxide cluster-surface modified titanium(IV) dioxide photocatalysts. Aust. J. Chem. 2012; 65: 624–632.

[8] Asahi R, Morikawa T, Ohwaki T, Aoki K, Taga Y: Visible-light photocatalysis in nitrogen-doped titanium oxides. Science 2001; 293: 269–271. DOI: 10.1126/science.1061051

[9] Khan S U M, Al-Shahry M, Ingler Jr. W B: Efficient photochemical water splitting by a chemically modified n-TiO_2. Science 2002; 297: 2243–2245. DOI: 10.1126/science.1075035

[10] Serpone N: Is the band gap of pristine TiO_2 narrowed by anion- and cation-doping of titanium dioxide in second-generation photocatalysts? J. Phys. Chem. B 2006; 110: 24287–24293. DOI: 10.1021/jp065659r

[11] Zhang H, Chen G, Bahnemann D W: Photoelectrocatalytic materials for environmental applications. J. Mater. Chem. 2009; 19: 5089–5121. DOI: 10.1039/B821991E

[12] Anpo M, Takeuchi M: The design and development of highly reactive titanium oxide photocatalysts operating under visible light irradiation. J. Catal. 2003; 216: 505–516. DOI: 10.1016/S0021-9517(02)00104-5

[13] Kitano M, Funatsu K, Matsuoka M, Ueshima M, Anpo M: Preparation of nitrogen-substituted TiO_2 thin film photocatalysts by the radio frequency magnetron sputtering deposition method and their photocatalytic reactivity under visible light irradiation. J. Phys. Chem. B 2006; 110: 25266–25272. DOI: 10.1021/jp064893e

[14] Liu G, Wang L, Yang H G, Cheng H-M, Lu G Q: Titania-based photocatalysts-crystal growth, doping and heterostructuring. J. Mater. Chem. 2010; 20: 831–843. DOI: 10.1039/B909930A

[15] Irie H, Shibanuma T, Kamiya K, Miura S, Yokoyama T, Hashimoto K: Characterization of Cr(III)-grafted TiO_2 for photocatalytic reaction under visible light. Appl. Catal. B. 2010; 96: 142–148. DOI: 10.1016/j.apcatb.2010.02.011

[16] Murakami N, Chiyoya T, Tsubota T, Ohno T: Switching redox site of photocatalytic reaction on titanium(IV) oxide particles modified with transition-metal ion controlled by irradiation wavelength. Appl. Catal. A. 2008; 348: 148–152. DOI: 10.1016/j.apcata.2008.06.040

[17] Yu H, Irie H, Shimodaira Y, Hosogi Y, Kuroda Y, Miyauchi M, Hashimoto K: J. Phys. Chem. C 2010; 114: 16481–16487. DOI: 10.1021/jp1071956

[18] Libera J A, Elam J W, Sather N F, Rajh T M, Dimitrijevic N M: Iron(III)-oxo centers on TiO_2 for visible-light photocatalysis. Chem. Mater. 2010; 22: 409–413. DOI: 10.1021/cm902825c

[19] Irie H, Miura S, Kamiya K, Hashimoto K: Efficient visible light-sensitive photocatalysts: Grafting Cu(II) ions onto TiO_2 and WO_3 photocatalysts. Chem. Phys. Lett. 2008; 457: 202–205. DOI: 10.1016/j.cplett.2008.04.006

[20] Tada H: Encyclopedia of Surface and Colloid Science. Ed. Hubbard A T, New York: Marcel Dekker, 2002. ISBN: 0-8247-0633-1

[21] Wang C-M, Heller A, Gerischer H: Palladium catalysis of O_2 reduction by electrons accumulated on TiO_2 particles during photoassisted oxidation of organic compounds. J. Am. Chem. Soc. 1992; 114: 5230–5234. DOI: 10.1021/ja00039a039

[22] Jin Q, Fujishima M, Tada H: Visible-light-active iron oxide-modified anatase titanium(IV) dioxide. J. Phys. Chem C 2011; 115: 6478–6483. DOI: 10.1021/jp201131t

[23] Fujishima M, Jin Q, Yamamoto H, Tada H, Nolan M: Tin oxide-surface modified anatase titanium(IV) dioxide with enhanced UV-light photocatalytic activity. Phys. Chem. Chem. Phys. 2012; 14: 705–711. DOI: 10.1039/C1CP22708D

[24] Tada H, Jin Q, Nishijima H, Yamamoto H, Fujishima M, Okuoka S-i, Hattori T, Sumida Y, Kobayashi H: Titanium(IV) dioxide surface-modified with iron oxide as a visible light photocatalyst. Angew. Chem. Int. Ed. 2011; 50: 3501–3505. DOI: 10.1002/anie.201007869

[25] Nosaka Y, Nosaka A. Nyumon: Hikarisyokubai (Introduction to Photocatalyst). Tokyo: Tokyo-Tosho; 2004, 68 p. ISBN: 4-489-00684-5 C3043

[26] Navio J A, Colon G, Litter M I, Bianco G N: Synthesis, characterization and photocatalytic properties of iron-doped titania semiconductors prepared from TiO_2 and

iron(III) acetylacetonate. J. Mol. Catal. A: Chem. 1996; 106: 267–276. DOI: 10.1016/1381-1169(95)00264-2

[27] Wang C-Y, Bahnemann D W, Dohrmann J K: A novel preparation of iron-doped TiO_2 nanoparticles with enhanced photocatalytic activity. Chem. Commun. 2000; 1539–1540. DOI: 10.1039/B002988M

[28] Serpone N, Lawless D, Disdier J, Hermann J-M: Spectroscopic, photoconductivity, and photocatalytic studies of TiO_2 colloids: Naked and with the lattice doped with Cr^{3+}, Fe^{3+}, and V^{5+} cations. Langmuir 1994; 10: 643–652. DOI: 10.1021/la00015a010

[29] Jin Q, Yamamoto H, Yamamoto K, Fujishima M, Tada H: Simultaneous induction of high level thermal and visible-light catalytic activities to titanium(IV) oxide by surface modification with cobalt(III) oxide clusters. Phys. Chem. Chem. Phys. 2013; 15: 20313–20319. DOI: 10.1039/C3CP54304H

[30] Jin Q, Ikeda T, Fujishima M, Tada H: Nickel(II) oxide surface-modified titanium(IV) dioxide as a visible-light-active photocatalyst. Chem. Commun. 2011; 47: 8814–8816. DOI: 10.1039/C1CC13096J

[31] Iwaszuk A, Nolan M, Jin Q, Fujishima M, Tada H: Origin of the visible-light response of nickel(II) oxide cluster surface modified titanium(IV) dioxide. J. Phys. Chem. C 2013; 117: 2709–2718. DOI: 10.1021/jp306793r

[32] Jin Q, Fujishima M, Iwaszuk A, Nolan M, Tada H: Loading effect in copper(II) oxide cluster-surface-modified titanium(IV) oxide on visible- and UV-light activities. J. Phys. Chem. C 2013; 117: 23848–23857. DOI: 10.1021/jp4085525

[33] Shi J, Chen J, Feng Z, Chen T, Lian Y, Wang X, Li C: Photoluminescence characteristics of TiO_2 and their relationship to the photoassisted reaction of water/methanol mixture. J. Phys. Chem. 2007; 111: 693–699. DOI: 10.1021/jp065744z

[34] Gewirth A A, Thorum M S: Electroreduction of dioxygen for fuel-cell applications: Materials and challenges. Inorg. Chem. 2010; 49: 3557–3566. DOI: 10.1021/ic9022486

[35] Fermández J L, Walsh D A, Bard A J: Thermodynamic guidelines for the design of bimetallic catalysts for oxygen electroreduction and rapid screening by scanning electrochemical microscopy. M-co (M: Pd, Ag, Au). J. Am. Chem. Soc. 2005; 127: 357–365. DOI: 10.1021/ja0449729

[36] Muramatsu Y, Jin Q, Fujishima M, Tada H: Visible-light-activation of TiO_2 nanotube array by the molecular iron oxide surface modification. Appl. Catal. B: Environ. 2012; 119–120: 74–80. DOI: 10.1016/j.apcatb.2012.02.012

[37] Jin Q, Fujishima M, Nolan M, Iwaszuk A, Tada H: Photocatalytic activities of tin(IV) oxide surface-modified titanium(IV) dioxide show a strong sensitivity to the TiO_2 crystal form. J. Phys. Chem. C 2012; 116: 12621–12626. DOI: 10.1021/jp302493f

[38] Boppana V B R, Lobo R F: SnOx-$ZnGa_2O_4$ photocatalysts with enhanced visible light activity. ACS Catal. 2011; 1: 923–928. DOI: 10.1021/cs200137h

[39] Kisch H: Semiconductor photocatalysis-mechanistic and synthetic aspects. Angew. Chem. Int. Ed. 2013; 52: 812–847. DOI: 10.1002/anie.201201200

[40] Kitano S, Murakami N, Ohno T, Mitani Y, Nosaka Y, Asakura H, Teramura K, Tanakia T,

[41] Tada H, Hashimoto K, Kominami H: Bifunctionality of Rh^{3+} modifier on TiO_2 and working mechanism of Rh^{3+}/TiO_2 photocatalyst under irradiation of visible light. J. Phys. Chem. C 2013; 117: 11008–11016. DOI: 10.1021/jp311801e

[42] Lin X, Li D-Z, Wu Q-P, Fu X-Z, Wang X-X: Chem. Photocatalytic activity and mechanism of heterojunction thin films. J. Chin. Univ. 2005; 26: 727–730.

[43] Nosaka Y, Takahashi S, Sakamoto H, Nosaka A: Reaction mechanism of Cu(II)-grafted visible-light responsive TiO_2 and WO_3 photocatalysts studied by means of ESR spectroscopy and chemiluminescence photometry. J. Phys. Chem. C 2011; 115: 21283–21290. DOI: 10.1021/jp2070634

[44] Abe R, Takami H, Murakami N, Ohtani B: Pristine simple oxides as visible light driven photocatalysts: Highly efficient decomposition of organic compounds over platinum-loaded tungsten oxide. J. Am. Chem. Soc. 2008; 130: 7780–7781. DOI: 10.1021/ja800835q

[45] Naya S, Niwa T, Negishi R, Kobayashi H, Tada H: Multi-electron oxygen reduction by a hybrid visible-light-photocatalyst consisting of metal-oxide semiconductor and self-assembled biomimetic complex. Angew. Chem. Int. Ed. 2014; 53: 13894–13897. DOI: 10.1002/anie.201408352

[46] Jin Q, Arimoto H, Fujishima M, Tada H: Manganese oxide-surface modified titanium(IV) dioxide as environmental catalyst. Catalysts 2013; 3: 444–454. DOI: 10.3390/catal3020444

[47] Ide Y, Hattori H, Ogo S, Sadakane M, Sano T: Highly efficient and selective sunlight-induced photocatalytic oxidation of cyclohexane on an eco-catalyst under a CO_2 atmosphere. Green Chem. 2012; 14: 1264–1267. DOI: 10.1039/C2GC16594E

[48] Higasihmoto S, Shirai R, Osano Y, Azuma M, Ohue H, Sakata Y, Kobayashi H: Influence of metal ions on the photocatalytic activity: Selective oxidation of benzyl alcohol on iron (III) ion-modified TiO_2 using visible light. J. Catal. 2014; 311: 137–143. DOI: 10.1016/j.jca.2013.11.013

[49] Tang J, Grampp G, Liu Y, Wang, B-X, Tao F-F, Wang L-J, Liang X-Z, Xiao H-Q, Shen, Y-M: Visible light mediated cyclization of tertiary anilines with maleimides using nickel(II) oxide surface-modified titanium dioxide catalyst. J. Org. Chem. 2015; 80: 2724–2732. DOI: 10.1021/jo502901h

2

Practical Design of Green Catalysts for PET Recycling and Energy Conversion

Arvin Sangalang, Seunghwan Seok and Do Hyun Kim

Abstract

The recycling of chemicals and generation of alternative energy are central topics in the efforts toward sustainable development. Among these, research on plastics recycling and fuel cells has received significant attention, with the aim of designing novel catalysts to improve yield and efficiency. We highlight our work on these areas focusing on the chemical depolymerization of polyethylene terephthalate (PET) to recover its constituent monomer and the development of high-performance anode catalysts for polymer electrolyte membrane fuel cells (PEMFC). We demonstrate various flexible yet practical synthesis strategies (e.g. ultrasound-assisted deposition and biopolymer coating) that were used to obtain catalytic properties optimized for these applications. The effectiveness and simplicity of these methods render the catalysts to be truly green — from synthesis up to process application.

Keywords: PET glycolysis, fuel cell, ultrasound, polydopamine, nanocomposites

1. Introduction

Depleting resources and waste management are the major driving factors for environmental conservation practices, such as recycling and development of alternative energy. Plastics recycling and generation of renewable energy, in particular, are among the direct approaches that aim to curb the consumption of oil-derived commodities. Research efforts on these areas can contribute to the realization of long-term environmental sustainability, aiming for increased productivities and minimal environmental impact. In these efforts, the development of effective catalysts to improve their performance remains a major goal in optimizing processes for resource conservation. Taking both catalytic activity and practical synthesis into account, we present catalyst design strategies that truly conform to principles of green catalysis.

1.1. Chemical recycling of poly(ethylene terephthalate)

Poly(ethylene terephthalate) or PET is currently the world's most widely recycled plastic. Due to its extensive use in various applications, largely in packaging, its worldwide consumption has increased tremendously, adding to the generation of plastic solid wastes. The global market for PET packaging was worth $48.1 billion in 2014, equivalent to about 16 million tons of PET. Furthermore, the projected demand for PET packaging material is estimated to increase further by 4.6% annually, reaching up to 20 million tons by 2019 [1]. The dramatic increase in PET consumption over the past decades has led to significant efforts to recycle the thermoplastic polyester.

The case of PET recycling is currently a logistically well-established, large-scale recycling industry. Bottle recollection rates, for example, have shown a steady increase worldwide, which accounted for at least 2.8 million tons of collected bottles in 2008 (Figure 1). Most of recycled PET flakes produced worldwide are utilized for fiber applications. PET recycling, at present, is largely dominated by physical methods or melt-phase recycling strategies, although various process routes have been studied and remain open to major advancements for better sustainability [2–4].

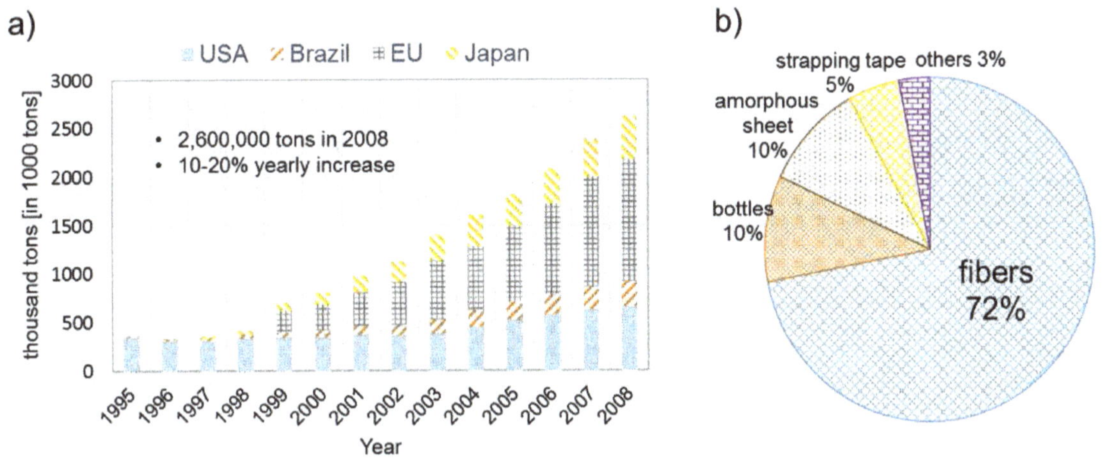

Figure 1. a) Increase in PET bottle recollection rates in various regions worldwide from 1995 to 2008. b) Use distribution of recycled PET flakes.

Significant efforts and renewed interest in chemical recycling routes have resurfaced recently, as stronger calls for long-term sustainability and new process developments in chemical recycling appeared. Extensive reviews on the progress and various aspects of PET chemical recycling are available elsewhere [2–7]. It has been regarded in the recycling literature that chemical recycling offers the most viable approach of sustainable recycling, allowing the recovery of raw materials and other useful chemicals while dealing with long-term waste disposal issues [5]. Figure 2 shows the monomer recovery routes in PET synthesis and chemical recycling. Among them, glycolysis is the simplest and well established, requiring low to moderate process conditions and capital investment (Table 1) [2].

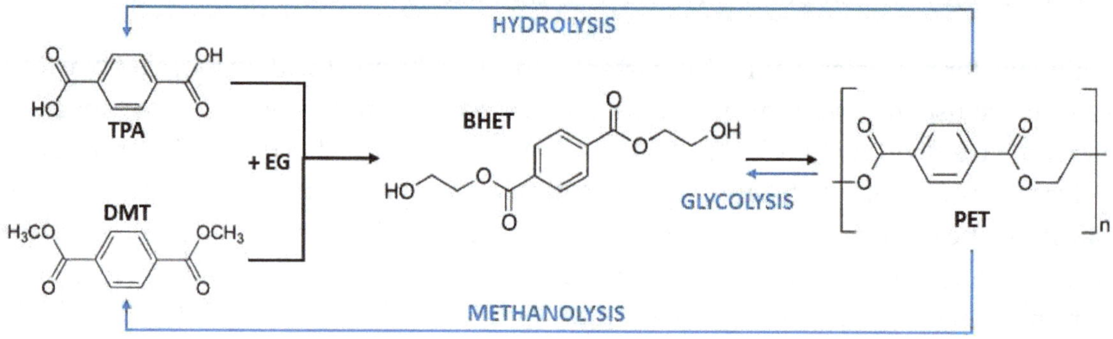

Figure 2. PET monomer recovery routes.

Recycling process	Qualitative plant cost	Economically competitive scale	Safety conditions
Hydrolysis	High	Large plants	High
Methanolysis	High	Large plants	High
Glycolysis	Medium-low	Small plants/batch reactor	Medium

Table 1. Process and economic considerations for PET monomer recovery routes [2]

The products obtained from glycolysis are the polycondensation monomer of PET, bis(2-hydroxyethyl terephthalate) or BHET (Figure 3) and low molecular weight oligomers, which are purified for subsequent use in PET manufacture or conversion into other chemicals. There have been significant issues, however, in process efficiency and product purification that led to recent efforts in the development of novel catalysts.

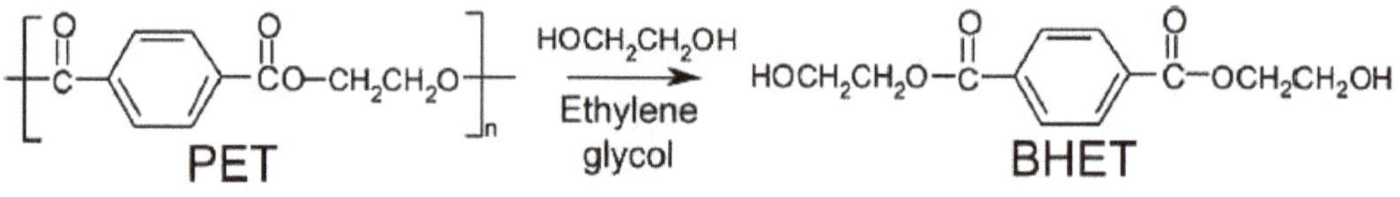

Figure 3. Glycolysis of PET chain releasing the monomer BHET.

Catalyst use is crucial in glycolysis and those used in earlier studies still required long reaction times despite catalyst application. These include metal acetates, alkalis, sulfates, and chlorides, some of which are those of toxic heavy metals [5]. Aside from efficiency, catalyst separation has been a major drawback in these catalysts together with environmental concerns. Despite chemical recycling being touted as the superior viable recycling route in the long-term, significant process developments are needed before it can truly be considered as a practical and competitive alternative to mechanical recycling. Results of life cycle assessments (LCA) show (Figure 4) that not only chemical recycling incurs greater process costs, but it also contributes to an overall negative impact in terms of a larger net consumption of non-renewable energy source and environmental effects [4]. Fortunately, recent developments in novel catalyst design may offer effective solutions to achieve the goal of a truly green and viable PET chemical recycling.

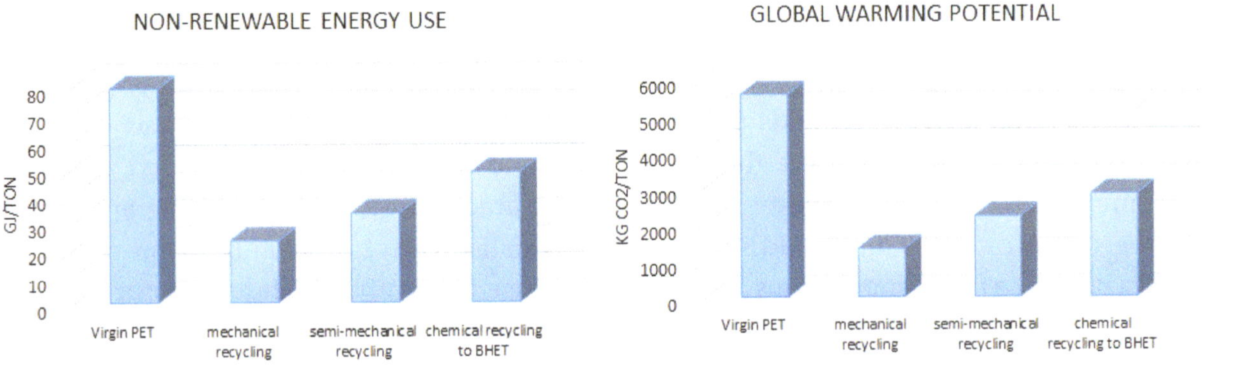

Figure 4. Comparison of life cycle assessment (LCA) results for PET manufacture and recycling in terms of non-renewable energy use and equivalent CO_2 generated per ton of PET fiber [4].

1.2. Hybrid catalysts for polymer electrolyte membrane fuel cells

The development of hybrid catalysts for polymer electrolyte membrane fuel cells (PEMFCs) has received attention recently due to advantageous qualities in its interaction characteristics and physical durability. The combination of conductivity enhancement and water management becomes widely used to improve PEMFC performance without external humidifying system. In this respect, requirements for next promising materials for PEMFCs include such categories as high proton conductivity, membrane electrode assemblies (MEA), and non-toxic and easy fabrication method. The humidifying ability of MEA in fuel cells is crucial for proton conductivity. As such, the approach to enhance conductivity and allow self-humidification at the same time has been considered recently to improve the performance of PEMFCs [8]. Given these requirements, the potential solution is platinum doping for conductivity and using silica as support for its hygroscopic properties. It is a challenge, however, to increase the platinum doping and enhance catalyst conductivity. With our bio-coating strategy for metal doping, we demonstrate a successful implementation of N-doped carbon composite through a silica-based templating technique [9]. In contrast to other reported methods that use chemical vapor deposition (CVD) and plasma treatment of various nitrogen precursors, our method remains on the practical side requiring no expensive equipment and complex treatment processes.

2. Synthesis and performance of metal oxide-based catalysts for PET glycolysis

Clearly, there is a need to consider various factors in designing PET glycolysis catalysts such as catalytic activity, ease of process applicability, and environmental impact. The coordination between a metal cation and the PET carbonyl oxygen is the widely exploited basis of catalytic activity in glycolysis, although other catalytic pathways do exist. Numerous metal compounds have been used as glycolysis catalysts with varying success [6]. Some of these made its way to commercial application through patented technologies [7]. Recently, a significant amount of work in the development of solid catalysts has appeared, potentially addressing the problem of product/catalyst separation.

The first solid catalysts used for glycolysis were zeolites, which showed catalytic performance of only moderate monomer yields (<70%). We have developed various glycolysis catalysts based on metal oxides that were thermally stable and recoverable and provided excellent monomer recovery yields. These are the ideal characteristics of industrially applicable heterogeneous catalysts that could solve issues in existing conventional glycolysis catalysts. We discuss in the following sections several methods used to synthesize them, their properties, and performance as stable and effective glycolysis catalysts. It was possible to obtain free and supported versions of these catalysts using flexible synthesis strategies [10–13] tailored to specific process applications.

2.1. Glycolysis reaction set-up and product analysis

Virgin PET pellets were obtained, mixed with dry ice in a grinder, and reduced to a fine powder with a particle size of <200 μm. The glycolysis reactions were carried out in a 10-mL stainless steel batch-type pressure reactor at 300°C and 1.1 MPa, unless otherwise specified. The reactant mixture of 0.3 g of PET, 1.1 g of EG, and a set amount of catalyst was loaded into the reactor. The reactor was placed in a furnace preheated at the reaction temperature, which, after the given reaction time, was taken out and quenched in cold water to stop the reaction. For quantitative determination of the monomer yield, the reaction products were dissolved in tetrahydrofuran (THF) and analyzed via HPLC. A reverse-phase Zorbax-C8 column and an ultraviolet (UV) detector set at 254 nm were used. The mobile phase used was a 50:50 (v/v) THF/H_2O solution at a flow rate of 1.0 ml/min. The molar yield of BHET was calculated based on the following equation:

$$\text{BHET Molar yield, \%} = \frac{\text{moles of BHET produced}}{\text{moles of PET units}} \times 100\%$$

The monomer BHET was separated from the glycolysis products and purified for qualitative analysis. Boiling water was added to the reaction mixture to dissolve BHET and the hot solution was immediately filtered. Repeated washing was done to extract residual BHET in the filter cake. The filtrate was stored in a refrigerator at 4°C for 24 h, after which white needle-like crystals formed. These were filtered and dried at 70°C for 12 h. The dried crystals were then used for various characterization steps, such as FT-IR, DSC, TGA, and NMR to verify the structure and properties of the recovered monomer, properly distinguishing BHET from its dimer and/or oligomers. Its chemical structure was analyzed by ^{1}H NMR and ^{13}C NMR using DMSO-d_6 as the solvent [10].

2.2. Supported metal oxide composites via ultrasound-assisted synthesis

Metal oxides find important application as catalysts, but agglomeration during their synthesis reduces the active surface area beneficial for catalysis. The combination of solid support and metal has been proposed, since composite materials can isolate the nanoparticle on the supporting material and effectively reduce its size. As a result, such nanocomposites can increase the surface area of metal oxide catalyst [14–16]. Silica, one of the most widely used catalyst supports, is synthesized in this study in the nanoscale and used as a metal oxide support material. Another material considered as support for nanocomposites is graphene oxide (GO). It is compatible with various organic/inorganic nanomaterials, taking advantage of possible chemical modification utilizing the oxygen-containing functional group on its sheet. In addition, it has high chemical stability and specific surface area [17–20]. The synthesis of silica and graphene oxide-based catalysts is presented, where ultrasound was used to reduce reaction time, simplify reaction steps, or perform synthesis in milder conditions.

2.2.1. Ultrasound-assisted catalyst synthesis

Ultrasound has been used in metal deposition, dish cleaning, chemical reactions, and particle dispersion in a solvent [21]. The cavitation phenomenon includes formation, growth, and collapse of bubbles generated in the aqueous solution under ultrasonic irradiation. The phenomenon causes both high temperature (~5000 K) and pressure (~1000 atm) during the sequence of bubble collapse. In a short span of time for bubble formation, it provides very high heating and cooling rate. This unique condition generates intense energy that converts water into H and OH radicals, promoting the formation of metal nanoparticles [22].

Our group applied the ultrasound irradiation method for the fabrication of metal-doped silica nanocomposites [8,11,23,24]. The acoustic cavitation phenomenon facilitates interparticle collision between metal and support material, inducing the binding between metal and support [25,26]. As a result, metal doped on supporting material can be prepared without surfactants or surface modification of support material in a short reaction time and mild reaction conditions.

2.2.2. Ultrasonic deposition of metal oxide catalyst on silica particles

Silica microparticles (SMPs) and silica nanoparticles (SNPs) were used as support for the catalysts. Silica microparticles with size 1–20 μm were purchased from Junsei Chemicals and used without any treatment. Nanosized silica particles were synthesized using the Stöber method, with some modifications. In a sealed round-bottomed flask, 8.0 ml of ammonium hydroxide (28 wt%) and 6.0 ml of deionized water were added to 100 ml of ethanol and stirred for 15 min. Then, 4.7 ml of the silica precursor tetraethyl orthosilicate (TEOS) were added to this solution and stirred at room temperature for 3 h. The resulting precipitate was centrifuged and washed with water and ethanol several times. The product was dried in an oven at 70°C for 8 h followed by calcination at 500°C for 12 h.

To synthesize the silica-supported manganese oxide or zinc oxide catalysts, a predetermined amount of silica support was added to a 1.0 M solution of the precursor [$Mn(NO_3)_2$ xH_2O or

$Zn(NO_3)_2 6H_2O$]. The metal oxide loading was set to be 15 wt%. Using a horn-type sonifier, the solution was sonicated for 45 min. A 0.1 M ammonia solution was added at the start of the sonication step in order to keep the pH at around 9.5. The ultrasound-assisted deposition process is illustrated in Figure 5. After separating the particles by centrifugation, they were washed with water and ethanol. The catalyst samples were then dried at 100°C for 8 h and calcined at 350°C for 3 h.

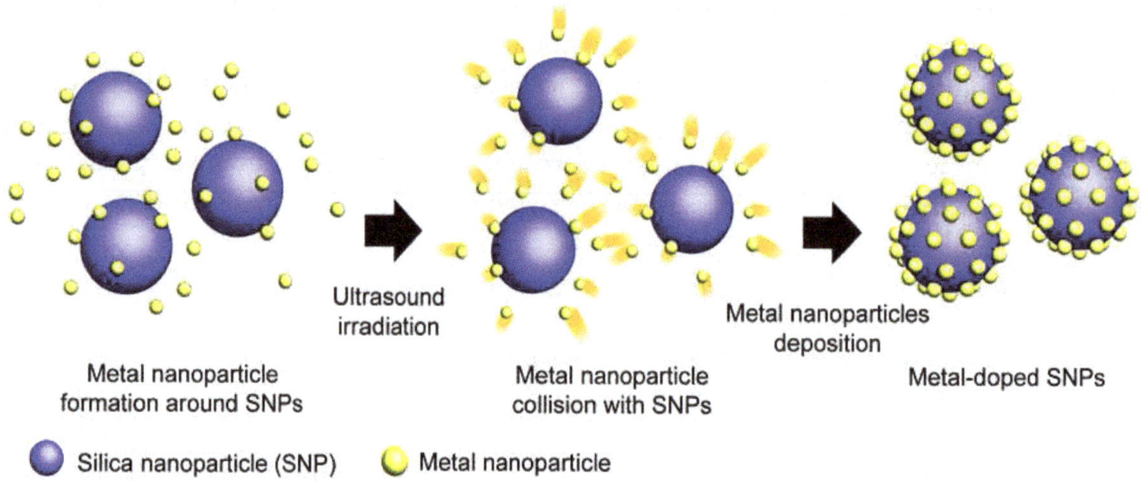

Figure 5. Overall synthesis procedure of metal-doped silica nanoparticle.

The properties of the synthesized silica-supported catalysts are given in Table 2. At 1.0 wt% catalyst-to-PET loading, glycolysis was performed at 300°C for 80 min, after which the product monomer was recovered. Shown in Figure 6 are the FT-IR spectra of the recovered BHET crystals and standard BHET sample. The matching spectra confirm that the product obtained using the silica-supported catalyst was indeed the monomer. The spectrum showed the presence of peaks corresponding to the functional groups in BHET: an -OH band at 3,424 and 1,128 cm^{-1}, an aromatic C-H at 1,456–1,502 cm^{-1}, C-O at 1,712 cm^{-1}, and alkyl C-H at 2,879 and 2,964 cm^{-1} [10].

From the monomer yield versus reaction time data in Figure 7, the order of catalytic activity can be determined as Mn_3O_4/SNPs>ZnO/SNPs>Mn_3O_4 SMPs>ZnO/SMPs. This trend follows the same order as the catalyst surface area and the pore volume given in Table 1. Among the four catalyst samples with different size of support and metal oxide doping, the large amount of active surfaces in the nanoparticle support and the activity of the Mn_3O_4 catalyst could be responsible for the fastest reaction rate and the maximum monomer yield [10]. Although it is evident that using a silica nanoparticle support could improve the catalytic performance, it also has its drawbacks in the practical and industrial perspective. Using a metal oxide catalyst is desired for easier purification of the glycolysis products, but efficient separation of a nanosized catalyst can be challenging. A trade-off between catalytic performance and practical applicability will be inevitable, unless an effective method to separate the catalysts is provided.

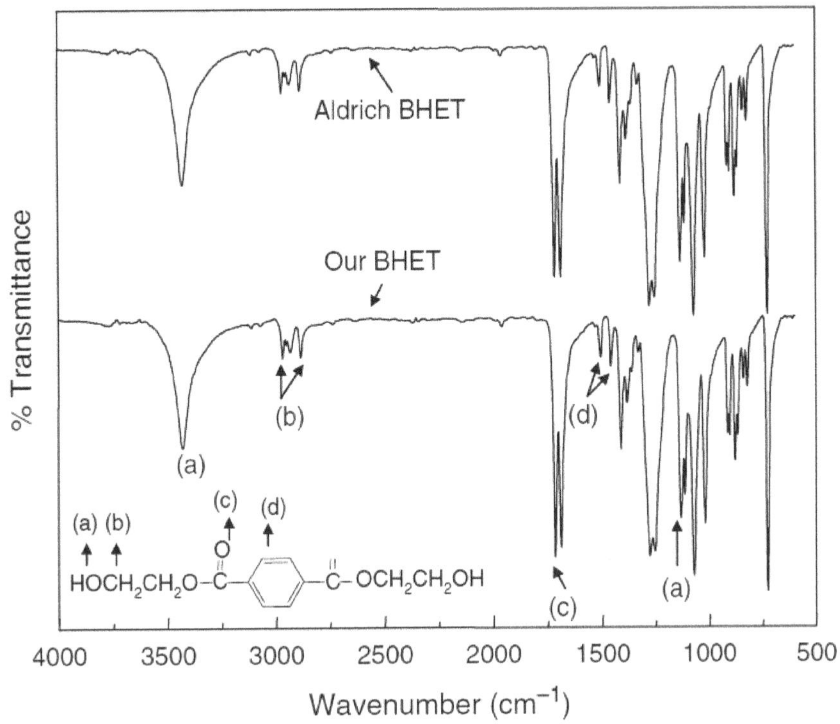

Figure 6. Comparison of FT-IR spectra of standard BHET and BHET recovered from catalyzed glycolysis. Figure from [10].

Catalyst	BET surface area (m^2/g)	Pore volume (cm^3/g)	Average pore diameter (nm)
ZnO/SMPs	2.49	0.014	21.4
Mn_3O_4/SMPs	3.38	0.020	24.4
ZnO/SNPs	22.44	0.154	30.2
Mn_3O_4/SNPs	45.09	0.214	18.9

Table 2. BET surface area, pore volume, and average pore diameter of silica-supported metal oxide catalysts [10]

2.2.3. Sonochemical synthesis of the GO-Mn_3O_4 composites

Mn_3O_4 and its nanocomposites have been utilized as highly effective catalysts for various applications, including PET glycolysis as demonstrated in our previous studies. Some of the known advantages of manganese oxides include excellent catalytic activity, low cost, abundance, and being environmental benign [27–30]. Depending on the oxidation states of manganese, there are several forms of MnO_x (e.g. MnO_2, MnO, Mn_2O_3, and Mn_3O_4), each of which has different applications. Hausmannite (Mn_3O_4), with both Mn^{2+} and Mn^{3+} ions in its crystal structure, has been widely used in catalytic applications. When loaded onto a support to provide a large surface area and prevent aggregation, it could significantly enhance the depolymerization of PET [10].

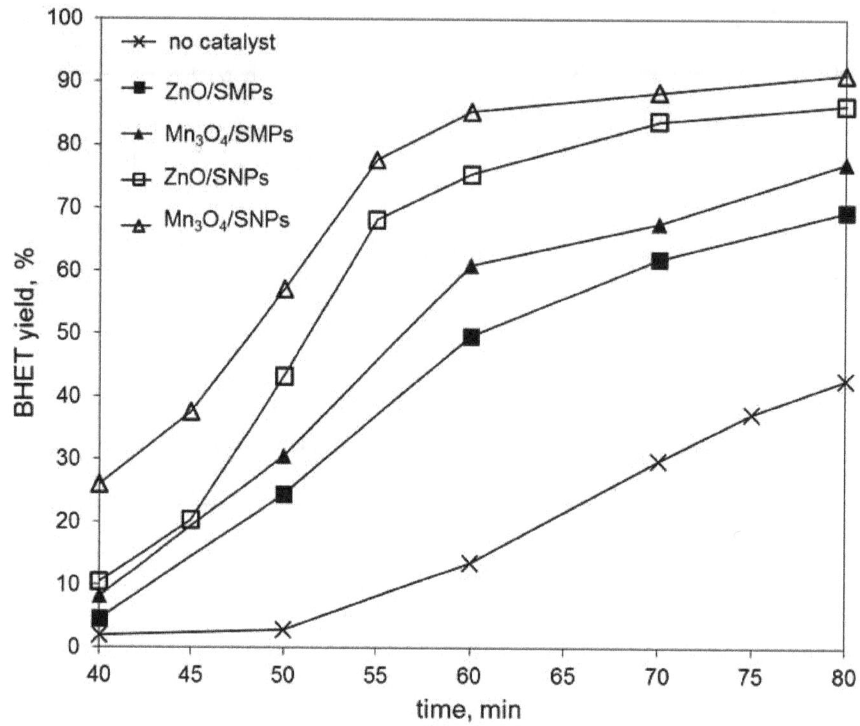

Figure 7. Molar yield of BHET at 300°C and 1.1 MPa using silica-supported catalysts. Figure from [10].

Simultaneous formation and direct deposition of MnO_x nanostructures have been reported by using redox reaction and electrodeposition [21,29–31]. One of the most favorable methods to do this is the reduction of permanganate ions into the insoluble manganese dioxide induced by carbon such as that in a graphene structure. The procedure is simple and the reaction has a self-limiting character [27,28,32]. This method can also be adapted to deposit Mn_3O_4 onto graphene through thermal reduction of MnO_2 over 1000°C [16]. This is an energy-intensive process, over which alternative methods of synthesis using milder conditions would be preferable [21,33]. In our work, sonochemical methods were used to facilitate mild conditions for synthesis and reduce reaction time involved in Mn_3O_4 deposition as illustrated in Figure 8.

Using the modified Hummers method, graphene oxide (GO) was prepared from graphite flakes [34]. One gram of graphite was added to 50 mL of concentrated H_2SO_4 in an ice bath. Then, 3.5 g of $KMnO_4$ were added and stirred for 2 h at 35°C. The suspension was then kept at 98°C to which deionized (DI) water was added dropwise. Then, 25 mL of 3% H_2O_2 aqueous solution were poured into the mixture and filtered with a 0.1 mm pore diameter Anodisc™ membrane. The product was washed with 10% HCl aqueous solution and DI water. By applying ultrasound to the filtered graphite oxide cake suspended in DI water, exfoliated GO was obtained and subsequently dried. Dispersion of GO in DI water at 0.5g/mL concentration was prepared. Mixtures of 10 mg/mL $KMnO_4$ and the GO dispersion at varied volume ratios of 0.01, 0.03, and 0.05 were used for the synthesis of GO-Mn_3O_4 composite samples A, B, and C, respectively. These are then subjected to ultrasonication at a power of 80 W/cm for 30 min using a horn-type sonicator. The resulting suspension was then filtered and washed with DI water and then ethanol [11].

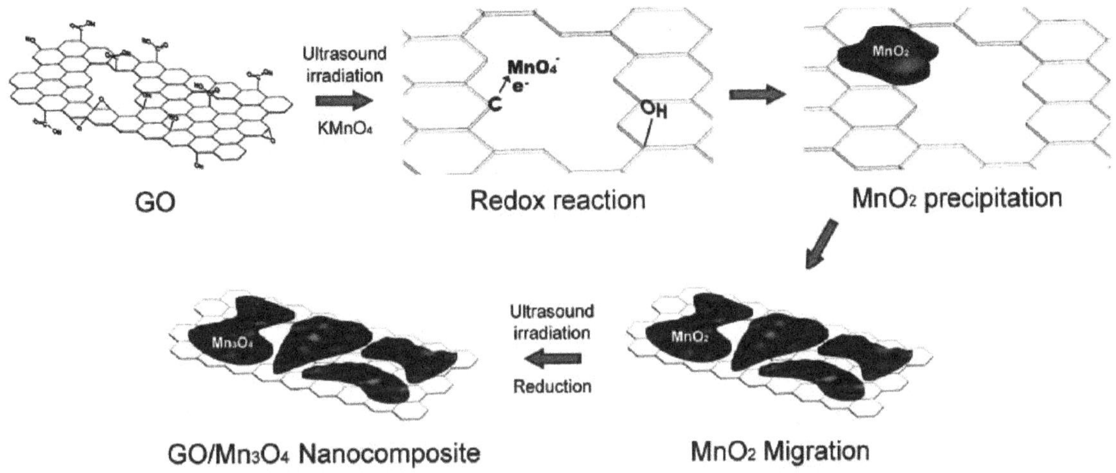

Figure 8. Schematic illustration of ultrasound-assisted synthesis of the GO-Mn_3O_4 composites. MnO_4 is first reduced to MnO_2 and precipitated onto the GO support by oxidizing carbon. The reduction of MnO_2 to Mn_3O_4 then takes place in the following steps.

The formation of GO-Mn_3O_4 nanocomposite was verified by various characterization methods such as XRD, XPS, and Raman spectroscopy. TEM images of pristine GO and the obtained composite are shown in Figures 9a and 9b, indicating the coverage of the GO surface. The high-resolution TEM image in Figure 9c shows the lattice fringes and diffraction pattern of the Mn_3O_4 crystal structure. Compared to the silica-supported composites and conventional metal salt catalysts [10,35], the monomer yield using the GO-Mn_3O_4 nanocomposite was comparable or higher, reaching more than 90% (Figure 10). The yields for the composite were all above 90%, showing improvement from that of bare Mn_3O_4 at 83%. However, the Mn_3O_4 without the support aggregated into micron scale. The GO support could prevent the aggregation of Mn_3O_4 and provide an enlarged and stable active sites [11].

2.3. Magnetically recoverable superparamagnetic γ-Fe_2O_3 nanocatalyst

Efficient recovery of the catalysts in PET glycolysis is an important aspect that several researchers have attempted to address recently. Among the recoverable catalysts studied are various ionic liquids, which were shown to provide molar yields up to 80% [36–39]. Aside from yield, there are several issues that should be considered in using ionic liquids as industrial glycolysis catalysts such as cost, stability, and robustness with respect to process variables such as moisture content, to which ionic liquids are very sensitive [10,11,40]. Urea was also reported to be a reusable glycolysis catalyst at mild temperatures [41]. Vacuum distillation was used to recover the catalyst, however, whose high energy requirements can be counterproductive.

We have studied magnetic nanoparticles as a recoverable glycolysis catalyst, among which γ-Fe_2O_3 was chosen as excellent candidate being known to have good performance in a number of reactions [42–44]. This was the first attempt to utilize magnetic nanomaterials in PET depolymerisation. Nanosized γ-Fe_2O_3 was selected due to its stability, high catalytic activity, and superparamagnetic property [42]. The superparamagnetic behavior allows recovery by application of a magnetic field yet allows good dispersion in the reaction medium, as it has

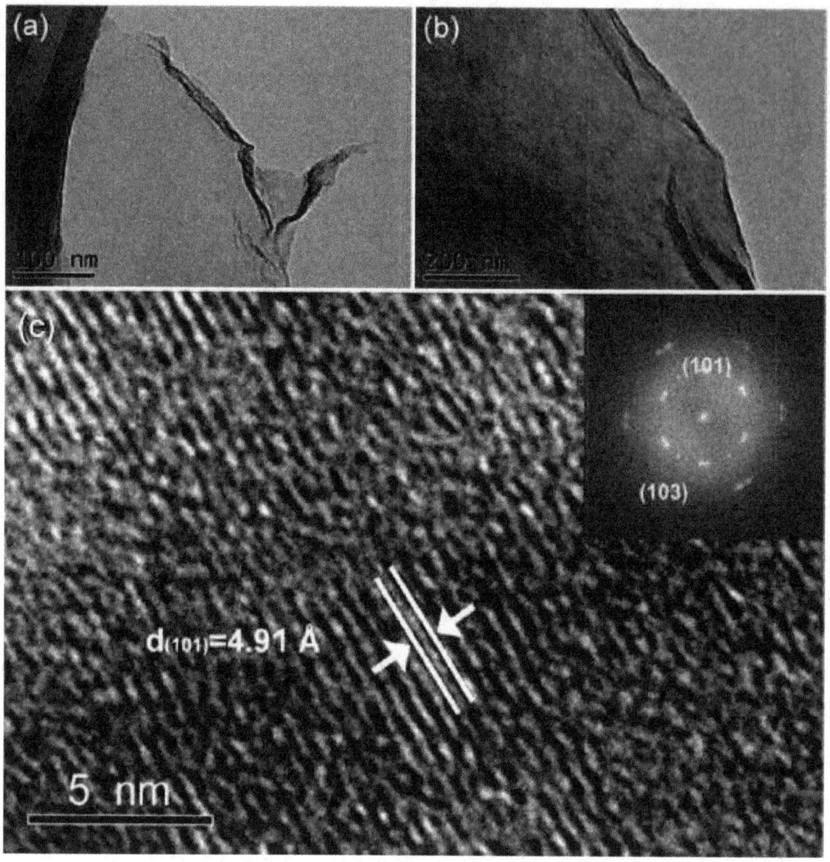

Figure 9. Low-magnification TEM images of (a) GO, (b) GO-Mn_3O_4, and (c) high-resolution TEM image of GO-Mn_3O_4 showing the d-spacing of the (101) crystal plane of Mn_3O_4 and its diffraction rings [11].

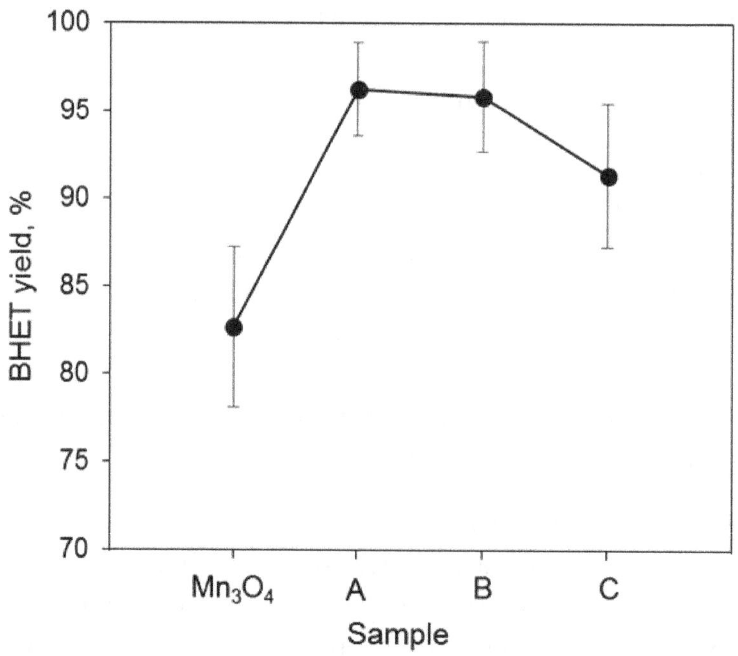

Figure 10. BHET yields for bare Mn_3O_4 and GO-supported Mn_3O_4 catalysts.

zero remanent magnetization. Iron oxides have further advantages being cheap, nontoxic, and abundant [12].

2.3.1. *Synthesis of superparamagnetic γ-Fe_2O_3 catalyst and characterization*

Slight modifications to the conventional co-precipitation method were done to obtain γ-Fe_2O_3 nanoparticles from Fe_3O_4 [45,46]. For the synthesis of Fe_3O_4 nanoparticles, predetermined amounts of $FeCl_2$ and $FeCl_3$ precursors were dissolved in an HCl solution (0.4 M, 25 mL) to prepare aqueous solutions of Fe^{2+}/Fe^{3+} with 1:2 molar ratio. Sodium hydroxide (1.5 M, 250 mL) was then added rapidly and stirred. The addition of NaOH solution instantly produced a black precipitate, characteristic of Fe_3O_4. The surfactant used to control the particle size was citric acid (0.2 M, 50 mL). The synthesis was performed in a nitrogen atmosphere. The formed black precipitate was separated by placing a magnet and decanting the solution. The product was washed with water four times and centrifuged at 4000 rpm for 4 min. The final washing step was done using a 0.01 M HCl solution in order to neutralize the anionic charges on the nanoparticle surface. Calcination of the dried Fe_3O_4 powder at 210°C for 3 h induced the phase transformation into γ-Fe_2O_3 [12].

The particle morphology of the synthesized catalyst was observed by a 200 kV transmission electron microscope (TEM). As shown in Figure 11a, the particles had a size distribution in the range of 8–14 nm, with mean size of 10 nm. The BET surface area measured was 147 m^2/g. The superparamagnetic property of the nanosized γ-Fe_2O_3 was confirmed by the magnetization curve obtained using a vibrating sample magnetometer (VSM). Figure 11b shows the magnetic behavior in the presence of a magnetic field, exhibiting a strong response with saturation magnetization reaching 47 emu/g. The curve does not have a hysteresis loop and retains no magnetization when the magnetic field is removed. The advantageous consequence of this property is redispersability of the catalyst particles when used in subsequent reactions [12].

The obtained X-ray diffraction (XRD) spectra of the catalyst suggest that the material is γ-Fe_2O_3. This is not conclusive, however, because the XRD patterns of Fe_3O_4 and γ-Fe_2O_3 are very similar. X-ray photoelectron spectroscopy (XPS) analysis of Fe2p cores was performed to distinguish the two phases. Higher binding energies before the calcination as shown in Figure 11c are indicative of Fe_3O_4. The shift to lower binding energies after the calcination step confirms the transformation of Fe_3O_4 to γ-Fe_2O_3 [12].

2.3.2. *Catalytic activity, recoverability, and stability with repeated use*

The catalytic performance of the synthesized γ-Fe_2O_3 nanoparticles was compared to previously studied silica nanoparticle-supported metal oxide catalysts [10,40]. Under the same reaction conditions and catalyst/PET weight ratio, γ-Fe_2O_3 delivered comparable performance (Figure 12). The BHET yield reached higher than 90% in 70 min at 1.0% catalyst-to-PET loading. As with the supported nanocatalysts, the excellent catalytic performance of the γ-Fe_2O_3 nanoparticles may be attributed to the high surface area and greater accessibility to active sites [12].

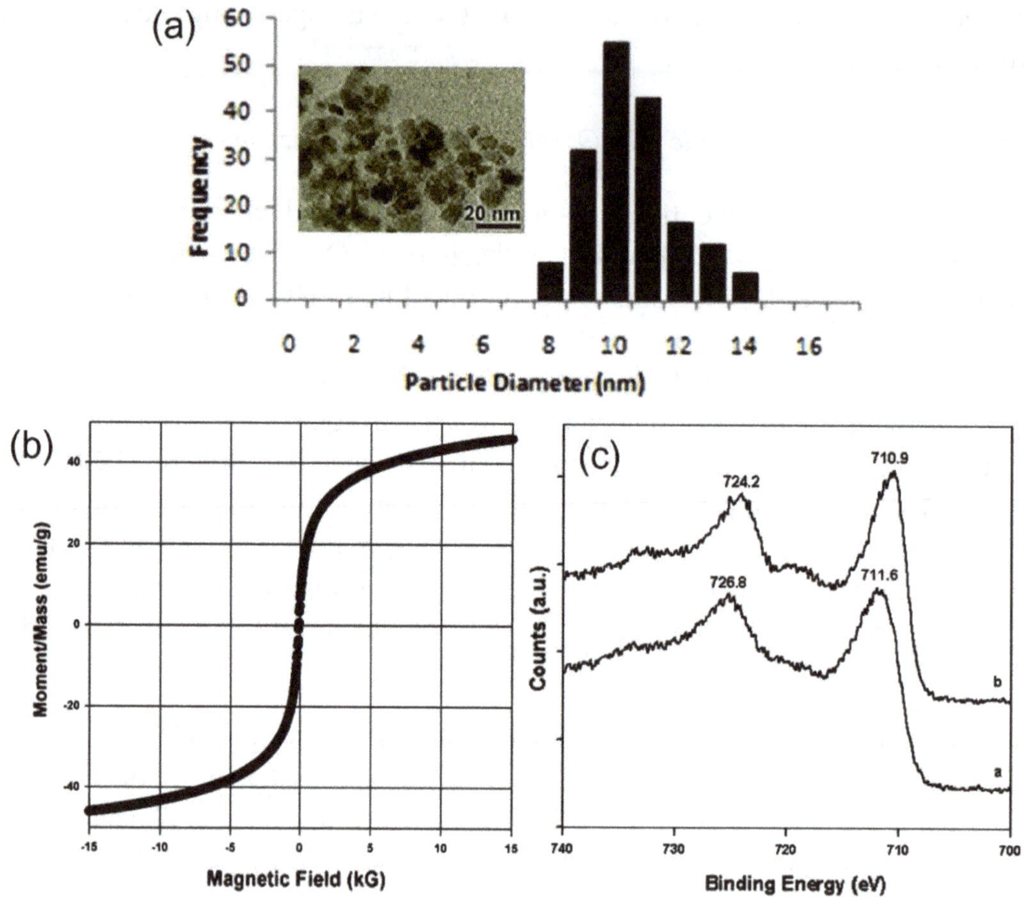

Figure 11. (a) Particle morphology, (b) superparamagnetic property, and (c) phase identification of the synthesized γ-Fe_2O_3 nanoparticles. Figure from [12].

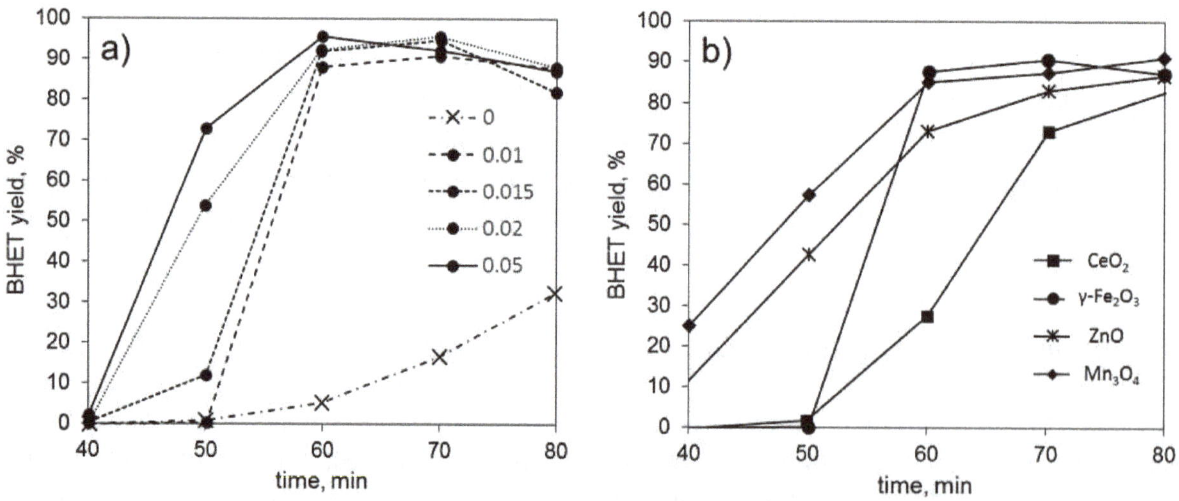

Figure 12. Catalytic performance of γ-Fe_2O_3 nanoparticles at various loadings and comparison to other metal oxide catalysts. Figure from [12].

The main potential of using γ-Fe_2O_3 as candidate for industrial glycolysis catalyst is the easy separation method and stability. In addition to the superparamagnetic property of γ-Fe_2O_3 nanoparticles that was beneficial to catalytic performance and separation, its stability with repeated use was also successfully demonstrated. The catalyst was reused in 10 reaction repeat cycles and did not significantly affect the BHET monomer yield, as shown in Figure 13. The thermal stability was supported by thermogravimetric analysis (TGA), while the XRD spectra of the used catalyst also proved phase stability [12]. These demonstrate the robust characteristics of the catalyst, withstanding repeated use at elevated temperature and pressure without deterioration in performance. Given comparable performance to a homogenous catalyst as shown in Figure 13b, the superparamagnetic catalyst provides a more practical approach to catalyst separation by application of a magnetic field.

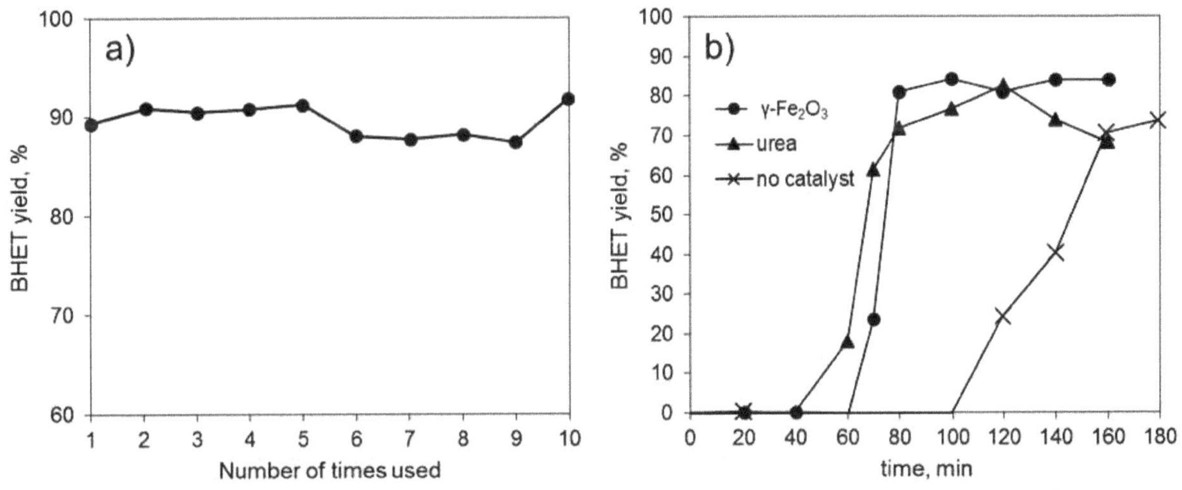

Figure 13. Assessment of catalyst stability with repeated use and comparison to a recoverable homogeneous catalyst at the same reaction conditions. Figure from [12].

2.4. Mesoporous spinel oxide catalysts

Due to several advantages such as high mechanical strength, possibility of regeneration, easier separation, and robust process integration, metal oxides can be considered superior to conventional PET glycolysis catalysts. Moreover, there are numerous possibilities to tailor their physical and chemical properties for the desired catalytic performance and functionality [47]. For the enhancement of catalytic activity, for example, altering the metal composition by introduction of another metal could result in higher catalytic activity [48–50]. The same principle is used to potentially enhance the performance of metal oxide catalysts for PET glycolysis. In this section, pure oxides and mixed-oxide spinel oxides of zinc, manganese, and cobalt were synthesised by simple precipitation or co-precipitation methods.

Simple precipitation method was used to synthesize the pure metal oxides. A 1.0 M solution of the salt precursors ($Mn(NO_3)_2\ xH_2O$, $Zn(NO_3)_2\ 6H_2O$, or $Co(NO_3)_2\ 6H_2O$) was mixed with 0.1 M ammonium hydroxide to set the pH at 9.0. Precipitates of the corresponding metal hydroxides formed, which were filtered, washed with water, and dried at 100°C for 8 h. The oxide form was obtained by calcination of the dried powder at 600°C for 4 h. For the mixed metal oxides, a modified co-precipitation method was implemented [51–53]. The molar ratios of the metal precursors were fixed to be 1:2. Similar to the previous synthesis, a 0.1 M ammonium hydroxide solution was stirred into the bimetallic precursor solutions, setting the pH value at 9.0. The same procedures for filtering, washing, drying, and calcining were performed on the mixed metal oxides [13]. The physical and chemical properties of the synthesized catalysts are summarized in Table 3. The analysis of atomic composition of the oxides via EDS was performed with results shown in Table 4.

Catalyst	BET surface area (m^2/g)	Pore volume (cm^3/g)	Average pore diameter (nm)	Acid site concentration (mmol/g)
ZnO	7.67	0.06	31.33	0.041
Co_3O_4	17.50	0.09	18.80	0.032
$ZnCo_2O_4$	21.65	0.08	12.23	0.045
Mn_3O_4	22.57	0.01	14.70	0.060
$CoMn_2O_4$	25.10	0.07	10.16	0.067
$ZnMn_2O_4$	32.40	0.14	14.50	0.088

Table 3. Surface area, pore dimensions, and acid site concentration of the pure and mixed oxide catalysts [53]

Catalyst		Atom percentage			Atomic ratio (Mixed oxides)		
Zn	Mn	Co	O	Mn/Zn	Mn/Co	Co/Zn	
ZnO	56.3	-	-	43.7	-	-	-
Mn_3O_4	-	61.4	-	38.6	-	-	-
Co_3O_4	-	-	75.9	24.1	-	-	-
$ZnMn_2O_4$	21.7	40.9	-	37.4	1.88	-	-
$CoMn_2O_4$	-	36.2	21.0	42.9	-	1.72	-
$ZnCo_2O_4$	24.1	-	38.9	37.0	-	-	1.61

Table 4. Atomic analysis of the synthesized oxide catalysts by EDX [53]

The catalytic activity of metal oxides in PET glycolysis is influenced by the interaction of the metal cation and the carbonyl oxygen in the polyester. The nature of the metal, oxidation state, and crystal structure affect this interaction. In spinel oxides, the metal cations can be located in tetrahedral and octahedral sites. The metal covalency of two or more different metals in the mixed oxides can result in a beneficial interaction that could enhance its redox properties and catalytic activity. In this study, the best catalytic activity for glycolysis was demonstrated by

$ZnMn_2O_4$ catalyst (Figure 14). This catalyst has the ion pair Zn^{2+}/Mn^{3+} in its crystal lattice compared to Co^{2+}/Mn^{3+} and Zn^{2+}/Co^{2+} in the other mixed spinels. The high activity of the catalyst was attributed to the nature of the manganese ions combined with structural effects in the spinel crystal [13].

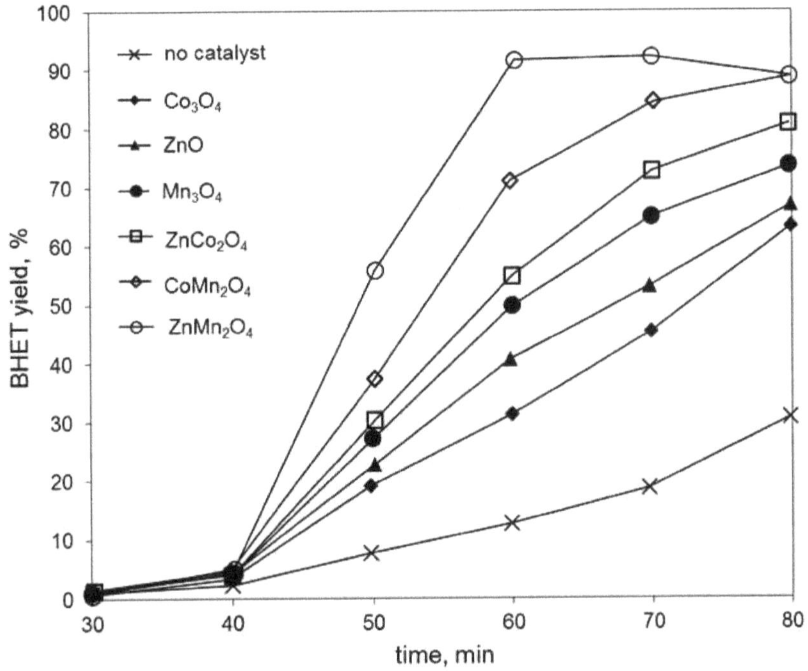

Figure 14. Comparison of BHET yields among the pure and mixed oxide catalysts [13].

2.5. Purity of recovered BHET monomer

For all the glycolysis reactions above using various catalysts, the recycled BHET was recovered via simple recrystallization. Several characterization methods have been employed to verify the structure and purity of the recovered monomer. The FT-IR spectra of the recycled BHET matched that of the standard sample [10], without extra peaks characteristic of contamination. As it is also possible that dimers and oligomers are not effectively separated, thermal and structure analyses have been performed using thermogravimetric analysis (TGA) and nuclear magnetic resonance spectroscopy (NMR). The proton and carbon NMR spectra identified peaks corresponding to distinct groups in the monomer and dimer backbones [13]. Along with TGA thermogram profiles, the structure characterization verified the good separation of the BHET from its oligomers.

3. Dopamine-induced carbon@silica hybrid as fuel cell anode catalyst

Aside from high proton conductivity and self-humidifying properties, principles of practical design for potential materials for polymer electrolyte membrane fuel cell (PEMFC) demand a fabrication method that is non-toxic and requires no complex apparatus for easy and large-scale production. Hybrid catalysts with carbon material and inorganic material have received attention from many researchers for PEMFC [54,55].

In hybrid material synthesis, a variety of methods are used to combine organic and inorganic materials. However, these fabrication methods require expensive equipment and complex processes, thereby limiting mass production [56–58]. To resolve this problem, our group has developed a simple and green fabrication of a hybrid catalyst for PEMFC. The hybrid composite consists of Pt nanoparticle, conductive dopamine coating layer, and silica support. Pt nanoparticle has been used as a typical catalyst for electrode catalyst for energy storage application. We also used Pt nanoparticle as the anode catalyst for PEMFC. To deposit Pt nanoparticles on the surface of silica, we applied dopamine coating as an adhesive layer [9]. The overall synthesis procedure for the metal and dopamine-induced carbon@silica nanocomposite is illustrated in Figure 15.

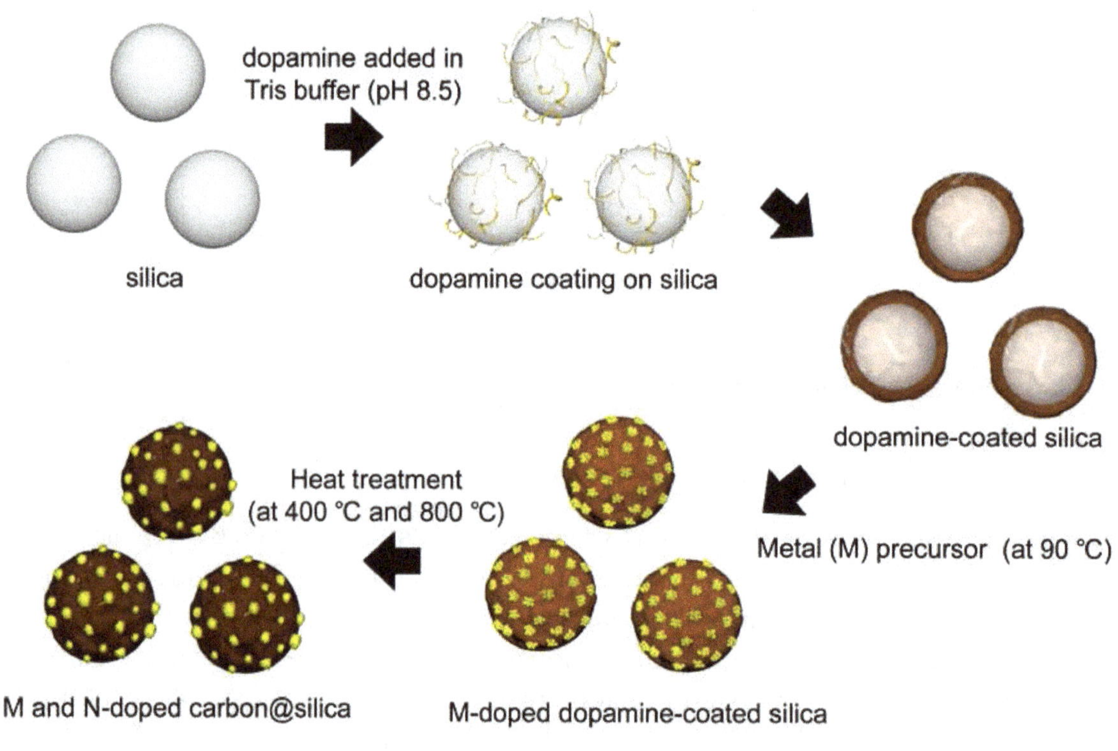

Figure 15. Overall procedure for synthesis of metal (M)- and nitrogen (N)-doped carbon@silica nanocomposite.

For the preparation of a hybrid anode catalyst, silica was fabricated by water-in-oil microemulsion method that produces a narrow size distribution. Silica is generally applied as support material because of its mechanical, chemical, and thermal stability. Its size and surface property can also be easily modified. Additionally, it has hygroscopic property that maintains the humidity needed for the proton conductivity in PEMFC. The humidity dependence of the membrane's conductivity typically required humidity control using an external equipment. However, the application of silica as support for the anode catalyst could provide both humidity and conductivity to the proton conductive membrane [8].

To deposit Pt nanoparticles on the surface of silica, we applied dopamine coating as an adhesive layer and self-reducing agent. Dopamine is a mussel-inspired adhesive based on a protein located at the end of the byssus of mussels. The protein allows mussels to strongly attach on rock surfaces or support in sea. The adhesion property of the protein helps mussels hold onto substrates in wet condition. Mussel-inspired dopamine is similar in chemical structure, with catechol and amine functional group as the protein [59]. Attempting to mimic the adhesive functionality, many researchers coated dopamine on various types of substrates. After dopamine coating, the substrate achieved a more hydrophilic surface and can be well dispersed in water. Dopamine coating also acts as a metal adhesive layer through the self-reducing ability of catechol group in the dopamine chemical structure [60]. Moreover, the coating is a simple and controllable dip-coating process in alkaline water solution (pH 8.5). The coating thickness can be tailored in nanoscale by adjusting the temperature, pH, time, and concentration of dopamine. Because of these advantages, dopamine coating has been applied to a variety of fields, such as surface modification, heavy metal removal, carbon material functionalization, and fabrication of conductive N-doped carbon sources [56–58,61].

We prepared Pt and N-doped carbon@silica with dopamine coating via metal decoration and carbonization steps. The dopamine coating procedure was performed on a Tris buffer solution at pH 8.5. Silica particles were dispersed in the solution. Then, 10 mM dopamine was added and mixed mildly for 6 h. After the coating step, the particle was washed several times with DI water. A thin layer of dopamine was observed to be coated well on silica with a thickness of 11.5 nm as shown in Figures 16c and 16d. The prepared dopamine-coated silica was re-dispersed in DI water. The dopamine-coated silica was then used to serve as catalyst support with metal adhesion properties. The Pt nanoparticles on the coated silica were deposited through the self-reducing ability of catechol groups in dopamine, as shown in Figure 17. This phenomenon releases electrons via the transformation of the R-OH group to a C=O group, which reduce Pt ions into Pt nanoparticles.

The dopamine coating layer naturally prohibits electron transfer (low conductivity). To enhance catalyst conductivity, we conducted carbonization of the Pt and dopamine-coated silica in an inert atmosphere at 800 °C. The carbonized dopamine layer contained nitrogen (N) and had other functional groups removed. N-doped carbon layer from dopamine coating enhanced both mechanical stability and conductivity of the material [57].

We applied Pt and dopamine-induced carbon@silica hybrid composite for polymer membrane electrolyte fuel cell as anode catalyst. The composite was evaluated and compared with commercial Pt/C catalyst under zero relative humidity. The carbonized dopamine coating and silica could be useful in PEMFC under non-humidified conditions. Pt and N-doped carbon@silica-based composites and commercial Pt/C were evaluated by galvanostatic polarization and power density curves as functions of current density for a single PEMFC at zero relative humidity (Figure 18). The Pt and dopamine-induced carbon@silica composite exhibited a maximum power density of 0.55 W cm^{-2}, which exceeds that of commercial Pt/C (0.45

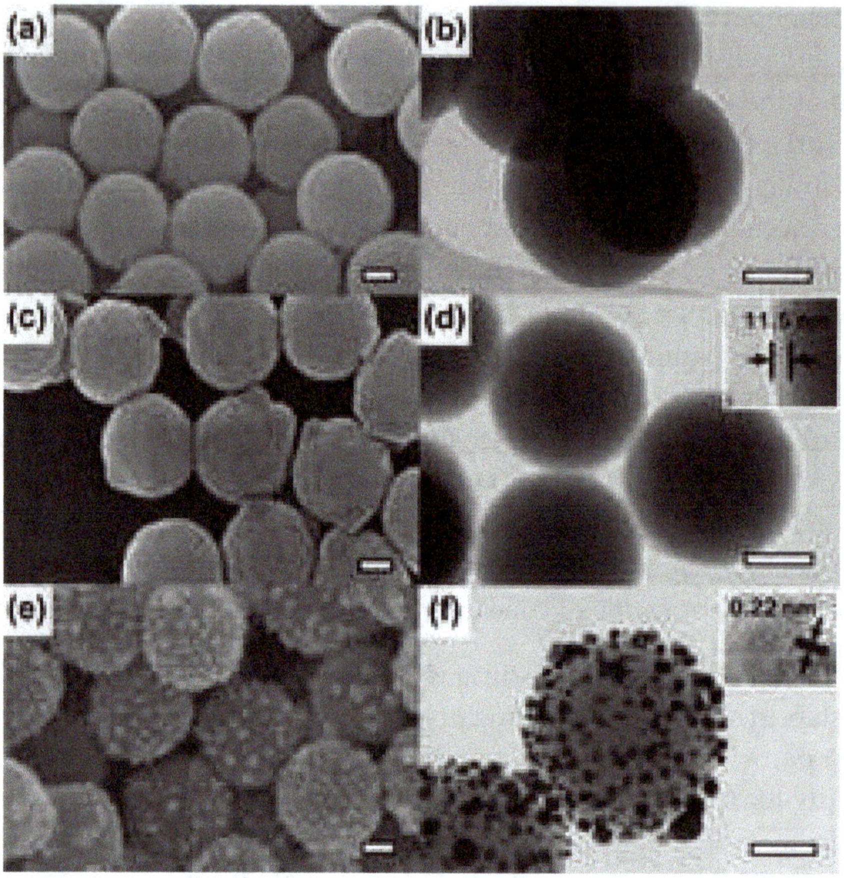

Figure 16. SEM and TEM images of (a and b) silica, (c and d) dopamine-coated silica, and (e and f) Pt and dopamine-induced carbon@silica. Scale bars are 100 nm.

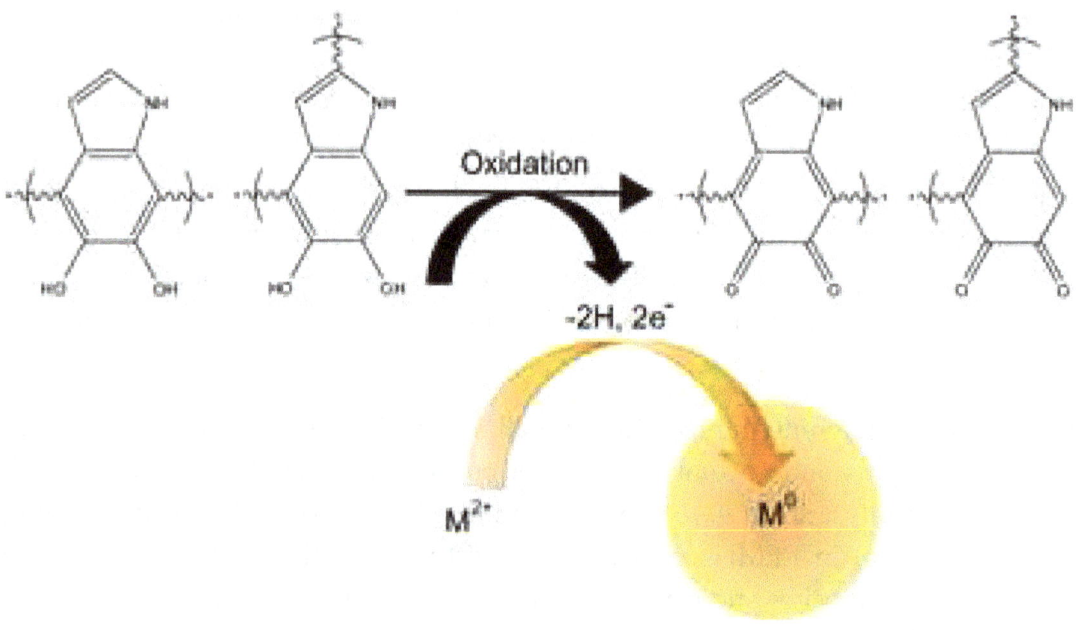

Figure 17. Reduction pathway from Pt ion into Pt nanoparticle using self-reducing ability of dopamine coating layer.

W cm^{-2}). The improved performance of the composite is attributed to the hybrid synergy effect of silica hygroscopic property and N-doped dopamine-induced carbon layer. Additionally, the N-doped carbon obtained after the carbonization step could make the composite more conductive with the enhancement of charge transfer at interfaces. As a result, these advantages of the Pt and dopamine-induced carbon@silica composite could contribute to enhance performance under low humidity conditions compared to the commercial Pt/C catalyst.

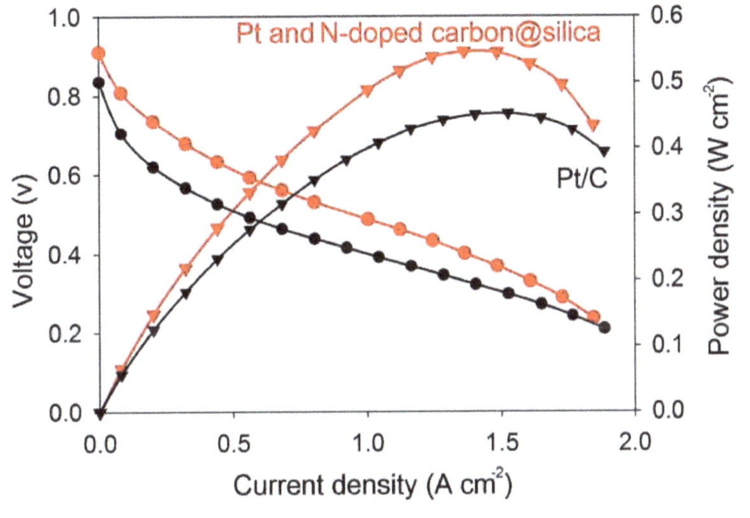

Figure 18. I-V polarization (left axis) and power density (right axis) curve for PEMFC single cell (H_2/O_2) with Pt and N-doped carbon@silica nanocomposites (red triangle) and commercial Pt/carbon (black circle) used as an anode catalyst under relative humidity at 0%.

4. Conclusion

We have presented various techniques to obtain novel catalysts that successfully enhance the efficiency and productivity of PET glycolysis for monomer recycling as well as the performance of polymer electrolyte membrane fuel cells. The catalysts can be considered a practical solution that addresses various issues in these processes, such as catalyst separability, reusability, and performance in limited operating conditions. The use of ultrasound-assisted synthesis provided an efficient alternative synthesis approach to obtain metal-oxide composites of silica and graphene oxide under mild conditions. The use of these catalyst composites offers the potential of industrial-scale use given their high activity in addition to thermal and chemical stability. Another practical and effective approach is the use of a magnetically recoverable catalyst, γ-Fe_2O_3. The superparamagnetic nanocatalyst, offering comparable performance and stability as the other solid catalysts, has further advantages of efficient separation and relatively simplicity of synthesis approach. For fuel cell development, we developed a non-toxic and easy chemical method to fabricate Pt and dopamine-induced carbon@silica composite without complex apparatus. The silica nanoparticles as support material acted as a self-humidifying material, while the dopamine coating played the role of a self-reducing agent and adhesive layer for anchoring the Pt nanoparticles. Additionally, the N-doped carbon from carbonization of dopamine coating enhanced electrical conductivity of the composite. This resulted to overall superior performance of the composite as anode catalyst even under low humidity conditions. The fabrication methods and alternative catalysis strategies presented will be valuable for the commercialization of green catalysts for PET monomer recycling and energy conversion systems.

Acknowledgements

This work was supported by the R&D Center for Valuable Recycling (Global-Top Environmental Technology Development Program) funded by the Ministry of Environment (Project No.: GT-11-C-01-250-0) and by Basic Science Research Program through the National Research Foundation of Korea (NRF) funded by the Ministry of Science, ICT & Future Planning (No. 2014R1A5A1009799).

Author details

Arvin Sangalang, Seunghwan Seok and Do Hyun Kim*

*Address all correspondence to: DoHyun.Kim@kaist.ac.kr

Department of Chemical and Biomolecular Engineering, KAIST, Daejeon, Republic of Korea

References

[1] Smithers P. Demand for PET packaging material in 2019 [Internet]. April 2014. Available from: http://www.smitherspira.com/news/2014/april/demand-for-pet-packaging-material-in-2019 [Accessed: 2015-10-03]

[2] Scheirs J, Long TE, editors. Modern Polyesters: Chemistry and Technology of Polyesters and Copolyesters. England: John Wiley & Sons Ltd; 2003. 750 p.

[3] Welle F. Twenty years of PET bottle to bottle recycling. Resources Conserv. Recycling. 2011;55(11):865–875. DOI: 10.1016/j.resconrec.2011.04.009

[4] Shen L, Worrell E, Patel MK. Open-loop recycling: a LCA case study of PET bottle-to-fibre recycling. Resources Conserv. Recycling. 2010;55(1):34–52. DOI: 10.1016/j.resconrec.2010.06.014

[5] Bartolome L, Imran M, Cho BG, Al-Masry WA, Kim DH. Recent developments in the chemical recycling of PET. In: Achilias D, editor. Material Recycling — Trends and Perspectives. INTECH; 2012. p. 65–84. DOI: 10.5772/33800

[6] George N, Kurian T. Recent developments in the chemical recycling of postconsumer poly(ethylene terephthalate) melts. Ind. Eng. Chem. 2014;53(37):14185–14198. DOI: 10.1021/ie501995m

[7] Lorenzetti C, Manaresi P, Berti C, Barbiroli G. Chemical recovery of useful chemicals from polyester (PET) waste for resource conservation: a survey of state of the art. J. Polym. Environ. 2006;14(1):89–101. DOI: 10.1007/s10924-005-8711-1

[8] Choi I, Lee KG, Ahn SH, Kim DH, Kwon OJ, Kim JJ. Sonochemical synthesis of Pt-deposited SiO_2 nanocomposite and its catalytic application for polymer electrolyte membrane fuel cell under low-humidity conditions. Catal. Commun. 2012;21:86–90. DOI: 10.1016/j.catcom.2012.02.005

[9] Seok S, Choi I, Lee KG, Choi BG, Park KJ, Park JY, Kwon OJ, Lee SJ, Kim DH. Dopamine-induced Pt and N-doped carbon@silica hybrids as a high-performance anode catalyst of polymer electrolyte membrane fuel cell. RSC Adv. 2014;4:42582–42584. DOI: 10.1039/C4RA06819J

[10] Imran M, Lee KG, Imtiaz Q, Kim BG, Han M, Kim DH. Metal oxide-doped silica nanoparticles for the catalytic glycolysis of polyethylene terephthalate. J. Nanosci. Nanotechnol. 2011;11(1):824–828. DOI: 10.1166/jnn.2011.3201

[11] Park G, Bartolome L, Lee KG, Lee SJ, Kim DH, Park TJ. One-step sonochemical synthesis of a graphene oxide-manganese oxide nanocomposite for catalytic glycolysis of poly(ethylene terephthalate). Nanoscale. 2012;4(13):3879–3885. DOI: 10.1039/C2NR30168G

[12] Bartolome L, Imran M, Lee KG, Sangalang A, Ahn JK, Kim DH. Superparamagnetic γ-Fe_2O_3 nanoparticles as easily recoverable catalyst for the chemical recycling of PET. Green Chem. 2014;16(1):279–286. DOI: 10.1039/C3GC41834K

[13] Imran M, Kim DH, Al-Masry WA, Mahmood A, Hassan A, Haider S, Ramay SM. Manganese-, cobalt-, and zinc-based mixed-oxide spinels as novel catalysts for the chemical recycling of poly(ethylene terephthalate) via glycolysis. Polym. Degrad. Stabil. 2013;98(4):904–915. DOI: 10.1016/j.polymdegradstab.2013.01.007

[14] Min BK, Santra AK, Goodman DW. Understanding silica-supported metal catalysts: Pd/silica as a case study. Catal. Today. 2003;85(2):113–124. DOI: 10.1016/S0920-5861(03)00380-8

[15] Juszczyk W, Karpiński Z. Characterization of supported palladium catalysts; II. $PdSiO_2$. J. Catal. 1989;117(2):519–532. DOI: 10.1016/0021-9517(89)90361-8

[16] Xie Z, Liu Z, Wang Y, Yang Q, Xu L, Ding W. An overview of recent development in composite catalysts from porous materials for various reactions and processes. Int. J. Mol. Sci. 2010;11(5):2152–2187. DOI: 10.3390/ijms11052152

[17] Stankovich S, Dikin DA, Dommett GHB, Kohlhaas KM, Zimney EJ, Stach EA, Piner RD, Nguyen ST, Ruoff RS. Graphene-based composite materials. Nature. 2006;442(7100):282–286. DOI: 10.1038/nature04969

[18] Stankovich S, Piner RD, Nguyen ST, Ruoff RS. Synthesis and exfoliation of isocyanate-treated graphene oxide nanoplatelets. Carbon. 2006;44(15):3342–3347. DOI: 10.1016/j.carbon.2006.06.004

[19] Muszynski R, Seger B, Kamat PV. Decorating graphene sheets with gold nanoparticles. J. Phys. Chem. C. 2008;112(14):5263–5266. DOI: 10.1021/jp800977b

[20] Bai H, Li C, Wang X, Shi G. A pH-sensitive graphene oxide composite hydrogel. Chem. Commun. 2010;46(14):2376–2378. DOI: 10.1039/C000051E

[21] Suslick KS, Price GJ. Applications of ultrasound to materials chemistry. Annu. Rev. Mater. Sci. 1999;29(1):295–326. DOI: 10.1146/annurev.matsci.29.1.295

[22] Okitsu K, Bandow H, Maeda Y. Sonochemical preparation of ultrafine palladium particles. Chem. Mater. 1996;8(2):315–317. DOI: 10.1021/cm950285s

[23] Park G, Lee KG, Lee SJ, Park TJ, Wi R, Kim DH. Synthesis of graphene-gold nanocomposites via sonochemical reduction. J. Nanosci. Nanotechnol. 2011;11(7):6095–6101. DOI: 10.1166/jnn.2011.4446

[24] Lee KG, Wi R, Park TJ, Yoon SH, Lee J, Lee SJ, Kim DH. Synthesis and characterization of gold-deposited red, green and blue fluorescent silica nanoparticles for biosensor application. Chem. Commun. 2010;46(34):6374–6376. DOI: 10.1039/C0CC00762E

[25] Doktycz SJ, Suslick KS. Interparticle collisions driven by ultrasound. Science. 1990;247:1067–1069. DOI: 10.1126/science.2309118

[26] Suslick KS. Sonochemistry. Science. 1990;247:1439–1445. DOI: 10.1126/science.247.4949.1439

[27] Wang H, Cui LF, Yang Y, Casalongue HS, Robinson JT, Liang Y, Cui Y, Dai H. Mn_3O_4-graphene hybrid as a high-capacity anode material for lithium ion batteries. J. Am. Chem. Soc. 2010;132(40):13978–13980. DOI: 10.1021/ja105296a

[28] Yan J, Fan Z, Wei T, Qian W, Zhang M, Wei F. Fast and reversible surface redox reaction of graphene-MnO_2 composites as supercapacitor electrodes. Carbon. 2010;48(13):3825–3833. DOI: 10.1016/j.carbon.2010.06.047

[29] Chen S, Zhu J, Wang X. From graphene to metal oxide nanolamellas: a phenomenon of morphology transmission. ACS Nano. 2010;4(10):6212–6218. DOI: 10.1021/nn101857y

[30] Yin J, Gao F, Wu Y, Wang J, Lu Q. Synthesis of Mn_3O_4 octahedrons and other manganese-based nanostructures through a simple and green route. CrystEngComm. 2010;12(11):3401–3403. DOI: 10.1039/C003551N

[31] Liu B, Thomas PS, Williams RP, Donne SW. Thermal characterisation of chemically reduced electrolytic manganese dioxide. J. Thermal Anal. Calorim. 2005;80(3):625–629. DOI: 10.1007/s10973-005-0704-8

[32] Huang H, Wang X. Graphene nanoplate-MnO_2 composites for supercapacitors: a controllable oxidation approach. Nanoscale. 2011;3(8):3185–3191. DOI: 10.1039/C1NR10229J

[33] Qiu LG, Li ZQ, Wu Y, Wang W, Xu T, Jiang X. Facile synthesis of nanocrystals of a microporous metal-organic framework by an ultrasonic method and selective sensing of organoamines. Chem. Commun. 2008;31:3642–3644. DOI: 10.1039/B804126A

[34] Offeman RE, Hummers WS. Preparation of graphitic oxide. J. Am. Chem. Soc. 1958;80:1339.

[35] López-Fonseca R, Duque-Ingunza I, De Rivas B, Arnaiz S, Gutiérrez-Ortiz JI. Chemical recycling of post-consumer PET wastes by glycolysis in the presence of metal salts. Polym. Degrad. Stabil. 2010;95(6):1022–1028. DOI: 10.1016/j.polymdegradstab.2010.03.007

[36] Wang H, Li Z, Liu Y, Zhang X, Zhang S. Degradation of poly (ethylene terephthalate) using ionic liquids. Green Chem. 2009;11(10):1568–1575. DOI: 10.1039/B906831G

[37] Wang H, Liu Y, Li Z, Zhang X, Zhang S, Zhang Y. Glycolysis of poly (ethylene terephthalate) catalyzed by ionic liquids. Eur. Polym. J. 2009;45(5):1535–1544. DOI: 10.1016/j.eurpolymj.2009.01.025

[38] Wang H, Yan R, Li Z, Zhang X, Zhang S. Fe-containing magnetic ionic liquid as an effective catalyst for the glycolysis of poly (ethylene terephthalate). Catal. Commun. 2010;11(8):763–767. DOI: 10.1016/j.catcom.2010.02.011

[39] Yue QF, Wang CX, Zhang LN, Ni Y, Jin YX. Glycolysis of poly (ethylene terephthalate) (PET) using basic ionic liquids as catalysts. Polym. Degrad. Stabil. 2011;96(4):399–403. DOI: 10.1016/j.polymdegradstab.2010.12.020

[40] Wi R, Imran M, Lee KG, Yoon SH, Cho BG, Kim DH. Effect of support size on the catalytic activity of metal-oxide-doped silica particles in the glycolysis of polyethylene terephthalate. J. Nanosci. Nanotechnol. 2011;11(7):6544–6549. DOI: 10.1166/jnn.2011.4393

[41] Wang Q, Yao X, Tang S, Lu X, Zhang X, Zhang S. Urea as an efficient and reusable catalyst for the glycolysis of poly (ethylene terephthalate) wastes and the role of hydrogen bond in this process. Green Chem. 2012;14(9):2559–2566. DOI: 10.1039/C2GC35696A

[42] Koukabi N, Kolvari E, Khazaei A, Zolfigol MA, Shirmardi-Shaghasemi B, and Khavasi HR. Hantzsch reaction on free nano-Fe_2O_3 catalyst: excellent reactivity combined with facile catalyst recovery and recyclability. Chem. Commun. 2011;47(32):9230–9232. DOI: 10.1039/C1CC12693H

[43] Garade AC, Bharadwaj M, Bhagwat SV, Athawale AA, Rode CV. An efficient γ-Fe_2O_3 catalyst for liquid phase air oxidation of p-hydroxybenzyl alcohol under mild conditions. Catal. Commun. 2009;10(5):485–489. DOI: 10.1016/j.catcom.2008.10.044

[44] Picasso G, Quintilla A, Pina MP, Herguido J. Total combustion of methyl-ethyl ketone over Fe_2O_3 based catalytic membrane reactors. Appl. Catal. B: Environ. 2003;46(1):133–143. DOI: 10.1016/S0926-3373(03)00219-4

[45] Kang YS, Risbud S, Rabolt JF, Stroeve P. Synthesis and characterization of nanometer-size Fe_3O_4 and γ-Fe_2O_3 particles. Chem. Mater. 1996;8(9):2209–2211. DOI: 10.1021/cm960157j

[46] Jeong, JR, Lee SJ, Kim JD, Shin SC. Magnetic properties of γ-Fe_2O_3 nanoparticles made by coprecipitation method. Phys. Stat. Solidi B Basic Res. 2004;241(7):1593–1596. DOI: 10.1002/pssb.200304549

[47] Henry CR. Catalysis by nanoparticles. In: Heiz U, Landman U, editors. Nanocatalysis. Heidelberg: Springer Berlin/Heidelberg; 2007. p. 245–268. DOI: 10.1007/978-3-540-32646-5

[48] Ghose J, Murthy K. Activity of Cu^{2+} ions on the tetrahedral and octahedral sites of spinel oxide catalysts for CO oxidation. J. Catal. 1996;162(2):359–360. DOI: 10.1006/jcat.1996.0293

[49] Guilhaume N, Primet M. Catalytic combustion of methane: copper oxide supported on high-specific-area spinels synthesized by a sol-gel process. J. Chem. Soc. Faraday Trans. 1994;90(11):1541–1545. DOI: 10.1039/FT9949001541

[50] Słoczyński J, Ziółkowski J, Grzybowska B, Grabowski R, Jachewicz D, Wcisło K, Gengembre L. Oxidative dehydrogenation of propane on $Ni_xMg_{(1-x)}Al_2O_4$ and $NiCr_2O_4$ spinels. J. Catal. 1999;187(2):410–418. DOI: 10.1006/jcat.1999.2626

[51] Niu X, Du W, Du W. Preparation and gas sensing properties of ZnM2O4 (M=Fe, Co, Cr). Sens. Actuators B: Chem. 2004;99(2):405–409. DOI: 10.1016/j.snb.2003.12.007

[52] Ferraris G, Fierro G, Jacono ML, Inversi M, Dragone R. A study of the catalytic activity of cobalt-zinc manganites for the reduction of NO by hydrocarbons. Appl. Catal. B: Environ. 2002;36(4):251–260. DOI: 10.1016/S0926-3373(01)00289-2

[53] Mokhtar M, Basahel SN, Al-Angary YO. Nanosized spinel oxide catalysts for CO-oxidation prepared via CoMnMgAl quaternary hydrotalcite route. J. Alloys Compounds. 2010;493(1):376–384. DOI: 10.1016/j.jallcom.2009.12.106

[54] Zhang L, Wang L, Holt CMB, Zahiri B, Li Z, Malek K, Navessin T, Eikerling MH, Mitlin D. Highly corrosion resistant platinum-niobium oxide-carbon nanotube electrodes for the oxygen reduction in PEM fuel cells. Energy Environ. Sci. 2012;5(3):6156–6172. DOI: 10.1039/C2EE02689A

[55] Long D, Li W, Qiao W, Miyawaki J, Yoon SH, Mochida I, Ling L. Partially unzipped carbon nanotubes as a superior catalyst support for PEM fuel cells. Chem. Commun. 2011;47(33):9429–9431. DOI: 10.1039/C1CC13488D

[56] Hong D, You I, Lee H, Lee S-G, Choi IS, Kang SM. Polydopamine circle-patterns on a superhydrophobic AAO surface: water-capturing property. Bull. Kor. Chem. Soc. 2013;34(10):3141–3142. DOI: 10.5012/bkcs.2013.34.10.3141

[57] Ryu S, Chou JB, Lee K, Lee D, Hong SH, Zhao R, Lee H, Kim SG. Direct insulation-to-conduction transformation of adhesive catecholamine for simultaneous increases of electrical conductivity and mechanical strength of CNT fibers. Adv. Mater. 2015;27(21):3250–3255. DOI: 10.1002/adma.201500914

[58] Lee M, Rho J, Lee D-E, Hong S, Choi SJ, Messersmith PB, Lee H. Water detoxification by a substrate-bound catecholamine adsorbent. ChemPlusChem. 2012;77(11):987–990. DOI: 10.1002/cplu.201200209

[59] Lee H, Dellatore SM, Miller WM, Messersmith PB. Mussel-inspired surface chemistry for multifunctional coatings. Science. 2007;318(5849):426–430. DOI: 10.1126/science.1147241

[60] Hong S, Lee JS, Ryu J, Lee SH, Lee DY, Kim D-P, Park CB, Lee H. Bio-inspired strategy for on-surface synthesis of silver nanoparticles for metal/organic hybrid nanomaterials and LDI-MS substrates. Nanotechnology. 2011;22(49):494020. DOI: 10.1088/0957-4484/22/49/494020

[61] Kang SM, Park S, Kim D, Ruoff RS, Lee H. Simultaneous reduction and surface functionalization of graphene oxide by mussel-inspired chemistry. Adv. Funct. Mater. 2011;21(1):108–112. DOI: 10.1002/adfm.201001692

3

Ag-Cu Nanoalloys as Oxygen Reduction Electrocatalysts in Alkaline Media for Advanced Energy Conversion and Storage

Adnan Qaseem, Fuyi Chen and Nan Zhang

Abstract

Silver-based nanoalloys owing to their cost, performance and stability are an attractive electrocatalyst system for oxygen reduction reaction (ORR) in the alkaline fuel cells and metal air batteries. A systematic computational and experimental approach has been adopted to investigate their performance for ORR in alkaline environment. Firstly, genetic algorithm (GA) based calculations have been performed to look for the stable compositions and structures of these nanoalloys. Later, density functional theory (DFT) is employed to simulate the working of those stable nanoalloys in actual working conditions. Finally, the most promising nanoalloys have been synthesized by physical and chemical routes to confirm their performance in real-life conditions. It has been found that the alloying of silver with copper enhances the catalytic performance of Ag nanoparticles. The enhancement in performance can be related to the modification of the electronic and physical structure of Ag due to copper doping. The superior performance of Ag–Cu nanocatalysts in alkaline fuel cells and metal air batteries along with their modest cost and long-term stability make them a promising candidate for deployment as a catalyst for ORR in alkaline media.

Keywords: Electrocatalyst, silver nanoalloys, metal air batteries, oxygen reduction reaction, rechargeable zinc air battery

1. Introduction

Oxygen reduction reaction (ORR) plays a vital role in the working of fuel cells and metal-air batteries. Both of these technologies utilize oxygen from the air to generate electrical energy. The ORR mechanism in acidic environment is accompanied with the formation of water.

$$O_2 + 4H^+ + 4e^- \rightarrow 2H_2O \tag{1}$$

For alkaline medium, hydroxyl formation takes place as

$$O_2 + 2H_2O + 4e^- \rightarrow 4OH^- \tag{2}$$

The current Li-ion battery technology is unable to offer the solutions for the long-range electric vehicles and energy storage grids. It is also postulated that the Li-ion battery system may reach its theoretical limit which will still be short of the demand for the long-range EVs. Metal air batteries such as Li-air (11,000 Wh/kg) and Zn–air systems (1,084 Wh/kg) offer much higher energy capacities [1–3]. A great deal of contemporary research is directed toward the realization of these high energy systems. Apart from being dense in energy, metal air batteries and fuel cells are green source of energy. The product of their working is free from toxic and harmful waste gases which damage the environment. These systems provide a valuable opportunity to cut the greenhouse gas emissions on a tremendous scale [4,5].

Electrocatalyst plays a crucial role in the working of metal air batteries and fuel cells. The ORR and OER mechanism are strongly related to the functionality of the electrocatalyst. Pt and Pt-based alloy catalysts are widely used for ORR, but prohibitive cost and catalytic poisoning are major drawbacks associated with Pt. Therefore, it is highly desirable to explore novel Pt-free cost-effective catalysts. Different non-platinum catalysts have been considered as a replacement of Pt. Silver being about 50 times cheaper than platinum is an attractive choice for catalyst in alkaline fuel cells. The pourbaix diagram reveals the superior stability of Ag over platinum in alkaline environment [6,7].

The ORR is accompanied with the formation of various adsorbed intermediates such as O, OH and OOH. Norkosov et al. evaluated the effect of potential on the free energy of various intermediates on Pt (111) by DFT calculations [8]. At high potential, the adsorbed oxygen was found to be stable. The ORR reaction was found to proceed only by lowering the potential, hence giving rise to overpotential. Bond energies of oxygen and hydroxyl on different metals were also calculated by DFT. The rate of ORR is limited by the removal of O, OH for metals which bind oxygen strongly where as in case of metals with poor oxygen binding, the rate is limited by the weak bonding of the adsorbed species. A volcano plot as shown in Figure 1 is obtained as a result of DFT calculations performed on various systems. Although platinum sits near the top of volcano plot developed by DFT calculations, alloying of metals can result in new materials with adsorption energies for the intermediates that are different from the constituent pure metals. Therefore, new generation of superior electrocatalysts can be developed by alloying of metals to yield optimum binding of adsorbates onto alloy surface.

Tremendous amount of research has been performed in the past decade to enhance the activity of ORR catalyst. A myriad of bimetallic and multi metallic alloy compositions have been developed in a variety of structures such as core shell, skin alloys, thin films, ordered intermetallics and solid solutions [9–12]. Skin alloys have been widely popular because of their

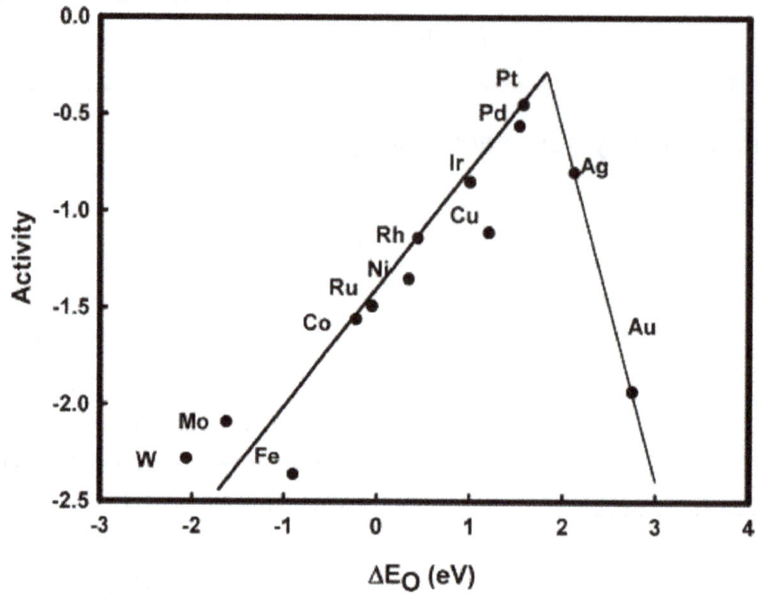

Figure 1. Variation of activity with oxygen adsorption energy for metals. Reproduced with permission from ref. 8. Copyright 2004, American Chemical Society.

superior activity as compared to the bulk alloys. This is accompanied with the reduction of cost of precious metal cost such as Pt in skin alloys by using a monolayer thick platinum on top of a non-precious metal/alloy core. As the skin is of Pt, therefore, stability in the corrosive chemical environment is still maintained in these alloys. The enhancement in the activity of these skin alloys is attributed to the geometric and ligand effect of the subsurface atoms on the skin. Recently, these structural form alloys have been reported to improve the ORR performance of the silver-based alloys. Some possible modifications of the surface electronic structure by the ligand mechanism are revealed in Figure 2.

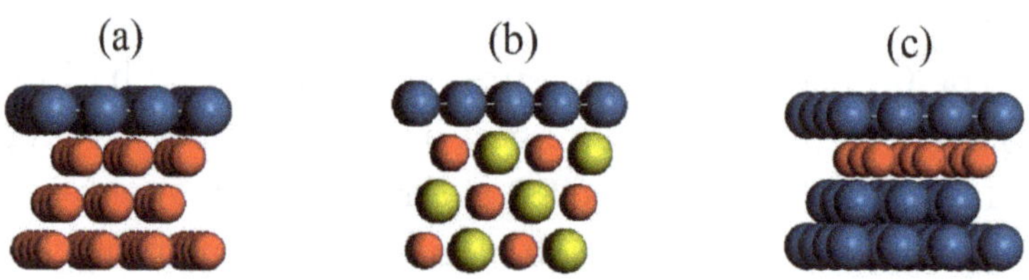

Figure 2. Modification of the surface geometric and electronic structure by subsurface configuration (a) skin on pure metal, (b) skin on alloy/compound, (c) subsurface ligand.

The rate-limiting step for ORR in case of Pt catalyst is the removal of adsorbed OH species. It is well established that the decrease in binding energy by 0.1–0.2 eV will optimize the performance of the Pt-based catalyst [13,14]. By using the Pt skin on a transition metal core, this reduction in binding energy can be achieved. This is attributed to the modification of the electronic structure of the surface atoms by the core. In case of transition metals such as Ti, Co, Ni, and Ru, charge transfers from the core to the dband of the surface atoms of platinum [15]. This occupancy of dband lowers the dband centre of the platinum atoms which in turns decreases the adsorption of the OH on the Pt skin. As a result of this mechanism, the activity of these alloy systems has been reported to be much superior than that of commercial Pt/C catalyst. The skin of platinum atoms can be either in compression or tension. This is because of the subsurface structure effect on the skin alloy. The compression of the Pt skin increases the overlap of the d orbitals which consequently increases the dband width. The result of this perturbation of the structure is the lowering of the dband center of the surface atoms [16, 17]. According to the dband theory, the lower the dband center of the surface, the lower its reactivity and vice versa. By the combination of geometric and ligand effects, the dband center of the surface can be tuned to the desired value so as to achieve the optimum adsorption of adsorbates such as O, OH and OOH on the surface of the metal. This is in accordance with the Sabatier principle which implies that for the catalytically induced chemical reactions catalyst for the reaction should have neither strong nor weak adsorption for the reaction species [18].

From Figure 1, it is evident that weak binding of oxygen onto silver is the cause of its shift from the volcano peak and weak ORR activity.

The oxygen reduction reaction for alkaline can occur by a direct four electron transfer method as in Eq. 2 or by indirect 2e transfers [19–21] as:

$$O_2 + H_2O + 2e^- \rightarrow HO_2^- + OH^- \tag{3}$$

$$H_2O + 2e^- + HO_2^- \rightarrow 3OH^- \tag{4}$$

In the indirect mechanism, second step, i.e., Eq. (4), is the rate-determining step. The oxygen reduction reaction by indirect mechanism seriously limits the performance of the cell. This is because of the fact that if the reaction 4 does not occur by direct mechanism, then the total electrons transferred during ORR reduce to two only. As the result, this lowers the total output voltage and energy density. Also the peroxide formed in Eq. (3) can undergo catalyst-induced conversion to O_2 and OH^- by Eq. (5).

$$HO_2^- \rightarrow O_2 + 2OH^- \tag{5}$$

Reaction 5 limits the catalytic activity of the catalyst and hence it is desirable that the ORR proceeds by direct four electron transfer mechanism. The effect of pH on the ORR mechanism has been studied by Blizanac et al. [22]. In case of ORR on Ag(111) in alkaline medium, four electron transfer was found to be the dominant mechanism at all overpotentials, but in case of low pH, i.e., in acidic solutions, 2 electron pathway was favored at low overpotentials. It was observed that the ORR on Ag (111) by 4 electron pathway could take place only at high overpotentials in the low pH electrolytes.

Owing to the stability and efficient ORR mechanism of silver-based catalysts in alkaline media, various research groups have focused on the synthesis of Ag-based catalysts. Holewinski et al. have investigated the effect of alloying on silver for the ORR performance [23]. Density functional theory (DFT) based calculations were performed on Ag alloy slabs with an Ag skin on top to investigate the effect of alloy core on the ORR performance. The 4 electrons transfer has been proposed by the following mechanism:

$$Ag + O_2 + e^{-1} \overset{H-OH}{\rightarrow} Ag\cdots OOH + OH^{-1} \tag{6}$$

$$Ag\cdots OOH + e^{-1} \overset{H-OH}{\rightarrow} Ag\cdots O + OH^{-1} \tag{7}$$

$$Ag\cdots O + e^{-1} \overset{H-OH}{\rightarrow} Ag\cdots OH + OH^{-1} \tag{8}$$

$$Ag\cdots OH + e^{-1} \overset{H-OH}{\rightarrow} Ag + OH^{-1} \tag{9}$$

A reaction coordinate diagram was developed for different alloys of silver on the basis of DFT calculation for the above mentioned reactions. From the theoretical calculations it was concluded that in case of silver-based catalysts, the rate-determining step for ORR is the initial adsorption of oxygen on the catalyst surface to form OOH adsorbate. In order to confirm the theoretical findings, the group synthesized Ag–Co alloys. After chemical etching of the alloy an Ag-skin with an Ag-Co core was developed. The resulting silver alloy showed phenomenal improvement in the ORR as compared to the unalloyed catalyst. Functionality tests on the Ag-skin /Ag-Co alloy core yielded an improvement in the area specific activity by a factor of 6 @ 0.8 V_{RHE}. This significant enhancement was attributed to the perturbation of the electronic structure of the surface silver atoms which resulted in the lowering of the activation energy barriers for the ORR.

Composites of silver/graphene oxide and silver/graphene oxide/carbon were developed to investigate their ORR performance in alkaline environment [24]. The composites were facially synthesized by the reduction of $AgNO_3$ with graphene oxide with or without the presence of Vulcan XC-72 carbon black. The average particle size of Ag/CO/C composites (ca d = 12.9 nm) was found to be almost twice of Ag/CO composite (ca d = 6.9). The composites were electro-chemically characterized which revealed the superior performance of Ag/CO/C for ORR as compared to Ag/CO composite. Rotating disc electrode (RDE) analysis revealed that the onset potential and the half wave potential shift positively for Ag/CO/C as compared to Ag/CO composite. This enhancement in ORR performance of Ag/CO/C composites as compared to Ag/CO composites was attributed to the 3D composite support which not only improves the electrical conductivity but also facilitated the mass transport in the catalyst layer. A similar beneficial effect of catalyst support was observed in the case of $Ag/Mn_3O_4/C$ catalysts [25]. The catalyst performance for ORR in alkaline media was found to be superior to the simple Ag/C catalyst. This improvement was ascribed to the Mn_3O_4 support which perturbed the electronic structure of the silver particles. Charge was transferred to Mn_3O_4 support from Ag which was manifested by the lowering of the binding energy of the Ag 3d electrons in XPS measurements. This was accompanied with the rise of the d-band center of Ag in the Ag/Mn_3O_4/C catalyst as compared to the Ag/C catalyst. This was attributed to the tensile strain which results in less overlap of d orbitals and a corresponding rise of the dband center which in turn favors the kinetics of ORR by O—O bond breakage. Figure 3 reveals the oxygen reduction polarization curves for Ag/C, Ag/Mn_3O_4/C and Pt/C at 1600 rpm in O_2-saturated 0.1 M KOH. The onset potential for Ag/Mn_3O_4/C was 0.92 V (vs. RHE) which is close to the onset potential for Pt/C catalyst. The limiting current incase of Ag/Mn_3O_4/C (c.a. I_l = 5.5 mA/cm^2) was also found to be very close to the limiting current of Pt/C (c.a. I_l = 5.62 mA/cm^2). Moreover, the electron transfer mechanism was found to proceed by four electrons from the Koutecky–Levich plots.

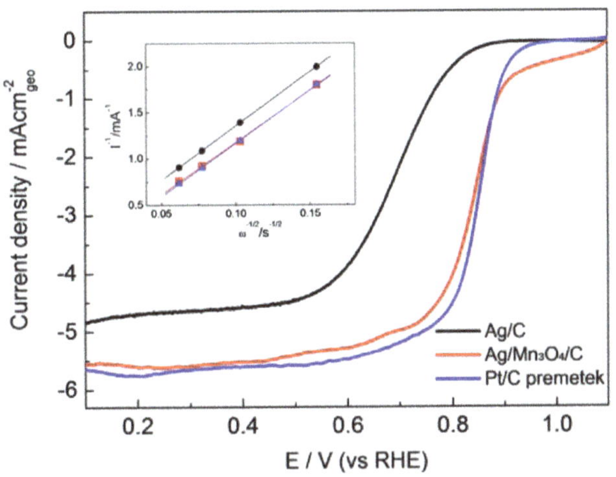

Figure 3. Oxygen reduction polarization curves for Ag/C, Ag/Mn3O4/C and Pt/C premetek at 1,600 rpm in O_2-saturated 0.1 M KOH at 10 mV s^{-1}, and (inset) Koutecky–Levich plots for ORR in the presence of Ag/C, Ag/Mn3O4/C and Pt/C premetek at 0.32 V vs. RHE. Reproduced with permission from ref. 25

The effect of morphology of the silver catalyst particles was studied by Ohyama et al. [26]. Silver particles with three distinct morphologies were investigated for their ORR performance in alkaline medium. Among the spherical, worm-like and the faceted particles, the maximum specific activity was observed in the worm-like particles with subsurface oxygen at surface defects. This was followed by the multifaceted particles with surface $AgCO_3$ layer and defects. The smooth and spherical particles had the least specific activity of the three types. This increase in activity is justified by the large number of defects on these irregular-shaped particles which increase the reactivity of the silver catalyst toward the O_2 during ORR.

2. Computational work

The computational work involves initially the search for the structure with global minimum energy by genetic algorithm. The structure with minimum energy is further optimized by density functional theory (DFT) calculations. The most optimum structures are employed for simulations of the ORR reactions by density functional theory calculations.

2.1. Structural optimization

The first step toward computational modeling of the nanoalloys is to search for the most stable geometry at absolute zero. This requires the modeling of the potential energy surface for the multi-element alloy. Gupta potential was used for the atomistic modeling of the nanoalloy [27, 28]. It is a semi-empirical method for the approximation of the potential energy surface. This method is based on the second moment approximation to the tight binding theory (SMATB). Total energy E for the nanoalloy A_xB_{N-x} is written as a sum of an attractive term E_j^b and a repulsive term E_j^r.

$$E = \sum_j (E_j^b + E_j^r) \tag{10}$$

where E_j^b and E_j^r are further defined as:

$$E_j^b = \sqrt{\sum_{i \neq j}^{N} \xi^2 e^{-2q\left(r_{ij}/r_0 - 1\right)}} \tag{11}$$

and

$$E_j^r = \sum_{i \neq j}^{N} A e^{-p\left(\frac{r_{ij}}{r_o} - 1\right)} \tag{12}$$

N is the number of atoms, r_{ij} is the distance between atoms i and j in the cluster and r_o is the nearest-neighbor distance. The parameters A, r_o, ξ, p and q for the pure species are fitted to several bulk experimental values, such as the cohesive energy, the lattice parameter and the elastic constants. The heteroatom interactions are fitted to the solubility energy of an impurity A into a B bulk. The Gupta potential parameters used for the Ag–Cu system are listed in Table 1 [29].

Parameters	$A(eV)$	ξ (eV)	p	q	$r_o(A^o)$
Ag–Ag	0.1031	1.1895	10.85	3.18	2.8291
Ag–Cu	0.0980	1.2274	10.70	2.805	2.72405
Cu–Cu	0.0894	1.2799	10.55	2.43	2.556

Table 1. Gupta potential parameters for Ag–Cu system

From the modeling of the potential energy surface the next step forward is to search for the global minimum (GM) by optimization. This is performed by the help of Birmingham cluster genetic algorithm (GA) [30]. From the initially randomly generated cluster population, the algorithm looks for the most stable cluster structure by mutation and cross over. For each generation, parent clusters are chosen with a probability depending on their fitness and offsprings are developed from parents by a cross-over scheme which is followed by a mutation step on the offspring to bring diversity in population. The whole population is ranked by fitness and the less fit, i.e. high-energy, clusters are replaced with more stable structures. The whole process is repeated till a predefined convergence criterion is met.

The 13 atom Ag–Cu bimetallic cluster was chosen for further geometric optimization by $Dmol^3$ module available in the materials studio software package [31,32]. The 13 atom cluster is a magic size owing to special stability and relative abundance in case of Ag–Cu alloy clusters [33,34]. Spin-polarized DFT calculations are performed in real space within the framework of DFT-based semi core pseudo potentials (DSPPs) with the double numerical plus polarization (DNP) function. Grid integration is performed with a global cutoff of 5.0 °A. Self-consistent field procedures are performed with a convergence criterion of 10^{-6} Hartree on the total energy and the electron density. The Perdew et al. generalized gradient approximation (PBE/GGA) is used for the exchange–correlation functional during the geometry optimization for the Ag cluster. The ascending order of stability for the pure 13 atom Ag cluster is icosahedron (Ih) with binding energy $E_b = -18.682$ eV, decahedron (Dh) with $E_b = -18.731$ eV and cuboctahedron (COh) with $E_b = -18.958$ eV, where cuboctahedron (COh) structure was found to be the most stable configuration for the 13 atom Ag cluster. In case of the single Cu surface-doped $Ag_{12}Cu$ cluster, the increasing order of stability is also icosahedron (Ih) with $E_b = -19.18727$ eV, decahedron (Dh) with $E_b = -19.40207$ eV and cuboctahedron (COh) with $E_b = -19.55135$ eV. So for both pure Ag_{13} and $Ag_{12}Cu_{surface}$ clusters, the most stable structural form was found to be cuboctahedron (COh).

2.2. ORR on 13 atom Ag-Cu clusters

Ma et al. performed first principle calculations for the ORR process in alkaline media on the 13 atom pure Ag and Cu doped Ag clusters [35]. 13 atom Ag-Cu nanoalloy clusters have been previously identified as a potential candidate for ORR catalyst [36]. Pure Ag_{13} and $Ag_{12}Cu_{surface}$ clusters with cuboctahedron (COh) symmetry were used for these calculations. The doping of copper significantly improves the ORR process. The ORR reaction was observed to occur by the efficient four electron transfer mechanism. Pure silver is a poor catalyst for the ORR because of its weak adsorption of oxygen. Doping of silver with copper atom on the surface improves the binding of the intermediates such as O, OH and OOH on the nanoalloy cluster. This optimum binding is critical for the efficient ORR. Binding energies of different adsorbates are described in Table 2.

From Table 2 it is evident that the binding energy of every adsorbate is more negative on the $Ag_{12}Cu_{surface}$ cluster as compared to pure Ag_{13} cluster. This stronger binding facilitates the electron transfer reactions in the ORR process. A schematic of the ORR mechanism is provided in Figure 4.

ORR pathway	$^{\cdot}O_2$	$^{\cdot}O_2^-$	$^{\cdot}OOH$	$^{\cdot}OOH^-$	$^{\cdot}OH + ^{\cdot}OH$	$^{\cdot}OH$
Ag_{13} (eV)	−0.216	−0.259	−1.144	−0.980	−4.594	−2.958
$Ag_{12}Cu$ (eV)	−1.019	−1.154	−1.697	−1.304	−5.405	−3.312

Table 2. Binding energy of different adsorbates on nanoclusters

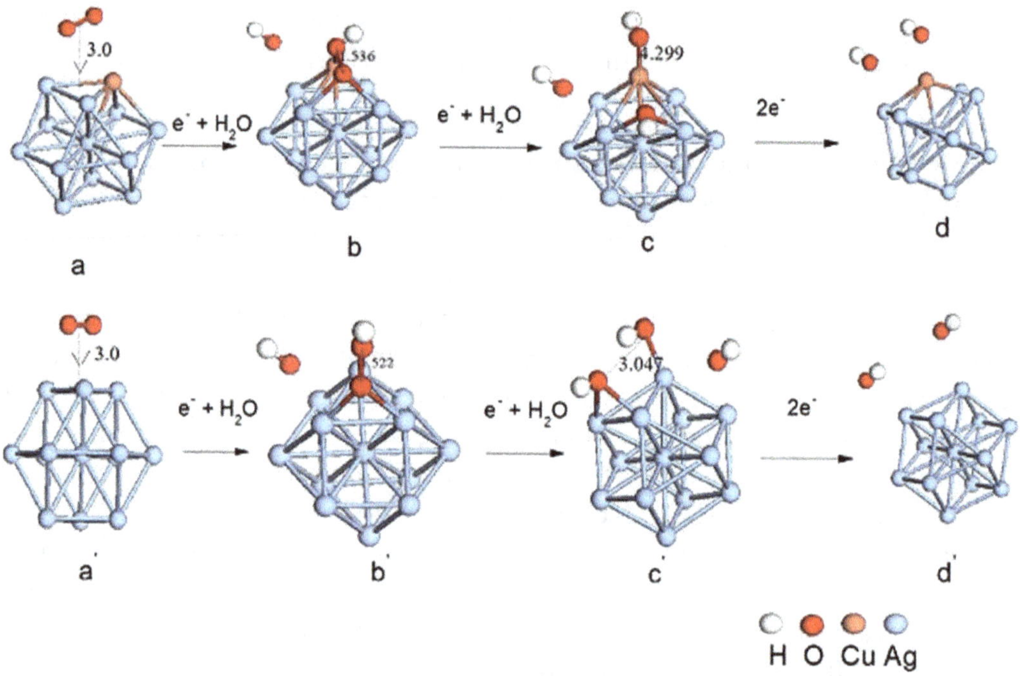

Figure 4. ORR pathway on Ag_{13}(a,b,c,d) and $Ag_{12}Cu$ (a', b', c',d'). Reprinted with permission from ref. 35.

The improvement of the ORR performance of the copper doped 13 atom nanocluster can be attributed to the modification of the dband. The dband center of the Ag_{13} cluster is at −3.078 eV. Doping with a copper atom at surface raises the dband center to −1.507eV. This can be explained by the dband theory of the reactivity of metal surfaces. According to that theory, the deeper the dband from the Fermi level, the lower is the surface reactivity. Alloying with copper raises the dband of the nanoalloy as compared to the pure metal cluster, which in turn raises the binding ability of the intermediates onto the nanoalloy cluster. The stronger binding of the intermediates is the reason behind the better ORR performance of $Ag_{12}Cu$ cluster as compared to the Ag_{13} cluster.

2.3. ORR on 38 atom Ag–Cu clusters

The 38 atom $Ag_{38-x}Cu_x$ cluster has been studied for ORR because of its relative stability by another group [37]. Truncated octahedron (TO) $Ag_{32}Cu_6$ alloy cluster was observed to perform as a better catalyst for ORR as compared to the TO $Ag_{32}Cu_6$ core-shell clusters. With the use of Gupta potential-based potential energy surface and genetic algorithm (GA) search for global minimum, polyicosahedron (PIh) $Ag_{32}Cu_6$ core-shell structure is found to be the most stable structure for $Ag_{32}Cu_6$ nanoalloy by Zhang et al.[38]. The stability of polyicosahedron (PIh) $Ag_{32}Cu_6$ core-shell structure exceeds that of truncated octahedron (TO) $Ag_{32}Cu_6$ core shell by 0.564 eV. For both pure Ag_{38} and Cu_{38} clusters truncated octahedron (TO) was the most stable geometry. Figure 5 shows that the minimum energy structure of the pure 38 atom silver took 35 iterations while more than 70 iterations were required for the $Ag_{32}Cu_6$ nanoalloy cluster.

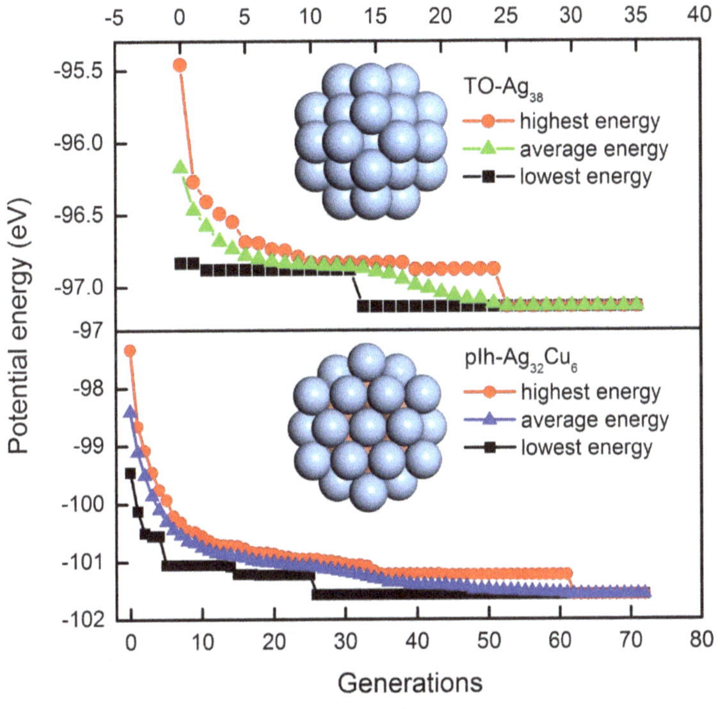

Figure 5. Generations to reach global minimum by genetic algorithm. Adapted from ref. 38.

The polyicosahedron (PIh) $Ag_{32}Cu_6$ core-shell structure was further investigated for ORR because of its stability. ORR was found to proceed more favorably by the dissociation mechanism as compared to the associative mechanism by 0.1 eV. Hence, dissociative ORR mechanism involving the scission of molecular oxygen to atomic form, i.e., $O_2 \rightarrow O + O$, was considered for computational analysis. Four non-equivalent sites were identified on the $Ag_{32}Cu_6$ core-shell structure as B1 to B4. The ORR was then followed with the bond fracture and subsequent adsorption of atomic oxygen at hollow sites marked H1, H2 and H3 as shown in Figure 6.

Of the four adsorption configurations, we notice that the adsorption energy on B4 site has a highest value of –0.149 eV, and also the maximum value of 1.209 eV for the dissociation barrier, and an energy release of 0.259 eV, dissociating to H2 and H3 sites. The B1 site, which has similar adsorption energy to B4 site, –0.146 eV, further dissociates to two H2 sites with barrier of 0.993 eV and exothermicity of 0.259 eV. The O_2 on B2 and B3 sites have bond-cleavage barriers of 0.715 and 1.134 eV and energy release of 1.088 and 0.368 eV, respectively. It is clear that the most favorable pathway for O_2 dissociation is B2 sites with a minimum value of activation energy barrier.

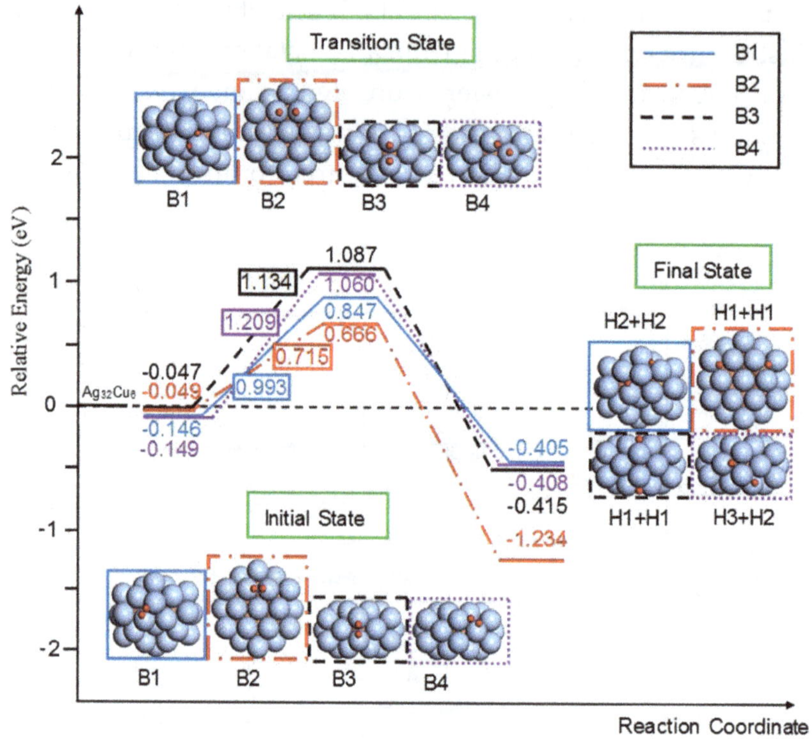

Figure 6. Reaction coordinate diagram. Adapted from ref. 38.

The interaction strength of atoms and molecules with metal surface is defined by the *d*-band center of the metal. In order to explain further that B2 site is most favorable to display a good catalytic behavior, the electronic structure of these four adsorption configurations was addressed and the position of the *d*-band center relative to the Fermi energy for these different sites was calculated as shown in Figure 7. The *d*band center of B2 site is –3.395 eV, which is closest to the Fermi level. By having the dband center closest to the Fermi level as compared to the other adsorption sites on the polyicosahedron (PIh) $Ag_{32}Cu_6$ core-shell structure, B2 site is the most conducive for ORR as it enhances the otherwise weak affinity of silver alloys for the reaction intermediates.

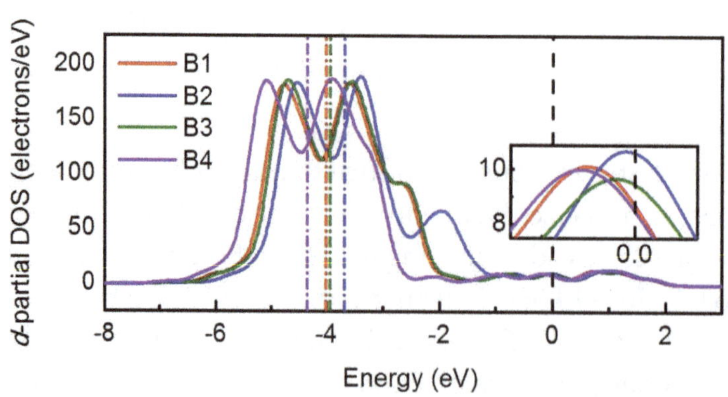

Figure 7. Partial density of states (PDOS). Adapted from ref. 38.

Moreover, frontier orbital theory describes that the states of a metal which are involved in electron transfer with the adsorbates are closest to the Fermi level [39]. Therefore, the density of states (DOS) at the Fermi level is an indicator of the chemical activity. We notice that the density of states at the Fermi energy level is maximal for B2 site. This further endorses the superior activity of the B2 site for ORR on $Ag_{32}Cu_6$ core-shell nanoalloy.

3. Synthesis of Ag–Cu nanoalloy

Bimetallic Ag–Cu nanoalloy particles were developed by both physical and chemical routes.

3.1. Pulse laser deposition

Pulse laser deposition (PLD) is widely used in the field because the approach is a feasible way to control the thickness of catalyst layer [40]. As the utilization of catalyst in the fuel cell is determined mainly by the surface area of catalyst with electrolyte, the reduction of the thickness of catalytic layer can lead to improvement of the catalyst utilization and reduction of the fuel cell cost [41]. Moreover, compared to chemical preparation techniques such as electrochemical deposition, chemical vapor deposition, reduction of salts and facile hydrothermal method, PLD method owns high repeatability and stability in process, making it to be a suitable route to obtain electrocatalyst with film state [42–44].

Ag–Cu alloyed catalyst was developed by PLD method in a vacuum chamber [45]. The target of Ag–Cu alloy with atomic ratio of 50:50 was irradiated with a nanosecond Q-switched Nd:YAG laser beam (EKSPLA, Lithuania). The wavelength was set to be 266 nm, and the pulse duration was ranging from 3 to 6 ns. The laser beam diameter was around 1 mm, with an energy density of 200 mJ/pulse. Both target and substrate (nickel foam) rotated at a speed of 5 rpm during deposition, and target was irradiated for 2 min at 10 Hz to clear away the oxide on the surface before deposition. The laser was operated at the frequency of 10 Hz. The deposition time is set as 90 min. The as-prepared product is $Ag_{50}Cu_{50}$ catalyst.

Figure 8 shows series of TEM analysis on $Ag_{50}Cu_{50}$ catalyst. According to Figure 8a, plenty of nanoparticles are distributed in a continuous film. The tiny nanoparticles with size under 5 nm dominate the film. Magnifying the blue rectangle area, the obtained HRTEM is shown in Figure 8b. It can be seen that they display two different states: few are amorphous, and the left are with crystallized state.

HAADF result shown in Figure 8c displays that contrast of the particles is brighter than the gap area between particles, demonstrating a higher atomic number Z for nanoparticles. The lower Z corresponding gap area then is attributed from Cu element. This is because Z of Cu ($Z = 29$) is smaller than Ag ($Z = 47$). Combining the amorphous state in gap area observed in Figure 8b, we can draw that $Ag_{50}Cu_{50}$ catalyst actually is Ag–Cu alloyed nanoparticles embedded in amorphous Cu film.

Electrochemical characterizations have been carried out on PLD synthesized $Ag_{50}Cu_{50}$ catalyst. Figure 9a shows RDE polarization curves of $Ag_{50}Cu_{50}$ catalyst with rotation rate 1,600 rpm in

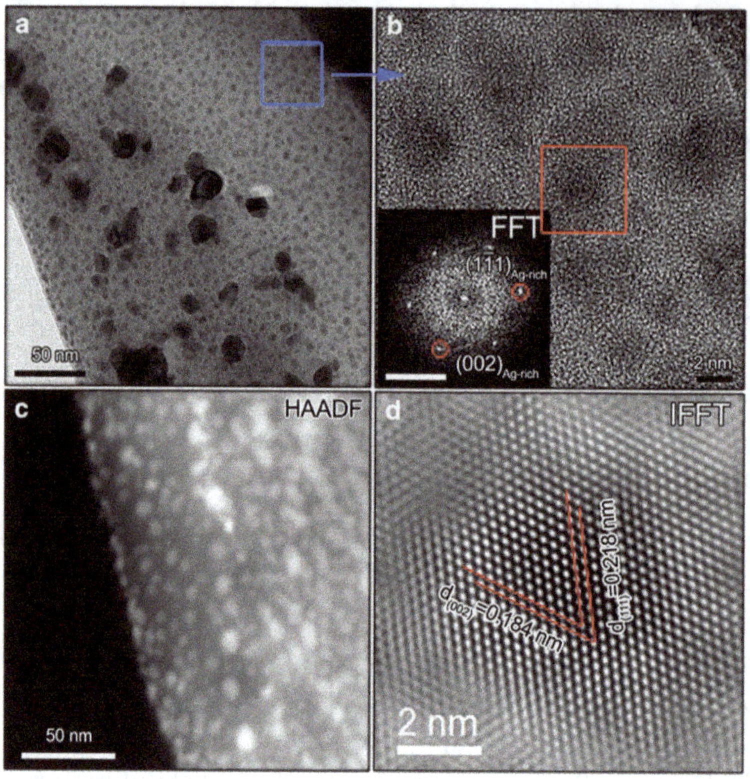

Figure 8. TEM and HAADF characterization of $Ag_{50}Cu_{50}$ catalysts. (a) bright field image, (b) HRTEM, (c) HAADF result and (d) IFFT image. Adapted from ref. 45.

N_2 and O_2 saturated 0.1 M KOH solutions. It can be seen that there is reduction in current density in O_2-saturated KOH solution, while that in N_2-saturated solution is flat. This shows that the catalyst indeed works on O_2. Figure 9b shows a set of RDE curves with rotation rates of 400, 800, 1,200 and 1,600 rpm. The Koutecky–Levich plots were then obtained from the limiting current density, as shown in Figure 9c. The plots show the inverse current density J^{-1} as a function of $\omega^{-1/2}$. From these plots the number of electrons transferred during ORR was found to be 3.76, 3.87, 3.85 and 3.97 when the potential was 0.5, 0.4, 0.3 and 0.2 V, respectively. Hence, four electrons route was found to be dominant for ORR in case of synthesized $Ag_{50}Cu_{50}$ catalyst. Finally, 9d gives a comparison of the performance of Ag, $Ag_{50}Cu_{50}$ and Pt/C(20 wt%).

The catalytic layer was used to assemble a zinc-air battery and results showed open-circuit voltage (OCV) of the cell was around 1.48 V close to the theoretical value, and the maximum power density is 67 mW cm^{-2} at 100 mA cm^{-2}. The resulting rechargeable zinc-air battery exhibits low charge–discharge voltage polarization of 1.1 V at 20 $mAcm^{-2}$ and high durability over 100 cycles in natural air.

Wu et al. deposited Ag–Cu nanoalloys on nickel foam by pulse laser deposition. Several Ag–Cu alloys with Ag/Cu atom ratios of 90:10, 50:50 and 25:75 are used as the target material with the high-purity nickel foam (99.97%) as substrate [46]. Ag–Cu nanoalloys were used as the catalyst layer of the air cathode for a single zinc-air battery, and found to exhibit good

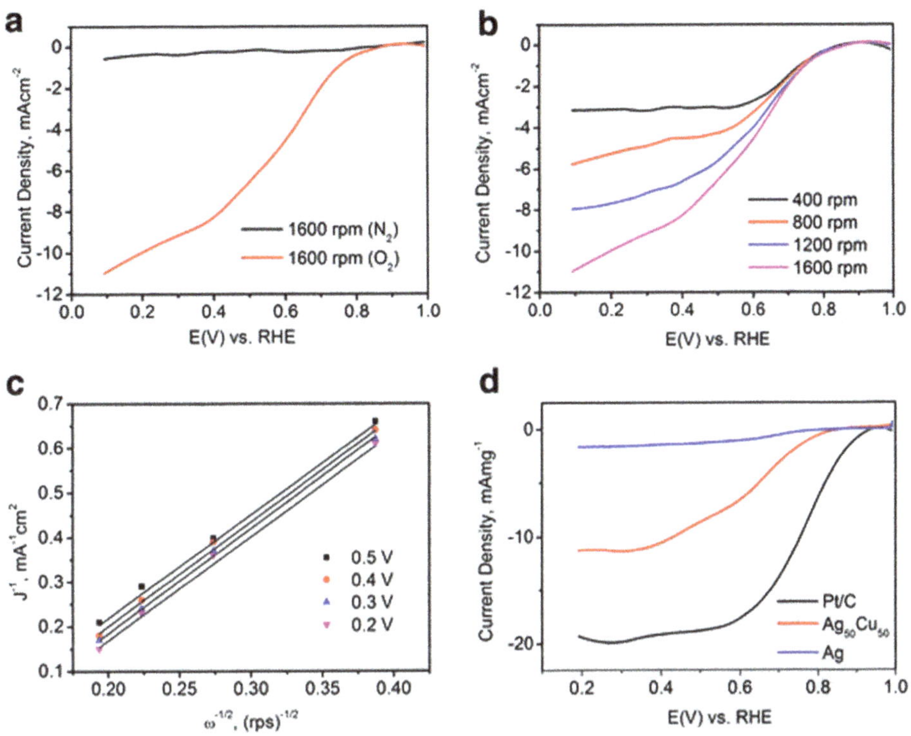

Figure 9. Electrochemical characterization of $Ag_{50}Cu_{50}$ catalyst. (a) The RDE curves of $Ag_{50}Cu_{50}$ catalyst in O_2- and N_2-saturated 0.1 M KOH solution; (b) the RDE curves at the rotation rates of 400, 800, 1,200 and 1,600 rpm; (c) the Koutecky–Levich plot of $Ag_{50}Cu_{50}$ catalyst; and (d) the ORR mass activity for Ag, $Ag_{50}Cu_{50}$ and Pt/C (20 wt%) catalysts. Adapted from ref. 45.

bifunctional catalytic performance. The effect of the Ag/Cu atom ratio on the average electron transfer numbers of the ORR was systematically investigated. This carbon-free binder-free bimetallic catalyst layer was found to possess both ORR and OER catalytic activity in the rechargeable zinc-air battery. Figure 10(a) shows representative microscopic images of the $Ag_{50}Cu_{50}$ alloys used in this work. The Ag–Cu nanoparticles were uniformly distributed in the substrate, and the electron diffraction pattern revealed the single phase and polycrystalline structure of the Ag–Cu alloy. Figure10 (b) shows a HRTEM image of more than 100 nanoparticles in the substrates. It is clear that the nanoparticles have an average size of 2.58 nm with a narrow size distribution between 1 and 5 nm. Figure 10(c,d) indicate that the nanoparticles are enriched with Ag atoms, with few doped copper atoms in them, and the films are enriched with Cu atoms.

Hence, it can be concluded that the nanocatalyst has crystalline Ag-enriched nanoparticles embedded in an amorphous Cu-enriched matrix. These copper-doped silver nanoparticles with composition $Ag_{50}Cu_{50}$ were observed to have superior catalytic performance for ORR as compared to pure silver as shown in Figure 11. The ORR was found to proceed via four electron transfer mechanism. It is for the first time the Ag-based electrocatalysts in amorphous films were created from the vapor phase under far-from-equilibrium condition by pulse laser ablation, previous works demonstrated that the face-centered cubic Ag–Cu solid solutions or completely amorphous Ag–Cu metal glasses were formed by rapid quenching from the liquid or vapor phase as corrosion-resistant non-equilibrium alloys and metastable phases.

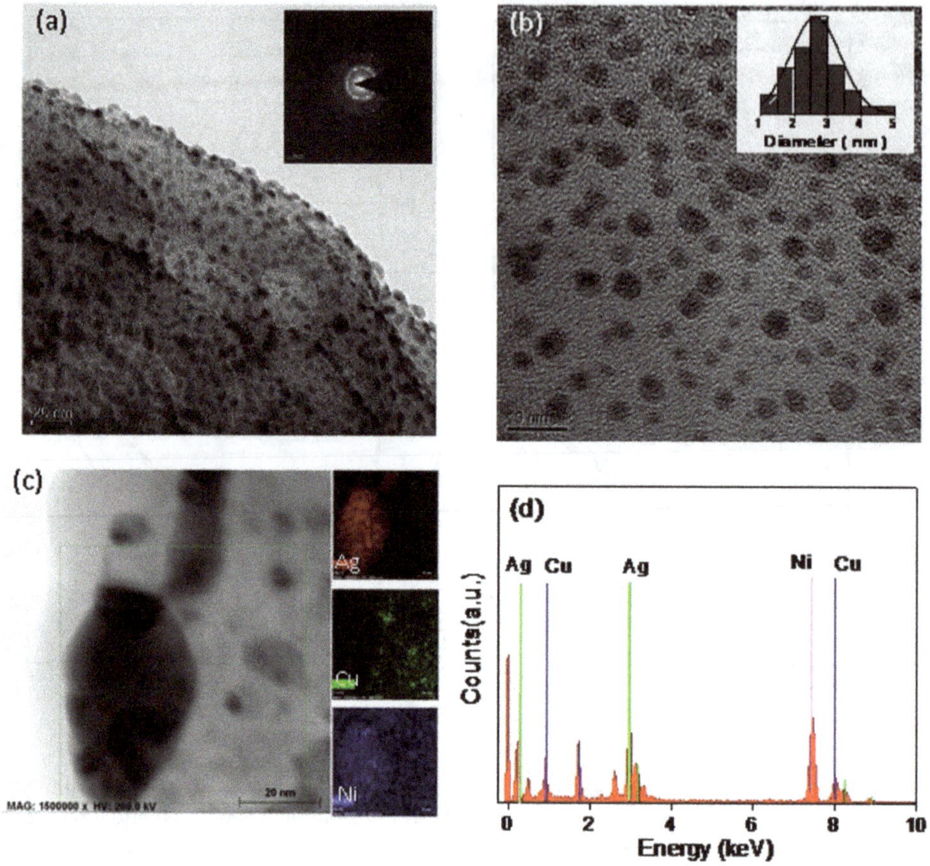

Figure 10. (a)TEM bright-field images and SAED patterns (inset) of the $Ag_{50}Cu_{50}$ film prepared by PLD. (b) HRTEM images and the particle size distribution (inlet) of Ag–Cu nanoparticles in the film. (c) TEM element mapping for Ag, Cu and Ni on a Ag–Cu nanoparticle deposited on nickel grid. (d) EDS of the $Ag_{50}Cu_{50}$ film on nickel grid. Reprinted with permission from ref. 46.

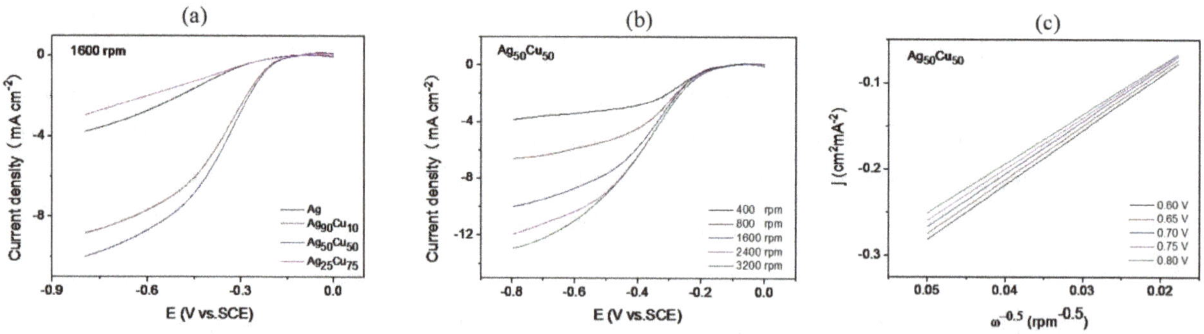

Figure 11. (a) RDE polarization curve of different compositions in O_2-saturated 0.1 M KOH solution. (b) RDE polarization curve for $Ag_{50}Cu_{50}$ same environment (c) Koutecky–Levich plots for (b). Reprinted with permission from ref. 46.

Zinc-air batteries were assembled using $Ag_{90}Cu_{10}$ and $Ag_{50}Cu_{50}$ catalysts in the air cathode, and the discharge performance is shown in Figure 12(a). The cell voltage decreases nearly with increasing current density, demonstrating that the cell performance shows a strong dependence on the resistance of the battery. For $Ag_{50}Cu_{50}$ and $Ag_{90}Cu_{10}$ catalysts, the open-circuit voltages of the single cell are about 1.42 V and 1.44 V, the peak power densities of the zinc-air batteries are 86.3 mW.cm^{-2} and 82.1 mW.cm^{-2}, and the current densities at a voltage of 1 V are 60 mA cm^{-2} and 50 mA cm^{-2}, which is higher than Ag/C, N-doped CNTs and silver-molybdate catalysts [47,48]. Figure 12 (b) records the change of cell voltage with time at a current density of 20 mA cm^{-2}. It is clear that $Ag_{90}Cu_{10}$ catalysts have a higher initial discharge voltage of 1.15 V than $Ag_{50}Cu_{50}$ catalysts (1.0V); however, after 30 h discharging, the discharge voltage of $Ag_{90}Cu_{10}$ catalysts gradually reduces to 1.11 V, while $Ag_{50}Cu_{50}$ catalysts gradually increase to 1.18 V. The discharge curve of the $Ag_{50}Cu_{50}$ catalysts gradually rises to a stable cell voltage and decrease by 16% as compared with the open-circuit potential, but for $Ag_{90}Cu_{10}$ catalysts, the curve gradually reduces to a stable voltage after about 30 h of discharge, showing a decrease of 20%. It can be concluded that the $Ag_{50}Cu_{50}$ catalyst has higher discharge voltage stability, being more stable than the $Ag_{90}Cu_{10}$ catalysts for applications in zinc-air batteries.

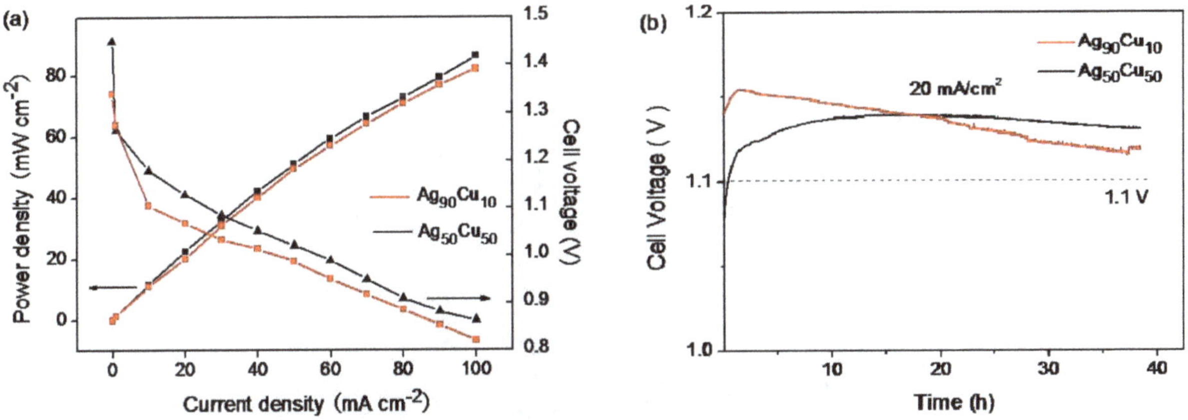

Figure 12. (a) The discharge polarization and power density curves for $Ag_{90}Cu_{50}$ and $Ag_{50}Cu_{50}$ catalyst layer of air-cathode in the primary zinc-air battery. (b) The single cell voltage and time curves at 20 mA cm^{-2} in the primary zinc-air battery. (In 6M KOH solution) Reprinted with permission from ref. 46.

3.2. Galvanic displacement synthesis

Ag–Cu catalysts were synthesized directly on Ni foams by galvanic displacement reaction, which is an environment-friendly and straightforward process [42]. In this method, the catalysts were directly grown on Ni foams, thereby freeing the catalytic layer from carbon and binder. The driving mechanism to grow the various catalysts in this work is the large difference of the redox potentials of Ni^{2+}/Ni (–0.25 V vs. SHE), Cu^{2+}/Cu (0.34 V vs. SHE) and $Ag+/Ag$ (0.799 V vs. SHE). The galvanic displacement reaction can be described by the following equations:

$$Ni(s) + Cu^{+2}(aq) \rightarrow Cu(s) + Ni^{+2}(aq) \quad (13)$$

$$Ni(s) + 2Ag^{+}(aq) \rightarrow 2Ag(s) + Ni^{+2}(aq) \quad (14)$$

$$Cu(s) + 2Ag^{+}(aq) \rightarrow 2Ag(s) + Cu^{+2}(aq) \quad (15)$$

Figure 13 shows the typical SEM images of (a) Ni foam, (b) Cu nanoparticles, (c) AgCu-10 and (d) Ag catalysts. Figure 13(a) shows the pure Ni foam smooth surface. Immersion into $CuSO_4$ solution for 3 h makes the surface of the foam rough (Figure 13(b)) along with coverage of octahedral copper nanoparticles (Figure 13(b) inset). The SEM of AgCu-10 catalyst prepared by two-step galvanic displacement reaction is shown in Figure 13(c). The catalyst has dendritic morphology. The dendrites shown in Figure 13(c) have a dense and uniform distribution, and the shape is complete. On the other hand, similarly dendritic morphologies are obtained for pure Ag catalyst (Figure 13(d)) prepared by directly immersing the as-prepared Ni foam into $AgNO_3$ solution. The dendrites of Ag catalyst are thinly distributed on the Ni foam compared to that in Figure 13(c). The difference of the dendrites between AgCu-10 and Ag catalysts can be because of the different sacrificial templates. For the AgCu-10 catalyst, the Ni foam was already covered by octahedral copper nanoparticles while the nickel surface was free from copper in case of pure Ag particles.

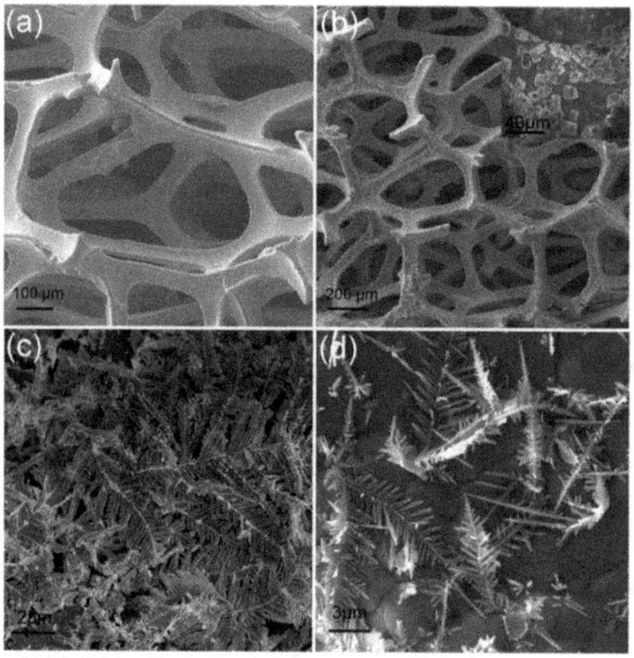

Figure 13. The FE-SEM images of (a) Ni foam. (b) Cu nanoparticles supported on Ni foam. The inset in (b) shows the high-magnification image of Cu nanoparticles. (c) Ag–Cu dendrites in AgCu-10 catalyst and (d) Ag dendrites supported on Ni foam. Reprinted with permission from ref. 42.

As shown in Figure 13(c), the one-dimensional dendrites prefer to form in a relatively high $AgNO_3$ concentration (10 mM). The AgCu-10 bimetallic catalyst possessed a hierarchical structure characteristic and highly rough surface which provided more catalytic active sites so it showed higher catalytic current during the RDE polarization measurements. The SEM elemental mapping analysis for AgCu-10 catalyst shown in Figure 14(a–d) clearly prove that Ag and Cu are uniformly distributed. HRTEM and SAED images are shown in Figure 14 (e, f). The clear lattice fringes are observed for AgCu-10 catalyst. The SAED pattern shows a typical polycrystalline structure with the sharp diffraction rings for AgCu-10 catalyst. The diffraction spots for Ag (111), (2 0 0), (2 2 0), (311) facets and Cu (111), (2 2 0) facets are observed. A fast

Fourier transform (FFT) diffraction pattern as shown in Figure 14 (g) is obtained on the area marked by the red rectangle in Figure 14 (e), exhibiting the two phases characteristic of AgCu-10 catalyst. The FFT image also reveals that Ag and Cu crystallites are in an epitaxial relationship relative to each other in parallel orientation. The grown orientation schematic is shown in Figure 14 (h).

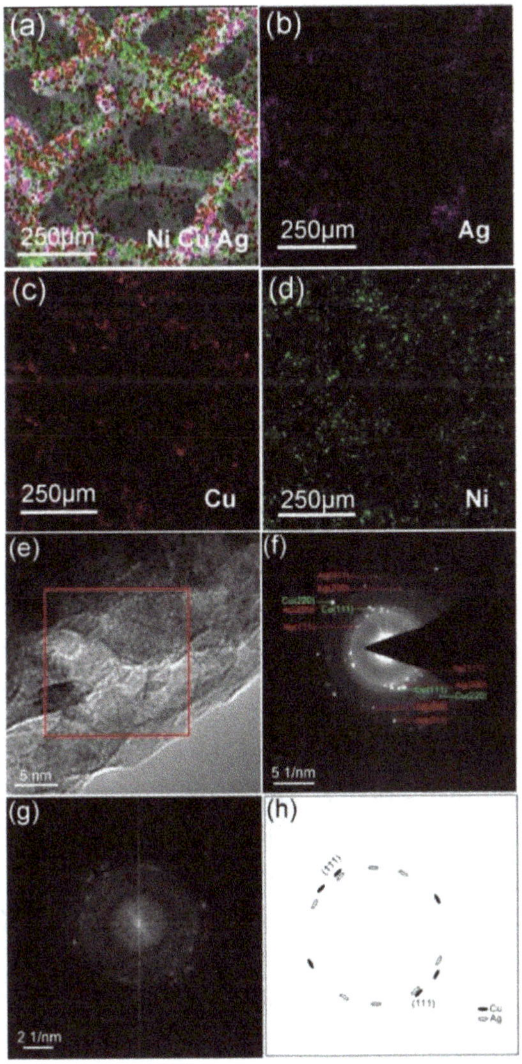

Figure 14. EDS elemental map of the AgCu-10 catalyst: (a) Overlay image. (b) Ag map. (c) Cu map. (d) Ni map. The TEM characteristic for AgCu-10 catalyst: (e) HRTEM image. (f) SAED pattern. (g) Fast Fourier transform (FFT) image corresponding to (e) and schematic drawing with index of reflections (h). Reprinted with permission from ref. 42.

Rotating disc measurements were performed on the catalysts made by galvanic displacement and comparison was made with pure silver. AgCu-10 performed efficiently as compared to the pure silver particles in terms of both onset potential and the limiting current. The Koutecky–Levich plots revealed the four electron transfer mechanism during ORR. Rechargeable zinc-air battery was also fabricated by as-prepared AgCu-10 catalyst-based air cathode, 6 M KOH solution with 0.2 M zinc acetate (zinc acetate was dissolved in KOH to form zincate to ensure reversible Zn electrochemical reactions at the anode) and pure zinc plate anode. A charge–discharge cycle experiment was performed with a short cycle period of 20 min and a long cycle period of 4 h at 20 mA cm^{-2}. At the short cycle period, the initial charge and discharge potentials of AgCu-10 based rechargeable zinc-air battery are 2.04 and 1.1 V, respectively. The round-trip efficiency corresponding to the first cycle is 53.9%. There is almost no apparent fluctuation for the charge and discharge potentials of the rechargeable zinc-air battery through all the cyclic process. The round-trip efficiency after 100 cycles is 53.08% compared to the initial 53.9% with a little decline of 0.82%. A further cycle performance study with a long cycle period of 4 h was carried out with the same rechargeable zinc-air battery after replacing the zinc anode and the electrolyte. The rechargeable zinc-air battery also shows high cycling stability at the long cycle period. The increase in charge and discharge potentials difference from the first to tenth cycle is as little as 0.06 V, which is comparable to the tri-electrode rechargeable zinc-air battery [49]. The cycling stability obtained on AgCu-10 catalyst based zinc-air battery is certainly appealing and significant for the large-scale application of metal-air batteries and fuel cells.

3.3. Electro-deposition of Ag–Cu nanoalloys

Ag–Cu catalysts were synthesized by the electrodeposition method under a potential of −0.4 V for a period of 50 s by using the conventional three-electrode cell system [50]. For synthesis, $AgNO_3$, $Cu(NO_3)_2\,3H_2O$ and 3 mM sodium citrate ($Na_3C_6H_5O_7$) are mixed by deionized water with the formula of Ag_xCu_{100-x} (labelled as $Ag_{25}Cu_{75}$, $Ag_{50}Cu_{50}$ and $Ag_{75}Cu_{25}$ for x = 25, 50 and 75 mM, respectively).

Figure 15(a) describes the TEM and SEM (the inset) images of the bimetallic Ag–Cu catalyst. The nanoplatelets have diameters of 40–50 nm. The HRTEM analysis in Figure 15(b) clearly shows lattice fringes, indicating good crystallinity. The lattice spacing obtained from the HRTEM image is 0.239 nm. The particle demonstrates a single crystal pattern in the SAED of Figure15(c), indicating that Ag–Cu catalysts grow larger by the oriented attachment from small Ag–Cu nanoparticles. The cell constant of the single crystal is 0.3986 nm, which is between the standard cell parameter of FCC-Ag (a = 0.4086 nm) and FCC-Cu (a = 0.3615 nm), suggesting that the Cu atoms are partially alloyed with the Ag atoms.

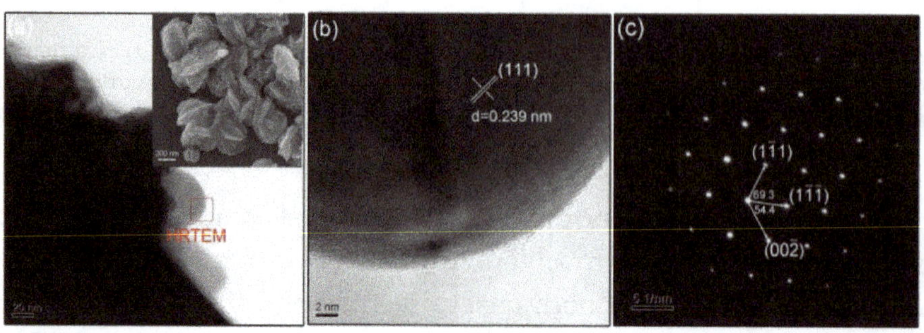

Figure 15. Images of the Ag–Cu catalyst (a) TEM (inset SEM). (b) HRTEM. (c) SAED. Reprinted with permission from ref. 50.

The EDX spectrum of the Ag–Cu nanoalloy in Figure 16(a) exhibits that the Ag–Cu deposits contain both Ag and Cu elements and the nominal atomic composition are 1.5:1, 5:1 and 10:1 for the $Ag_{25}Cu_{75}$, $Ag_{50}Cu_{50}$ and $Ag_{75}Cu_{25}$ samples, respectively.

The survey spectrum of XPS for the Ag–Cu catalyst is shown in Figure 16(b), which shows clear Ag and Cu peaks. The binding energies of the $Ag3d_{3/2}$ and $Ag3d_{5/2}$ orbits observed from the high-resolution spectrum (Figure 16(b) inset) are 374.4 and 368.4 eV. This result indicates that the Ag atoms are zero-valent (Ag^0) metals [51,52]. The binding energies of the $Cu2p_{1/2}$ and $Cu2p_{3/2}$ orbits are 952.6 and 932.7 eV (Figure 16(f)). An analysis of the Auger electron spectrum for Cu LMM (Figure 16(f) inset) is also conducted to determine the specific valence of Cu; an apparent kinetic energy peak at 918.6 eV is seen, corresponding to the zero-valent Cu [53]. Thus, the Ag and Cu atoms exist as elemental metallic substances in Ag–Cu catalyst.

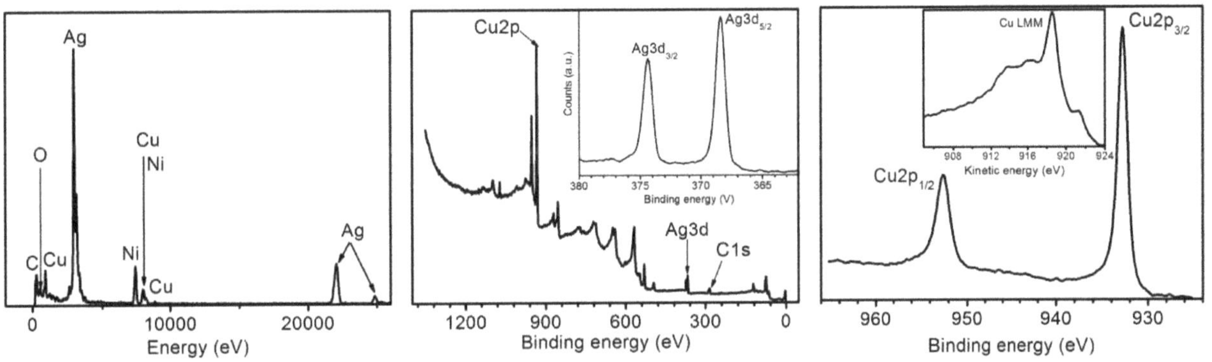

(a) (b) (c)

Figure 16. (a) EDX. (b) XPS survey spectrum, inset: high-solution spectrum of Ag3d. (c) High-solution spectrum of Cu2p, inset: Cu LMM. Reprinted with permission from ref. 50.

As shown in Figure 17(a), a reduction current peak is observed from the CV curves of the Ag–Cu catalyst in O_2-saturated 0.1 M KOH at −0.3 V (vs. SCE) but not in N_2-saturated solution, indicating that Ag–Cu catalyst has catalytic activity for ORR. To study the catalysis kinetics of Ag–Cu catalyst for ORR, an RDE experiment is performed in O_2-saturated 0.1 M KOH solution (Figure 17(b)). The number (n) of electrons transferred on the Ag–Cu catalyst during ORR, which determines the catalytic efficiency, is calculated by the Koutecky–Levich plots (see inset). The result with $n = 3.8$ and 3.7 at −0.7 and −0.8 V, respectively, indicates that the ORR catalyzed by the Ag–Cu catalyst occurs through a four-electron pathway, which is more efficient than a two-electron pathway.

The performance of primary battery fabricated from $Ag_{50}Cu_{50}$-based air cathode was evaluated as shown in Figure 18(a). The open-circuit voltage (OCV) and maximum power density are 1.49 V and 87 $mWcm^{-2}$, respectively, which have significant improvements [54,55]. After the primary zinc–air battery undergoes 10 discharging cycles, the OCV and power density decrease slightly. This result may be attributed to the polarization of the zinc anode caused by the zincate produced during the discharge process. Replacing the zinc anode and electrolyte revitalizes the battery performance. The battery has no obvious voltage loss compared with the first cycle, suggesting that the $Ag_{50}Cu_{50}$-based air cathode is stable in alkaline solution and

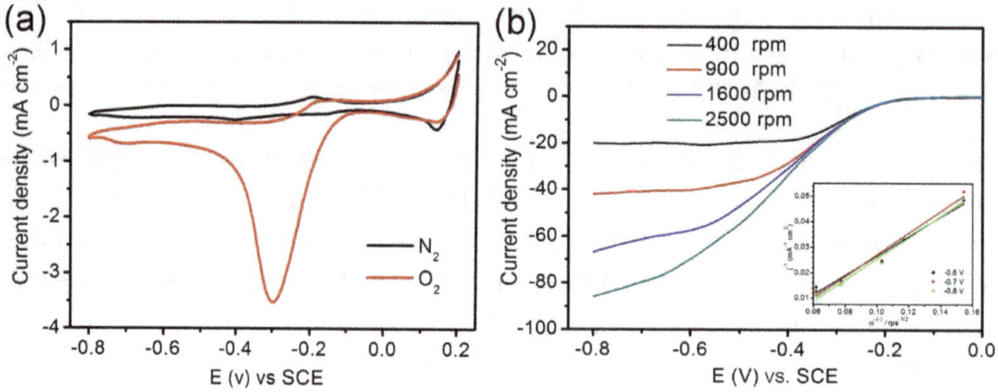

Figure 17. (a) CV curves for Ag–Cu catalyst in O_2- and N_2-saturated 0.1 M KOH solutions at a scan rate of 10 mV s^{-1}. (b) RDE polarization curves at different rotation rates in O_2-saturated 0.1 M KOH solution at a scan rate of 10 mV s^{-1}, inset: Koutecky–Levich plots. Reprinted with permission from ref. 50.

can be a potential candidate for rechargeable zinc-air batteries. Further analysis of discharging performance of this primary zinc-air battery is conducted at different constant current densities as in Figure 18(b). The cell voltages are stable during the whole discharge period of 20 h, except for a slight drop of approximately 0.05 V at the current density of 50 mA cm^{-2}. This voltage drop can be attributed to the electrode polarization under high current density.

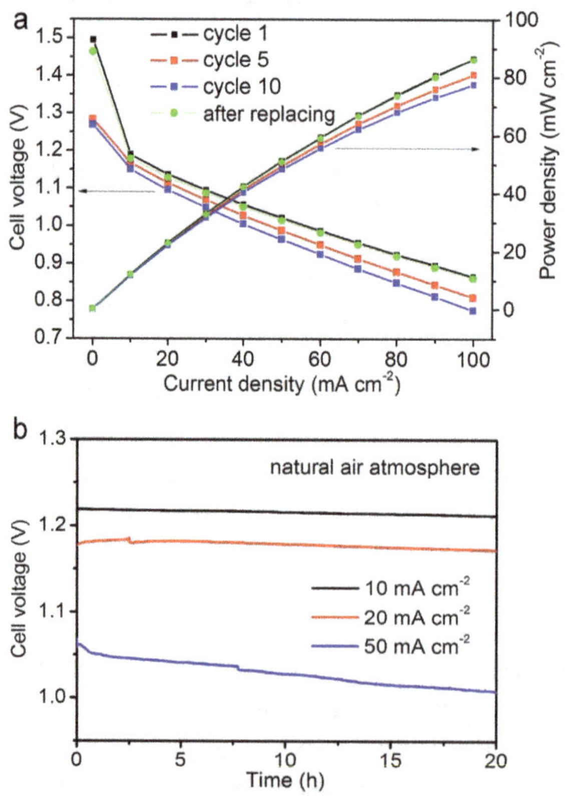

Figure 18. Performance of the primary zinc-air battery fabricated with an $Ag_{50}Cu_{50}$ catalyst-based air cathode: a) Cell voltage and power density polarization curves of the battery over 10 cycles and the polarization curve after replacing Zn anode and electrolyte; b) discharge voltage curves at different current densities. Reprinted with permission from ref. 50.

Figure 19(a) shows the charge and discharge polarization curves of the secondary zinc-air battery. An abrupt polarization occurs when the current densities increase from 0 to 10 mA cm^{-2} because of the activation polarization and anode polarization [56]. However, once the zinc-air battery begins its function, the polarization exhibits a steady increase with varying current densities from 10 to 100 mA cm^{-2}. The charge–discharge voltage gap (i.e., the overpotential) at 20 mA cm^{-2} is 0.9 V, which is lower than that of Co_3O_4-based rechargeable zinc-air batteries [57,58]. The cycle performance with different cycle periods is shown in Figure 19(b). As seen in the bottom of Figure 19(b), the secondary zinc-air battery undergoes 100 charge and discharge cycles at 20 mA cm^{-2} with 20 min per step. The difference between the charge and discharge potentials is 0.9 V, and the overpotential shows no apparent fluctuation through all 100 cycles. The round-trip efficiency is up to 56.4%, which is a considerable improvement. A more violent charge and discharge cycle experiment is conducted with a cycle period of 4 h for the same rechargeable zinc-air battery after replacing the zinc anode and electrolyte. As shown in the top of Figure 19(b), the charge and discharge voltages are still stable even with the long cycle period; this result is comparable to the result of the tri-electrode rechargeable zinc-air battery.

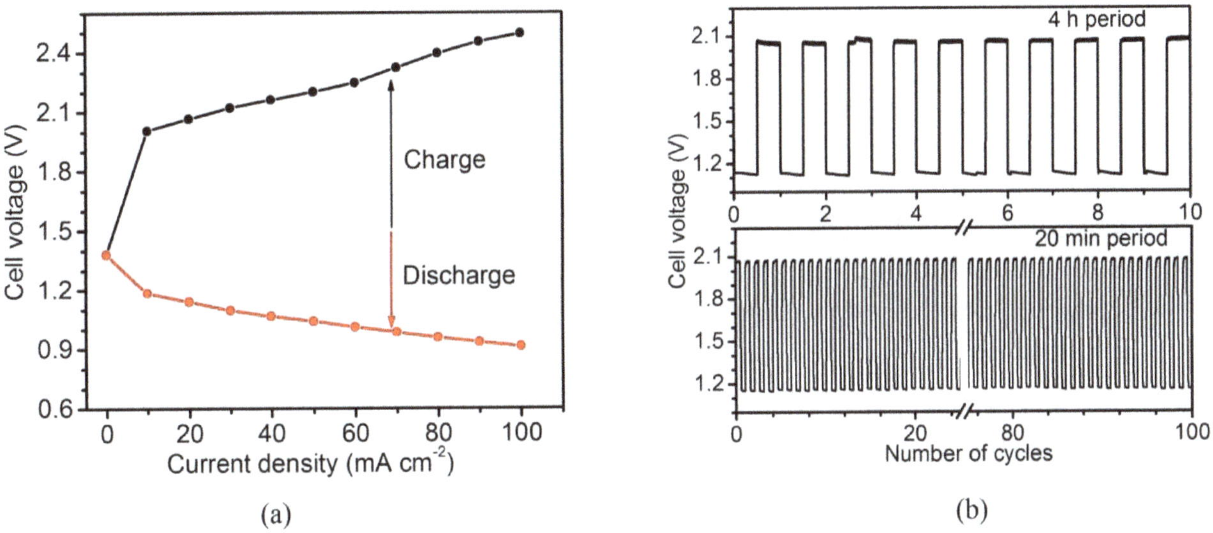

Figure 19. (a) Charge–discharge polarization curves for the rechargeable zinc air battery. b) Cycle performance for the rechargeable zinc-air battery at 20 mA cm–2 with a 20 min cycle period for 100 cycles and a 4 h cycle period for 40 h. Reprinted with permission from ref. 50.

4. Conclusions

Silver copper Ag–Cu nanoalloy particles have been investigated for prospective application as an electrocatalyst for oxygen reduction reaction in alkaline fuel cells and metal air battery systems. A holistic approach has been adopted incorporating density functional theory simulations along with synthesis of potential candidate compositions of Ag–Cu nanoalloys. Following conclusions can be drawn from this work:

1. Alloying of silver with copper significantly enhances the catalytic activity of silver for ORR. The enhancement of this catalytic activity is attributed to the modification of the surface electronic structure of the Ag–Cu nanoalloy catalyst.

2. Our DFT-based calculations confirm the rise of the d-band center of Ag–Cu nanoalloy catalyst toward Fermi level which in turn increases the adsorption strength of ORR intermediates on the catalyst surface during oxygen reduction. It has been well established that the optimum binding reaction intermediates such as O, OH, OOH are crucial for the fast kinetics of ORR.

3. The doping of a single Cu atom on the surface of 13 atom silver cluster improves the adsorption strength of ORR intermediates on the catalyst surface. We report that in case of 38 atom nanoalloy clusters, polyicosahedron (PIh) $Ag_{32}Cu_6$ core-shell structure is found to be the most stable structure by genetic algorithm calculations. The most active catalytic site on the $Ag_{32}Cu_6$ nanoalloy was also identified by our DFT calculations.

4. Ag–Cu nanoalloys have been successfully synthesized on Ni foams by pulse laser deposition as well as by chemical routes such as galvanic displacement and electrodeposition. Characterization of these nanoalloys by electron microscopy techniques (TEM, HRTEM, SEM) endorse the alloying of Ag and Cu atoms at the nanoscale.

5. Electrochemical characterization by rotating disc electrode methods (RDE) reveals the improvement of the ORR performance of the Ag–Cu nanoalloys due to the synergistic role of silver and copper in the Ag–Cu nanoalloys. Out of various compositions produced by PLD $Ag_{50}Cu_{50}$ was observed to be the most active toward ORR in alkaline media. A twofold increase in limiting current density along with more positive onset potential was recorded in case of AgCu-10 nanoalloy compared to pure silver. We report four electron transfer pathway to be dominant during ORR in alkaline conditions on the surface of Ag–Cu nanoalloys.

6. Ni foam is reported here to be a suitable support for Ag–Cu nanoalloy catalysts which enables the elimination of binder-based carbon supports.

5. Future outlook

Ag–Cu nanoalloy catalysts offer an attractive alternative for the otherwise costly Pt-based alloys. The performance of these Ag–Cu nanoalloy catalysts was found to be reasonable while the stability in alkaline media is much superior compared to the Pt-based alloy catalysts. The realization of high energy density metal air batteries and alkaline fuel cells for stationary and mobile applications demand swift ORR kinetics at cathode which can be achieved by highly active silver-based nanoalloy electrocatalysts. Very few Ag-based Ag/transition metal nanoalloy electrocatalysts have been reported for ORR. Recently, the number of publications on silver-based nanoalloy electrocatalyst for ORR in alkaline media has been on the rise. DFT-based calculations show that the surface electronic structure of the nanoalloy catalyst is sensitive to doping. This opens the window for carrying out investigations on the ternary Ag–

Cu–M catalyst systems with different morphologies to further improve the catalytic activity. Moreover, the effect of support such as CNT, MWCNT, RGO is still to be investigated for the Ag–Cu nanoalloys. These advance catalyst supports have been previously reported to be highly beneficial for Ag-based systems. The highest amount of catalytic activity in Ag-based systems is reported in the Ag composites owing to the ensemble effect. Incorporation of Ag–Cu nanoalloys into composite structures can provide a big leap forward toward improving the catalytic activity of these nanoalloys. At present, the activity of silver nanoalloys in alkaline media is limited because of weak adsorption of ORR intermediates. Highly active silver nanoalloy compositions can be identified by DFT calculations which when combined with active supports can rival or even surpass commercial Pt/C in alkaline media. A universal approach incorporating first principle simulations and experimental analysis is pivotal for the realization of highly active and stable silver-based nanoalloy electrocatalysts for ORR in alkaline media.

Acknowledgements

This study was supported by the National Natural Science Foundation of China (grant nos. 51271148 and 50971100), the Aeronautic Science Foundation Program of China (grant no. 2012ZF53073), the Science and Technology Innovation Fund of Western Metal Materials (grant no. XBCL-2-11) and the Doctoral Fund of Ministry of Education of China (grant no. 20136102110013).

Author details

Adnan Qaseem, Fuyi Chen* and Nan Zhang

*Address all correspondence to: fuyichen@nwpu.edu.cn

State Key Laboratory of Solidification Processing, Northwestern Polytechnical University, Xian, China

References

[1] Nobuyuki Imanishi and Osamu Yamamoto. Rechargeable lithium–air batteries: characteristics and prospects. Materials Today. 2014;17(1):24–30. DOI: 10.1016/j.mattod.2013.12.004

[2] Md. Arafat Rahman, Xiaojian Wang and Cuie Wen. A review of high energy density lithium–air battery technology. Journal of Applied Electrochemistry. 2014;44(1):5–22. DOI: 10.1007/s10800-013-0620-8

[3] Nuria Garcia-Araez and Petr Novák. Critical aspects in the development of lithium–air batteries. Journal of Solid State Electrochemistry. July 2013; 17 (7):1793–1807. DOI: 10.1007/s10008-013-1999-1

[4] Mark K. Debe. Electrocatalyst approaches and challenges for automotive fuel cells. NATURE. 2012;486(7401):43–51. DOI: 10.1038/nature11115

[5] İbrahim Dinçer and Calin Zamfirescu. Hydrogen and fuel cell systems. In: Sustainable Energy Systems and Applications. New York: Springer US; 2012. pp. 519–632. DOI: 10.1007/978-0-387-95861-3_13

[6] Marcel Pourbaix. Atlas of Electrochemical Equilibria in Aqueous Solutions. 2nd ed. US: National Association of Corrosion; 1974. 644 p. DOI: 0915567989

[7] Edward McCafferty. Thermodynamics of Corrosion: Pourbaix Diagrams. In: Edward McCafferty, editor. Introduction to Corrosion Science. Springer New York; 2010. pp. 95–117. DOI: 10.1007/978-1-4419-0455-3_6

[8] J. K. Nørskov, J. Rossmeisl, A. Logadottir and L. Lindqvist. Origin of the overpotential for oxygen reduction at a fuel-cell cathode. The Journal of Physical Chemistry B. 2004;108 (46):17886–17892. DOI: DOI: 10.1021/jp047349j

[9] Daojian Cheng, Xiangguo Qiua and Haiyan Yua. Enhancing oxygen reduction reaction activity of Pt-shelled catalysts via subsurface alloying. Physical Chemistry Chemical Physics. 2014;16(38): 20377-20381. DOI: 10.1039/c4cp02863e

[10] Bongjin Simon Mun, Guofeng Wang, Philip N. Ross, Christopher A. Lucas, and Nenad M. Marković. Improved Oxygen Reduction Activity on $Pt_3Ni(111)$ via Increased Surface Site Availability. Science. 2007;315 (5811):493–497. DOI: 10.1126/science.1135941

[11] Gongwei Wang, Bing Huang, Li Xiao, Zhandong Ren, Hao Chen, Deli Wang, Héctor D. Abruña, Juntao Lu, and Lin Zhuang. Pt Skin on AuCu Intermetallic Substrate: A Strategy to Maximize Pt Utilization for Fuel Cells. Journal of the American Chemical Society (JACS). 2014;136(27):9643–9649. DOI: 10.1021/ja503315s

[12] O A Petrii. Electrosynthesis of nanostructures and nanomaterials. Russian Chemical Reviews. 2015;84(2):159–193. DOI: 10.1070/RCR4438

[13] J. Greeley, I. E. L. Stephens, A. S. Bondarenko, T. P. Johansson, H. A. Hansen, T. F. Jaramillo, J. Rossmeisl, I. Chorkendorff and J. K. Nørskov. Alloys of platinum and early transition metals as oxygen reduction electrocatalysts. Nature Chemistry. 2009;1:552–556. DOI: 10.1038/nchem.367

[14] Hongliang Xin, Adam Holewinski, and Suljo Linic. Predictive Structure- Reactivity Models for Rapid Screening of Pt-Based Multimetallic Electrocatalysts for the Oxygen Reduction Reaction. ACS Catalysis. 2012;2:12–16. DOI: 10.1021/cs200462f

[15] Mitsuru Wakisaka, Satoshi Mitsui, Yoshikazu Hirose, Katsura Kawashima,Hiroyuki Uchida, and Masahiro Watanabe. Electronic structures of Pt-Co and Pt-Ru alloys for

CO-tolerant anode catalysts in polymer electrolyte fuel cells studied by EC-XPS. The Journal of Physical Chemistry B. 2006;110(46):23489–23496. DOI: 10.1021/jp0653510

[16] B. Hammer and J.K. Nørskov. Theoretical surface science and catalysis—calculations and concepts. In: Bruce C. Gates, Helmut Knozinger, editors. Advances in Catalysis. 45th ed. Academic Press; 2000. pp. 71–129. DOI: 10.1016/S0360-0564(02)45013-4

[17] Hongliang Xin, Adam Holewinski, Neil Schweitzer, Eranda Nikolla and Suljo Linic. Electronic structure engineering in heterogeneous catalysis: identifying novel alloy catalysts based on rapid screening for materials with desired electronic properties. Topics in Catalysis. 2012;55 (5-6):376–390. DOI: 10.1007/s11244-012-9794-2

[18] Paul Sabatier. Hydrogénations et déshydrogénations par catalyse. Berichte der deutschen chemischen Gesellschaft. 1911;44(3):1984–2001. DOI: 10.1002/cber.19110440303

[19] Ernest Yeager. Dioxygen electrocatalysis: mechanisms in relation to catalyst structure. Journal of Molecular Catalysis. 1986; 38(1–2):5–25. DOI:10.1016/0304-5102 (86)87045-6

[20] F. Bidault, D.J.L. Brett, P.H. Middleton and N.P. Brandon. Review of gas diffusion cathodes for alkaline fuel cells. Journal of Power Sources. 2009;187(1):39–48. DOI: 10.1016/j.jpowsour.2008.10.106

[21] Vladimir Neburchilov, Haijiang Wang, Jonathan J. Martin and Wei Qu. A review on air cathodes for zinc–air fuel cells. Journal of Power Sources. 2010;195(5):1271–1291. DOI: 10.1016/j.jpowsour.2009.08.100

[22] B.B. Blizanac, P.N. Ross and N.M. Markovic. Oxygen electroreduction on Ag(1 1 1): The pH effect. Electrochimica Acta. 2007;52 (6):2264–2271. DOI: doi:10.1016/j.electacta.2006.06.047

[23] Adam Holewinski, Juan-Carlos Idrobo and Suljo Linic. High-performance Ag–Co alloy catalysts for electrochemical oxygen reduction. Nature Chemistry. 2014;6(9):828–834. DOI: 10.1038/nchem.2032

[24] Lizhi Yuan, Luhua Jiang, Jing Liu, Zhangxun Xia, Suli Wang and Gongquan Sun. Facile synthesis of silver nanoparticles supported on three dimensional graphene oxide/carbon black composite and its application for oxygen reduction reaction. Electrochimica Acta. 2014;135:168–174. DOI: 10.1016/j.electacta.2014.04.137

[25] Shin-Ae Park, Hankwon Lim and Yong-Tae Kim. Enhanced oxygen reduction reaction activity due to electronic effects between Ag and Mn3O4 in alkaline media. ACS Catalysis. 2015;5(6):3995–4002. DOI: 10.1021/acscatal.5b00495

[26] Junya Ohyama, Yui Okata, Noriyuki Watabe, Makoto Katagiri, Ayaka Nakamura, Hidekazu Arikawa, Ken-ichi Shimizu, Tatsuya Takeguchi, Wataru Ueda and Atsushi Satsuma. Oxygen reduction reaction over silver particles with various morphologies and surface chemical states. Journal of Power Sources. 2014;245(1):998–1004. DOI: doi:10.1016/j.jpowsour.2013.07.034

[27] Raju P. Gupta. Lattice relaxation at a metal surface. Physical Review B. 1981;23(12): 6265–6270. DOI: http://dx.doi.org/10.1103/PhysRevB.23.6265

[28] Weiyin Li and Fuyi Chen. A density functional theory study of structural, electronic, optical and magnetic properties of small Ag–Cu nanoalloys. Journal of Nanoparticle Research. 2013;15(7):1809. DOI: 10.1007/s11051-013-1809-9

[29] F. Baletto, C. Mottet, and R. Ferrando. Growth simulations of silver shells on copper and palladium nanoclusters. Physical Review B. 2002;66(15):155420. DOI: http://dx.doi.org/10.1103/PhysRevB.66.155420

[30] Roy L. Johnston. Evolving better nanoparticles: Genetic algorithms for optimising cluster geometries. Dalton Transactions. 2003;(22):4193–4207. DOI: 10.1039/B305686D

[31] B. Delley. An all-electron numerical method for solving the local density functional for polyatomic molecules. The Journal of Chemical Physics. 1990;92(1):508. DOI: 10.1063/1.458452

[32] B. Delley. From molecules to solids with the DMol3 approach. The Journal of Chemical Physics. (2000);113(18): 7756. DOI: 10.1063/1.1316015

[33] J. Oviedo and R. E. Palmer. Amorphous structures of Cu, Ag, and Au nanoclusters from first principles calculations. The Journal of Chemical Physics. 2002;117(21):9548. DOI: 10.1063/1.1524154

[34] Jonathan P. K. Doye and David J. Wales. Global minima for transition metal clusters described by Sutton–Chen potentials. New Journal of Chemistry. 1998,;22(7):733-744. DOI: 10.1039/A709249K

[35] Wenqiang Ma, Fuyi Chen, Nan Zhang and Xiaoqiang Wu. Oxygen reduction reaction on Cu-doped Ag cluster for fuel-cell cathode. Journal of Molecular Modeling. 2014; 20:2454. DOI: 10.1007/s00894-014-2454-7

[36] Kihyun Shin, Da Hye Kim, Sang Chul Yeo and Hyuck Mo Lee. Structural stability of AgCu bimetallic nanoparticles and their application as a catalyst: A DFT study. Catalysis Today. 2012;185(1):94–98. DOI: 10.1016/j.cattod.2011.09.022

[37] Kihyun Shin, Da Hye Kim and Hyuck Mo Lee. Catalytic Characteristics of AgCu Bimetallic Nanoparticles in the Oxygen Reduction Reaction. ChemSusChem. 2013;6(6): 1044–1049. DOI: 10.1002/cssc.201201001

[38] N. Zhang, F. Y. Chen and X.Q. Wu. Global optimization and oxygen dissociation on polyicosahedral Ag32Cu6 core-shell cluster for alkaline fuel cells. Scientific Reports (2015) 5, Article number: 11984. DOI: 10.1038/srep11984

[39] Rutger a. Van Santen and Matthew Neurock. Concepts in Theoretical Heterogeneous Catalytic Reactivity. Catalysis Reviews: Science and Engineering. 1995 ;37(4):557–698. DOI: 10.1080/01614949508006451

[40] Waldemar Mróz, Bogusław Budner, Wojciech Tokarz, Piotr Piela and Michael L. Korwin-Pawlowski. Ultra-low-loading pulsed-laser-deposited platinum catalyst films for

polymer electrolyte membrane fuel cells. Journal of Power Sources. 2015;273(1):885–893. DOI: 10.1016/j.jpowsour.2014.09.173

[41] Sadesh Kumar Natarajan and Jean Hamelin. High-performance anode for polymer electrolyte membrane fuel cells by multiple-layer Pt sputter deposition. Journal of Power Sources. 2010;195(22):7574–7577. DOI: 10.1016/j.jpowsour.2010.06.027

[42] Yachao Jin and Fuyi Chen. Facile preparation of Ag-Cu bifunctional electrocatalysts for zinc-air batteries. Electrochimica Acta. 2015;158:437–445. DOI: 10.1016/j.electacta.2015.01.151

[43] M. J. Montenegro and M. Döbeli. Pulsed laser deposition of La0.6Ca0.4CoO3(LCCO) films. A promising metal-oxide catalyst for air based batteries. Physical Chemistry Chemical Physics. 2002;4(12):2799–2805. DOI: 10.1039/B200120A

[44] Shaomin Zhu, Zhu Chen, Bing Li, Drew Higgins, Haijiang Wang, Hui Li and Zhongwei Chen Nitrogen-doped carbon nanotubes as air cathode catalysts in zinc-air battery. Electrochimica Acta. 2011;56(14):5080–5084. DOI: 10.1016/j.electacta.2011.03.082

[45] Yimin Lei, Fuyi Chen, Yachao Jin and Zongwen Liu. Ag-Cu nanoalloyed film as a high-performance cathode electrocatalytic material for zinc-air battery. Nanoscale Research Letters. 2015;10:197. DOI: 10.1186/s11671-015-0900-9

[46] Xiaoqiang Wu, Fuyi Chen, Yachao Jin, Nan Zhang and Roy L. Johnston. Silver–copper nanoalloy catalyst layer for bifunctional air electrodes in alkaline media ACS Application of Materrial Interfaces. 2015;7(32):17782–17791. DOI: 10.1021/acsami.5b04061

[47] Jia-Jun Han, Ning Li and Tian-Yun Zhang. Ag/C nanoparticles as an cathode catalyst for a zinc-air battery with a flowing alkaline electrolyte. Journal of Power Sources. 2009;193(2):885–889. DOI: 10.1016/j.jpowsour.2009.02.082

[48] Ying Wanga,Yu Liuc and Xiaojian Lu. Silver-molybdate electrocatalysts for oxygen reduction reaction in alkaline media. Electrochemistry Communications. 2012;20:171–174. DOI: 10.1016/j.elecom.2012.05.004

[49] Yanguang Li, Ming Gong and Yongye Liang. Advanced zinc-air batteries based on high-performance hybrid electrocatalysts. Nature Communications. 2013;4:1805. DOI: 10.1038/ncomms2812

[50] Yachao Jin, Fuyi Chen, Yimin Lei and Xiaoqiang Wu. A Silver-copper alloy as an oxygen reduction electrocatalyst for an advanced zinc-air battery. Chemical Catalysis Catalytical Chemitry. 2015;7(15):2377–2383. DOI: 10.1002/cctc.201500228

[51] Sang Woo Han, Yunsoo Kim and Kwan Kim. Dodecanethiol-derivatized Au/Ag bimetallic nanoparticles: TEM, UV/VIS, XPS, and FTIR analysis. Journal of Colloid and Interface Science. 1998;208(1):272–278. DOI: 10.1006/jcis.1998.5812

[52] Xianfeng You, Feng Chen, Jinlong Zhang and Masakazu Anpo. A novel deposition precipitation method for preparation of Ag-loaded titanium dioxide. Catalysis Letters. 2005;102(3-4):247–250. DOI: 10.1007/s10562-005-5863-5

[53] C. C. Chusuei, M. A. Brookshier and D. W. Goodman. Correlation of relative X-ray photoelectron spectroscopy shake-up intensity with CuO particle size. Langmuir. 1999;15(8):2806–2808. DOI: 10.1021/la9815446

[54] Po-Chieh Li, Chi-Chang Hu, Tai-Chou Lee, Wen-Sheng Chang and Tsin Hai Wang. Synthesis and characterization of carbon black/ manganese oxide air cathodes for zinc-air batteries. Journal of Power Sources. 2014;269:88–97. DOI: 10.1016/j.jpowsour.2014.06.108

[55] Ting-Hao Yang, Shanmuganathan Venkatesan, Chien-Hung Lien, Jen-Lin Chang and Jyh-Myng Zen. Nafion/lead oxide–manganese oxide combined catalyst for use as a highly efficient alkaline air electrode in zinc–air battery. Electrochimica Acta. 2011;56 (17):6205–6210. DOI: 10.1016/j.electacta.2011.03.094

[56] Jia Wei Desmond Ng, Maureen Tang and Thomas F. Jaramillo. A carbon-free, precious-metal-free, high-performance O_2 electrode for regenerative fuel cells and metal–air batteries. Energy & Environmental Science. 2014;7(6):2017–2024. DOI: 10.1039/C3EE44059A

[57] Dong Un Lee, Ja-Yeon Choi, Kun Feng, Hey Woong Park, and Zhongwei Chen. Advanced Extremely Durable 3D Bifunctional Air Electrodes for Rechargeable Zinc-Air Batteries. Advanced Energy Materials. 2014;4(6):1301389. DOI: 10.1002/aenm.201301389

[58] Guojun Du, Xiaogang Liu, Yun Zong, T.S. Andy Hor, Aishui Yu and Zhaolin Liu. Co3O4 nanoparticle-modified MnO_2 nanotube bifunctional oxygen cathode catalysts for rechargeable zinc–air batteries. Nanoscale. 2013;5(11):4657–4661. DOI: 10.1039/C3NR00300K

4

Iron-based Nanomaterials in the Catalysis

Boris I. Kharisov, Oxana V. Kharissova, H.V. Rasika Dias,
Ubaldo Ortiz Méndez, Idalia Gómez de la Fuente, Yolanda Peña and
Alejandro Vázquez Dimas

Abstract

Available data on catalytic applications of the iron-containing nanomaterials are reviewed. Main synthesis methods of nZVI, nano-sized iron oxides and hydroxides, core-shell and alloy structures, ferrites, iron-containing supported forms, and composites are described. Supported structures include those coated and on the basis of polymers or inert inorganic materials (*i.e.*, carbon, titania or silica). Description of catalytic processes includes the decomposition reactions (in particular photocatalytic processes), reactions of dehydrogenation, oxidation, alkylation, C–C coupling, among a series of other processes. Certain attention is paid to magnetic recovery of catalysts from reaction systems and their reuse up to several runs almost without loss of catalytic activity.

Keywords: Iron nanomaterials, ferrites, catalysis, bimetallic nanoparticles, core-shell nanoparticles

1. Introduction

1.1. General information on the iron-containing nanostructures

Nanomaterials on iron basis mainly include zero-valent iron (ZVI and nZVI (nano zero-valent iron) are nowadays classic terms), iron-based nanoalloys or core-shell nanoparticles, iron(II and III) oxides, and ferrites, among others. Metallic iron is normally covered with iron(II) and iron(III) oxides [1]. The iron oxides (iron oxide nanoparticles are also referred in several reports to as superparamagnetic iron-oxide nanoparticles (SPIONs) although SPIONS have inducible magnetic properties) [2, 3], belong to the most technologically important oxides of transition metals. The collective term "iron oxides" is also used for oxides, hydroxides, and oxy-hydroxides containing Fe(II) and/or Fe(III) cations and OH^- and/or O^{2-} anions. In total, sixteen pure iron oxide phases, *i.e.*, oxides, hydroxides or oxy-hydroxides, are currently known. These compounds are $Fe(OH)_3$, $Fe(OH)_2$, $Fe_5HO_8 \cdot 4H_2O$, Fe_3O_4, FeO, five polymorphs of FeOOH and four of Fe_2O_3. In these oxide compounds, which are generally low soluble and possess brilliant colors, the iron is present in the form of Fe(III). The extremely important advantages of nanostructured iron, in comparison with other nanomaterials, are its relatively low toxicity and capacity to be biodegradable. This metal, in addition, is non-expensive and commonly widespread material [4].

Particle diameters of nZVI are normally in the range from 10 to 100 nm [5], exhibiting a classic core-shell structure. Their core contains metallic iron phase, meanwhile the oxidation products of zero-valent iron form mixed valent [*i.e.*, Fe(II) and Fe(III)] oxide shell. If stabilizers in excess are present, these core-shell nanoparticles could be protected against further oxidation [6]. Among such stabilizers, a series of organic compounds can be used for nZVI functionalization to stabilize nZVI aqueous dispersions, inhibiting strongly their further agglomeration. Such compounds can be used to satisfy this purpose, for example, PEG, polyacrylic acid, 4-butane-diphosphonic acid, and methoxyethoxyethoxyacetic acid (MEEA) [7].

The magnetite (Fe_3O_4) and maghemite (γ-Fe_2O_3) are of a particular interest talking about iron oxides (SPIONs). The magnetite structure corresponds to an inverse spinel ferrite. The oxygen ions are the part of a close-packed cubic lattice, containing the iron ions between two different interstices, tetrahedral sites (A), and octahedral sites (B). In a chemical point of view, the magnetite/maghemite can be represented by the following formula: Fe^{3+} $[Fe^{2+}{}_{1-y}$ $Fe^{3+}{}_{1-y}$ $Fe^{3+}{}_{1.67y}\square_{0.33y}]O_4$, where y=0 for pure magnetite and y=1 for pure maghemite (completely oxidized magnetite). From room temperature up to Curie temperature (T_c=860 K), the A sites are filled by Fe^{3+} ions and the B sites are filled by Fe^{3+} and Fe^{2+} ions in equal quantity. Although the lepidocrocite (γ-FeOOH) dehydration transforms into γ-Fe_2O_3, industrial fabrication of maghemite is based on a multistep process (1):

$$(\alpha \text{ and/ or } \gamma)\text{ - FeOOH}(\text{oxidation}) \rightarrow \alpha\text{ - Fe}_2\text{O}_3(\text{reduction}) \rightarrow \\ \rightarrow \text{Fe}_3\text{O}_4(\text{controlled oxidation}) \rightarrow \gamma\text{ - Fe}_2\text{O}_3 \quad (1)$$

In addition to the nZVI and SPIONs, a variety of composite inorganic iron-based nanomaterials have been discovered, in particular core-shell Fe(or Fe_xO_y)/Au or more complex trimetallic nanoparticles such as $(Fe_{60}Co_{49})_{core}/Au_{shell}$ [8]. These nanoparticles were classified [9] on the basis of on their complexity levels: 1) the nanostructures based of an iron-containing material with magnetic properties *different from iron oxide*; 2) the nanostructures with a non-spherical morphology (*e.g.* hollow structure); 3) the nanostructures with multi-material composition, *i.e.* each of them is constructed ≥2 more domains of joined together different inorganic materials.

2. Main synthesis methods

A number of currently used methods, shown below, are nowadays used for preparation of Fe-containing nanomaterials. At the same time, some well-known conventional *wet chemistry routes* have not been forgotten and are applied for nZVI synthesis [10] (for instance, by a borohydride reduction in laboratory scale) [11], Fe_2O_3 (sol-gel technique [12] or electrochemical deposition [13]), or Fe_3O_4 (urea- and NaOH-assisted hydrolysis of Fe^{3+} and Fe^{2+} salts and further *ultrasonic treatment* of FeO(OH)/$Fe(OH)_2$). Particle sizes and morphologies of the formed nanomaterials, synthesized by distinct methods, can vary depending on the synthesis conditions. For example, the Fe_3O_4 nanoparticles [14], obtained by *radio frequency nitrogen plasma* technique, represent a regular spherical form, meanwhile the nanoparticles synthesized by wet chemistry synthesis were seen as well shaped cubic form. In case of the plasma prepared iron oxides, their size distribution was wider resulting small (25–80 nm) and larger (>100 nm) particles.

2.1. nZVI and Fe-M nanoalloys and core-shell nanostructures

To produce nanopowders, the method of *electrical explosion of wire* was applied [15, 16], among other physical methods. Intermetallic phases can also be obtained by the *arc discharge* technique, for example Fe-Sn bimetallic nanoparticles [17]. Thus prepared nanoparticles have a core/shell structure consisting of a SnO_2 shell (5–10 nm in thickness) and a core containing polycrystalline intermetallic compounds. The intermetallic compounds $FeSn_2$ and Fe_3Sn_2 were shown to be generated; they coexist with the Sn phase as a single nanoparticle. *Microwave irradiation* (which is used to fabricate normally inorganic materials and composites, lesser organic- or organometallic-based materials), is an alternative method in comparison with classic heating of product precursors. Metal transformations in microwave field are extensively reviewed in a comprehensive recent book [18]. Elemental metal nanoparticles in various forms can also be fabricated using this method. For example, iron-based nanoparticles were obtained applying microwave–polyol route in ethyleneglycol at 100 and 150 °C with additives of polyvinyl pyrrolidone and dodecyl amine [19]. Also, pulsed excimer *laser radiation* (248 nm) was used to ablate a feedstock of permalloy (~2 μm, Ni 81%:Fe 19%) under both normal atmospheric conditions (in air) and in other gases, as well as under pressures [20]. α-Fe nanoparticles were also obtained with use of a modified *metal-membrane incorporation* method applying diffusing metal ions through a dialysis membrane [21] (the diffusion-time ≤15 min).

To get nZVI in laboratory conditions [22], the classic and usual synthesis technique is the reduction of Fe(II or III) salts using $NaBH_4$, $NaAlH_4$ or $LiAlH_4$ as reductants. Thus, nZVI was synthesized (reaction 2) [23] in ethanol medium by the method of reduction of FeX_n (X= Cl, OH, OR, CN, OCN, SCN) using sodium borohydride under atmospheric conditions [24].

$$2FeCl_3 + 6NaBH_4 + 18H_2O \rightarrow 2Fe^0 + 6NaCl + 6B(OH)_3 + 21H_2 \tag{2}$$

A patent [25] describes a route to metal nanoparticles by *thermal decomposition* of iron acetate $Fe(OOCCH_3)_2$, placed in a reaction vessel with a passivating solvent such as a glycol ether. Discussing the *pyrolysis* method, it should be noted the preparation of iron nanoparticles (embedded in a carbon matrix) from metal phthalocyanine as precursor [26] and carbon-encapsulated iron nanoparticles (size 5–20 nm) *via* a picric acid-*detonation-induced pyrolysis* of ferrocene as precursor; this route has such peculiarities as self-heating and extremely fast process [27]. Also, the *"greener" techniques* [28, 29, 30, 31, 32] have been applied for nanoparticle fabrication. Use of plant extracts and other natural products on polyphenole basis in these syntheses as reductants and capping agents at the same time for obtaining nZVI and several other Fe-containing nanoparticles is intriguing [33] as well. For instance, the *herbal tea extracts* were applied to reduce iron(III) chloride to elemental iron nanoparticles (50 nm) [34].

In addition, a variety of general physico-chemical methods have been applied for the production of as *Fe-containing bi- and polymetallic alloys* as core-shell nanostructures. For instance, high entropy Nd-Fe-Co-Ni-Mn alloy nanofilms were prepared [35] by *electrodeposition* at r.t. After preliminary preparation of alumina nanotemplates, Fe, Fe-Ni, and Fe-Pd nanowires were successfully electrodeposited within their porous structure. Also, the Fe-Pt nanocrystalline magnetic films (200 nm of thickness) with planar texture were obtained with use of *magnetron sputtering* and crystalline annealing in magnetic field [36].

2.2. Supported and coated iron nanoparticles

A number of publications are devoted to *carbon-supported* ZVI nanomaterials [37]. This type of protective carbon-cage encapsulation of iron nanoparticles can result hybrid core-shell nanomaterials with unique properties [38]. This way, carbon encapsulated iron core-shell nanoparticles (15–40 nm in size) were obtained *via* confined *arc plasma method* [39]. Resulting nanoparticles possessed a clear core-shell structure. The core (16 nm in diameter) of the particles corresponded to a BCC iron structure, and the shell (thickness 6–8 nm) was shown to be disorder carbon phase. A closely related *arc discharge technique* [40] is also frequently used for obtaining a variety of nanomaterials, in particular the iron containing ones. For instance, a simple, inexpensive and one-step arc discharge synthesis technique to prepare metal-containing carbon nanocapsules in aqueous solution is known [41]. It was established that iron nanoparticles can be *in situ* encapsulated in carbon shells when the arc discharge process was carried out in aqueous solutions of $FeSO_4$.

Combustion synthesis of iron oxide/iron-coated carbons such as cellulose fiber, anthracite, and activated carbon was reported a classic microwave oven with inverter technology [42]. The size of the iron oxide/iron nanoparticle-coated samples were determined to be in the range of 50–400 nm. It was found that iron oxide/iron nanoparticles exist in 4 main phases: γ-Fe_2O_3, α-Fe_2O_3, Fe_3O_4, and Fe, some of them had significant arsenic adsorption. In addition, carbon-coated iron nanoparticles with well-developed quasi-spherical shape were prepared with $Fe(NO_3)_3\cdot 9H_2O$ and starch as carbon source [43, 44]. This is an efficient approach for the mass production of nanocage structures under mild conditions, which needs to be further explored for preparing various carbon coated metal nanomaterials. *Radiation methods* are being applied more widely last 10–15 years, in that case for preparation of Fe-carbon-supported nanostructures. This way, 2 types of amorphous carbon films (15 at.% iron containing film) were deposited onto Si substrates by a sputtering method and further exposed to an electron flow, where the energy and dose rate were much smaller compared to the electron beam in a TEM. In this case, graphitic structures were observed in amorphous matrix at temperatures up to 450 K. It was established that the graphitization progressed more intensively during the electron irradiation than in annealing at 773 K. This was attributed to thermal and catalytic effects, strongly related to grain growth of metal clusters.

2.3. Free and supported iron oxides and ferrites

Zeolites and closely related supporting materials represent an ideal basis for iron oxide nanocomposites. This way, the zeolite loading with nanoiron oxide by a simple chemical process was described [45]. Final crystallite sizes of the doped nanomaterials were in the range of 4–6 nm. It was shown that the zeolites become to have magnetic properties after being doped with nanoiron oxide. Mesoporous nanocomposites "iron oxide/silicate" Fe_2O_3-SBA-15 (SBA-15 is an abbreviation for hexagonally ordered mesoporous silica) with iron loadings of 1.2–35.8 wt.% were prepared hydrothermally [46]. It was revealed that these composites contain well-dispersed iron oxide nanoclusters in the walls of ordered mesoporous silica and high surface area. Certain number of composite nanomaterials based on Fe_3O_4 is known, for instance core/shell Fe_3O_4 coated gold nanoparticles (diameter 50–100 nm) [47]. Their possible formation mechanism was proposed as follows: pH-sensitive polymer owing to a shrunken or stretched structure of polyethyleneimine (PEI), led to the aggregation of the Fe_3O_4-gold seed nanoparticles, then gold reduces onto the surface of Fe_3O_4-gold seed nanoparticles. It was concluded that these core/shell multifunction nanomaterials will not only have external magnetic separation by the core of Fe_3O_4 but also detect the large biological molecules using the shell of gold. In addition, iron phthalocyanine prepolymer/Fe_3O_4 nano hybrid magnetic material [48] can be applied as high temperature-resistant polymer magnetic composite material. At last, ferrites having different sizes, from ultrasmall (2 nm) to 50 nm, can be fabricated by distinct techniques [49] mainly co-precipitation method (CPM), sometimes without using any capping agents/surfactants.

3. Catalytic applications

3.1. nZVI and supported Fe^0 nanocomposites

3.1.1. Catalyzed removal or decomposition of pollutants

nZVI has been extensively reported to resolve a series of environmental problems, related with destruction, adsorption, precipitation, reduction or oxidation of heavy metals, salt anions, hydrocarbons and halogenated organic pollutants (Fig. 1) [50], leading to their conversion to final non-hazardous products. In these processes, iron nanoparticles have shown high efficiency and practically no damage for the environment because of absence of toxicity. Thus, the efficiencies of *nitrate removal* from aqueous solution by single (TiO_2 and nZVI), and composite (nano-TiO_2–Fe^0 composite, NTFC) system under UV illumination were studied [51]. Among the three systems, both nZVI and NTFC can effectively remove nitrate. However, only NTFC can achieve satisfactory transformation of nitrate to N_2. Reactive materials for catalytic *degradation of chlorinated organic compounds* in water at ambient conditions have been prepared on the basis of silica-supported Pd-Fe nanoparticles [52]. Nanoscale Fe-Pd particles were synthesized inside porous silica supports using $(NH_4)_3[Fe(C_2O_4)_3]$ and $[Pd(NH_3)_4]Cl_2$ or Pd acetate as reaction precursors. The reduction of these supported precursors using hydrogen led to materials, revealed high activity in the processes of perchloroethene (PCE) degradation and 2-chlorobiphenyl (2-ClBP) dechlorination. It was established that highly dispersed amorphous Fe-Pd bimetallic nanoparticles on silica support possess superior catalytic activity against PCE dechlorination, comparing with the free-standing Fe-Pd nanoparticles. It was also established that the addition of vitamin B12 (it is known to be an effective electron mediator, having strong synergistic effects with nZVI for reductive dehalogenation reactions) can significantly enhance the reductive dechlorination of PCE by nZVI [53]. A remarkable reductive dechlorination of PCE (0.25 ± 0.01 h^{-1}) was observed in nZVI suspension (0.05 g/24 mL) with 0.5 mM vitamin B12 in 6 h, while no significant reductive dechlorination of PCE was observed in the nZVI suspension without vitamin B12. Similar composite material based on deposition of nZVI particles and cyanocobalamine (vitamin B12) on a diatomite matrix was also used for catalytic transformation of PCE and other organic contaminants in water [54]. The composite material rapidly degrades or transforms completely a large spectrum of water contaminants, including halogenated solvents like TCE, PCE, and *cis*-DCE, pesticides like alachlor, atrazine and bromacyl, and common ions like nitrate, within minutes to hours. In a related publication [55], iron nanoparticles were applied for remediation of PCB-contaminated soil, taking into account a maximization of PCB destruction in each treatment stage. The efficiency of PCB destruction during the first step treatment (mixing of soil and iron nanoparticles in water) can be improved by increasing the water temperature. The PCB destruction efficiency of minimum 95% can be achieved. In air at 300 °C, Fe_2O_3 is also a good catalyst for remediating PCB-contaminated soils. In addition, photo-Fenton like method using nZVI/UV/H_2O_2 was applied [56] for removing total *petroleum hydrocarbons* (TPH) and determining the optimal conditions using Taguchi method. The removal rate in optimal conditions was between 95% and 100%. The nZVI particles can be reused in a magnetic field. This process may enhance the rate of diesel degradation in polluted water and could be used as a pretreatment step for the biological removal of TPH from diesel fuel in the aqueous phase. Among other degradation applications, we note the use of nZVI/AC for catalytic wet peroxide oxidation of phenol [57]. The catalytic activity of phenol degradation was improved by application of nZVI/AC catalysts compared to that of Fe/AC. For the range 150–1,000 mg/L, the phenol conversion ≥90% can be achieved using these nanocatalysts during 15 min of the reaction in presence of the stoichiometric hydrogen peroxide for complete mineralization. At last, as an example of classic organic synthesis, we emphasize the conversion of synthesis gas to C2 through C4 olefins with up to 60% selectivity by carbon [58], using catalysts which comprise Fe promoted nanoparticles (5–30 nm in diameter) homogeneously dispersed on weakly interactive α-alumina or carbon nanofiber supports.

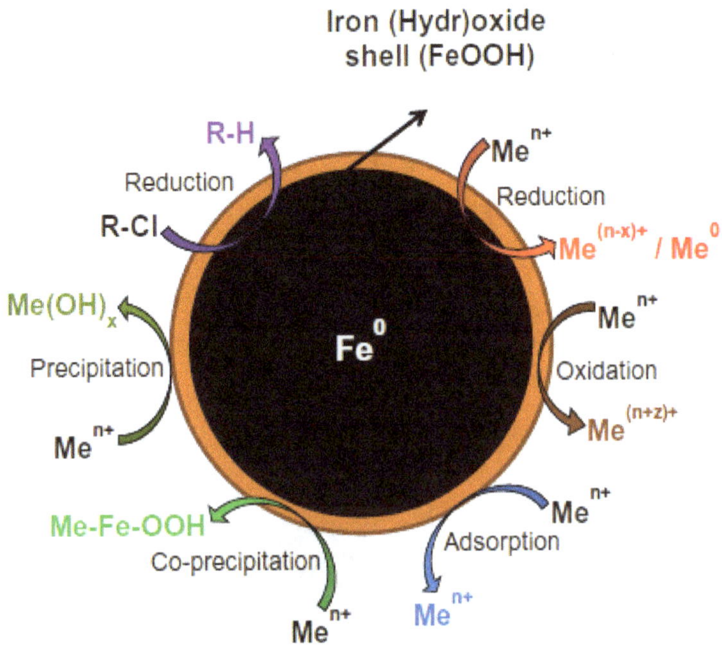

Figure 1. Core–shell structure of nZVI depicting various mechanisms for the removal of metals and chlorinated compounds. Adapted from Li et al. 2006 with permission.

Degradation of dyes with use of nZVI is carried out mainly on carbon-based nanocomposites. Thus, nZVI/activated carbon (nZVI/AC) was investigated as heterogeneous Fenton catalyst in 3D electrode system for methyl orange (MO) degradation [59]. The mineralization of MO was significantly improved by 20–35% compared to 2D AC system at the optimum conditions. A possible mechanism for decolorization and mineralization was proposed, which was attributed to the combination of adsorption, anodic oxidation, and Fenton oxidation in 3D nZVI/AC system. As an example of application of CNTs-based system, an ozone catalyst capable of working on acidic solution environments (labeled as CNTs-Fe^0), which was prepared by immobilizing nZVI onto the surface of MWCNTs and used for decomposition of methylene blue (MB) by formed hydroxyl radicals (HO·) [60]. At pH 3, the production of HO· was found to be considerably accelerated in the presence of CNTs-Fe^0 about 80 times in comparison with results using plain ozone. In the process of CNTs-Fe^0 catalytic ozonation, the CNTs support was analyzed to perform as effective "promoter" allowing the fast surface-mediated reactions, owing to the combination of its surface-active nature, conductivity, and chemical stability. All these and above technologies perfectly fit into advanced water treatment technologies, whose additional representative example is as follows. Thus, the *in situ* synthesis of air-stable nZVI embedded in cellulose fibers led to the assembly of highly reactive magnetic filter papers (membrane nanocomposites) [61]. This nZVI@FP nanocomposite (Fig. 2) showed high activity towards the removal of hexavalent chromium as well as an excellent catalytic ability to convert phenols into catechols, by simple filtration processes of the contaminated water solutions.

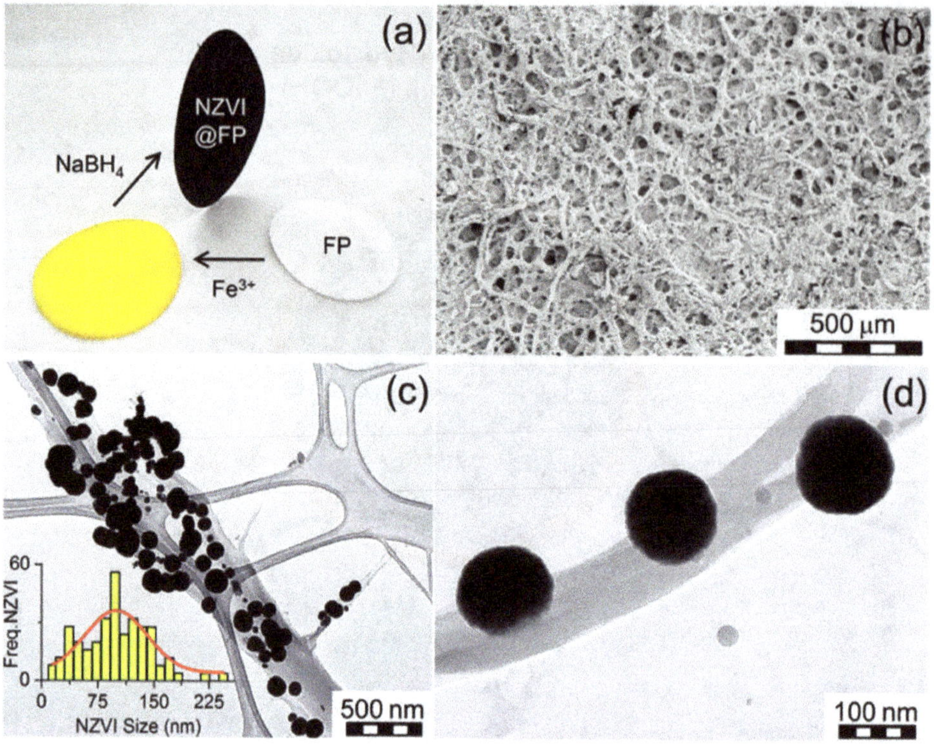

Figure 2. (a) Optical image showing the 2-step assembly of the magnetic nZVI@FP nanocomposite. (b) Scanning electron micrograph of nZVI@FP loaded with 5% in weight of nZVI. (c) TEM micrograph of NZVI entrapped over a cellulose fibre of FP. The inset shows the particle size distribution, as estimated from TEM (Freq.-nZVI *vs.* nZVI-size), together with the fitting analysis (red-line). (d) TEM micrograph of a magnified region of (c) revealing the intimate interaction between NZVIs and the fiber surface. FP = filter paper. Adapted from Datta et. al. 2014 with permission.

3.2. Fe–M nanoalloys, bimetallic NPs and core–shell nanostructures

Degradation of pollutants. Iron-based alloys, core-shell, and bimetallic nanoparticles, especially with noble metals, have been extensively applied in the catalysis and reviewed [62, 63, 64], so we will present in this section only their most representative recently reported examples. As well as nZVI described above, bimetallic Fe-containing nanoparticles (Fe with Pt, Ru, Rh, Ni, Co, Au, Cu, Ag) are used for the catalytic elimination of environmental pollutants. Reactions between them (pollutants and nanoparticles) can be mainly divided in 4 types [65]: a) catalytic replacement reactions for removal of heavy metals, b) hydrodehalogenation reactions (in case of halogenated hydrocarbons), c) azo and nitro hydrogenation reactions (for azo and nitro) compounds, and d) hydrodeoxygenation reactions (for oxyanions). In comparison with monometallic iron nanoparticles, the bimetallic iron NPs have considerable capacity to be separated and catalytic ability for degradation of non-biodegradable pollutants. Among them, Fe-Pt NPs are of an especial interest. Thus, Pt-Fe application as heterogeneous Fenton-like catalysts was reported for hydrogen peroxide decomposition and the decolorization of methylene blue [66]. FePt (and also Fe_3O_4, see more information below) NPs were prepared and tested as heterogeneous Fenton-like catalysts to evaluate and compare their efficiency toward the decolorization of MB dye in solution. Both FePt and Fe_3O_4 exhibited high activity toward the MB decolorization reaction though FePt exhibited a reaction rate that was 100 times faster for a 5 ppm catalyst concentration. Both FePt and Fe_3O_4 NPs are superparamagnetic and thus can be easily separated with a magnet and reused for subsequent catalytic cycles. The same objective was reached using core-shell NPs on the nZVI basis. Thus, a series of nanocomposites consisting of nZVI encapsulated in SiO_2 microspheres were applied for the degradation of organic dyes was investigated using MB as the model dye in the presence of H_2O_2 [67]. The degradation efficiency and apparent rate constant of the degradation reaction were significantly enhanced with increased nZVI encapsulated in SiO_2 microspheres, whereas the dosage of H_2O_2 remarkably promoted degradation rate without affecting degradation efficiency.

It should be also noted that Fe-containing *thin films* can be applied for dye degradation. Thus, 2D nano-TiO_2 and Fe-doped nano-TiO_2 thin films with large sizes were obtained [68] at low temperature in an aqueous system via molecular self-assembly approach. Degradation of methyl orange solution under action of UV and visible light radiation was applied for evaluation of the photocatalytic activity. The doped iron presence was shown to improve the TiO_2 photocatalytic activity. The degradation yields of methyl orange were 98.62% and 89.24%, respectively, under illumination by UV lamp and using visible light. As an example of another degradation process, we note that the longevity and reactivity of nZVI and palladized bimetallic particles (BNP) were evaluated in batch and column experiments for remediation of a trichloroethene (TCE)-contaminated plume within a clayey soil [69]. The particle behavior was found to be severely affected by clay sediments. Results of butch studies testified that TCE degradation in ORR clayey soil corresponds to a pseudo-first-order kinetic model with reaction rate constants (k) of 0.05–0.24 day^{-1} at varied iron-to-soil ratios. Despite of elevated reactivity in water phase, the BNP were less effective in the site-derived clay sediment resulting calculated TCE removal efficiencies of 98.7% and 19.59%, respectively.

Transformations of ketones using stabilized nZVI and bimetallic Fe NPs are known. Among these catalytic processes, we note that porous Pt-Fe bimetallic nanocrystals were found to effectively facilitate the manufacturing of 2-propanol from acetone [70]. It was suggested that the high reactivity is strictly related to the interface consist of bimetallic Pt-Fe alloy and the Fe_2O_{3-x}. As an example of reduction of substituted aromatic ketones to alcohols, iron nanoparticles (size 9 nm, 14 nm, and 17 nm), stabilized by polyethylene glycol (PEG), carboxymethyl cellulose (CMC), and poly N-vinyl pyrrolidone (PVP), were used as catalysts in the hydrogenation reaction of various substituted aromatic ketones to alcohols with $NaBH_4$ [71]. Fe-PEG NPs were found to act as better catalyst than Fe-CMC NPs and Fe-PVP NPs. The trend in the catalytic activity among metals falls in the line of decreasing size effect of the nanoparticles i.e., the order of the nanoparticle sizes increase as Fe-PEG < Fe-CMC < Fe-PVP. Also, effects of substituents in the aromatic ring of ketones revealed that +I substituents are better catalyzed than –I substituents. In addition, bimetallic ruthenium-iron nanoparticles constitute a magnetically recoverable heterogeneous catalyst for transfer hydrogenation with a pronounced selectivity for ketones over aldehydes and nitro groups [72]. The nanoparticles are recyclable up to five times without significant decrease in activity or leaching.

Other processes. An Fe group ternary nanoalloy FeCoNi (NA) catalyst (its synthesis see Fig. 3) enabled selective electrocatalysis towards CO_2-free power generation from highly deliverable ethylene glycol (EG) [73]. This FeCoNi nanoalloy catalyst exhibited the highest selectivities toward the formation of C_2 products and to oxalic acid, *i.e.*, 99% and 60%, respectively, at 0.4 V vs. the reversible hydrogen electrode (RHE), without CO_2 generation. The key feature was the formation of an atomically mixed FeCoNi alloy to enhance the synergetic effect of the Fe group elements on anti-self-oxidation and selective oxidation of EG to oxalic acid. Reduction processes also required polymetallic particles. Thus, trimetallic core/shell Pd/FePt NPs were applied in oxygen reduction reaction (ORR) catalysis [74]. The uniform FePt shell was formed by controlled nucleation of $Fe(CO)_5$ in the presence of a Pt salt and Pd NPs at designated reaction temperatures.

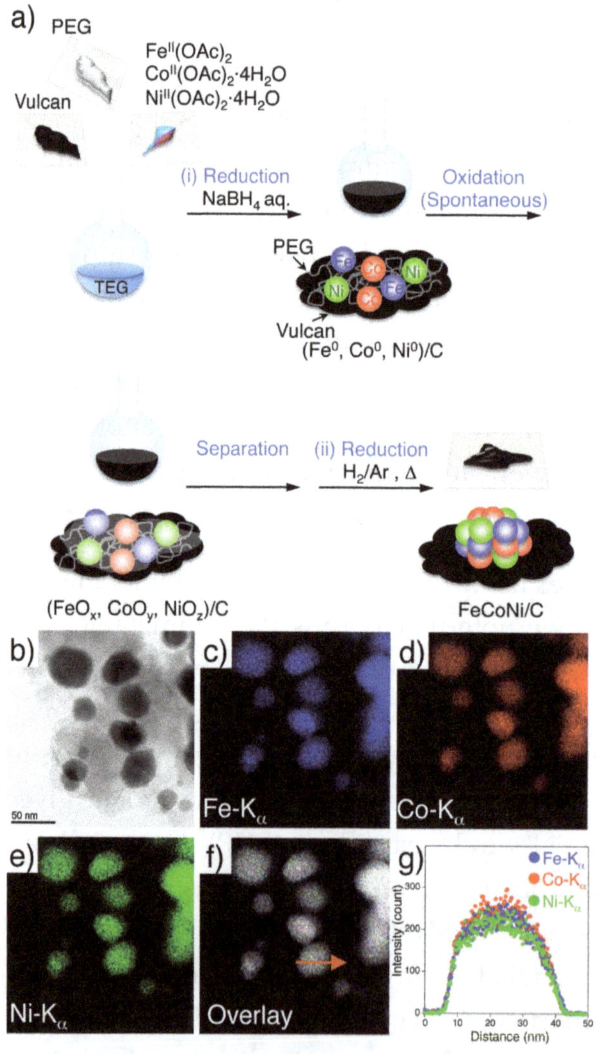

Figure 3. Synthetic scheme for the preparation of a FeCoNi nanoalloy catalyst supported on carbon (FeCoNi/C). Metallic Fe, Co and Ni form in the presence of polyethylene glycol (PEG) and a carbon support (vulcan) after the addition of an aqueous solution of $NaBH_4$. The metallic species are oxidised spontaneously, with production of an oxide mixture composed of Fe_3O_4, Co_3O_4, NiO, and so on is produced. FeCoNi/C was prepared by hydrogen reduction of the oxide mixture. Adapted from Matsumoto et al., 2014 with permission.

Iron-iron oxide core–shell nanoparticles were used as a catalyst for the *hydrogenation of olefins and alkynes* (reaction 3) under mild conditions in ethanol and in an aqueous medium [75]. The system is active in respect of a row of substrates and considerably selective for alkenes and alkynes over aromatic and carbonyl groups. The authors supposed that the presence of an oxide shell does not decrease its activity and provides a certain protection against oxidation by oxygen and water. In addition, highly active and well-defined AuPt nanoalloys, supported on the surface of ellipsoidal Fe@SiO_2 nanoparticles, were prepared by a method involving the loading of Pt NPs on the Fe_2O_3@SiO_2 nanocapsules *via* Sn^{2+} linkage and reduction, then *in situ* fabrication of Au nanoparticles by the galvanic replacement reaction between Au and Pt, and finally calcination and reduction to convert the nonmagnetic Fe_2O_3 to Fe core with high saturation magnetization [76]. The obtained Fe@SiO_2/AuPt samples exhibited a remarkably higher catalytic activity in comparison with the supported monometallic counterparts toward the *reduction of 4-nitrophenol to 4-aminophenol* by $NaBH_4$. The catalyst can be reused for several cycles with convenient magnetic separation.

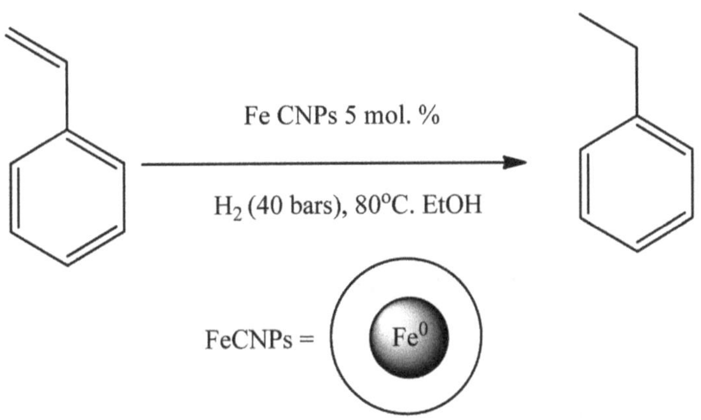

(3)

Hydrogenation of olefin catalyzed by Fe CSNPs.

3.3. Nano-Fe_2O_3 phases and their composites

Destruction of dangerous substances and pollutants. As well as Fe(0)-containing nanocomposites, oxidated iron forms are successfully used for degradation of several substances, both inorganic and organic. Thus, the decrease in temperature of *decomposition of ammonium perchlorate* in the presence of nano-ferric oxide was investigated [77] (see also section below on ferrites). It was shown that addition of nanometer-sized ferric oxide leads to a significant decrease in higher decomposition temperature of ammonium perchlorate. The catalytic activity of a colloidal catalyst (based on iron(III) oxides and obtained by hydrolysis followed by peptization of $FeCl_3 \cdot 6H_2O$ salt in water in the presence of 1% ethanol) in *decomposition of H_2O_2* was studied [78]. The obtained catalyst is mainly composed of α-Fe_2O_3 crystals with an admixture of other crystalline structures of iron oxides, as well as carbon-containing compounds. Its activity with respect to H_2O_2 decomposition varies nonlinearly and nonmonotonically and its particle size grows starting from 1–3 nm with an increase in the initial concentration of $FeCl_3 \cdot 6H_2O$ used to synthesize the catalyst. In addition, thermal decomposition of silver acetate at 200 °C in the presence of iron oxide microspheres in diphenyl ether led to the formation of iron oxide@Ag core–shell nanoparticles, exhibiting superparamagnetic behavior with a blocking temperature of about 42 K [79]. Their good catalytic activity and magnetic recovery was demonstrated by using two reactions, namely, *reduction of 4-nitrophenol* and *reduction of methylene blue* in aqueous solution. An especial case is the use of iron oxide / TiO_2 nanoparticles [80]. Thus, photocatalytic oxidation with TiO_2 nanoparticles (6–20 nm) was investigated as a promising water-treatment process [81, 82]. TiO_2 nanoparticles, after UV irradiation, are able to adsorb and degrade a huge variety of organic contaminants present in the environment. For example, in case of *organic arsenic species* (monomethylarsonic [MMA] and dimethylarsinic [DMA] acids), the strong affinity between the TiO_2 nanoparticles surface results a covalent bonding between MMA or DMA and the nanoparticle surface via formation of bidentate (AsMMA-Ti 3.32 Å) and monodentate (AsDMA-Ti 3.37 Å) inner-sphere complexes, respectively. Dopation of TiO_2 nanoparticles with Fe^{3+} ions at 0.1–0.5% may significantly increase the photocatalytic activity. The doped ions act as charge separators of the photoinduced electron–hole pair and enhanced interfacial charge transfers. Finally, an interesting application is known for *adsorption of elemental sulfur*. Thus, photoinduced sulfur desorption from the surfaces of Au Nps loaded on a series on metal oxides, in particular Fe_2O_3, was studied [83]. Elemental sulfur S_8 was selectively adsorbed on the Au Nps surfaces of Au/metal oxides in an atomic state. This phenomenon is applicable to the low temperature cleaning of sulfur-poisoned metal catalysts.

Organic synthesis. Free iron oxide(III) NPs and its nanocomposites have found numerous application in organic synthesis. Thus, hematite α-Fe_2O_3 NPs (diameters in the 7–18 nm range, synthesized by thermolysing a PVA-$Fe(OH)_3$ gel matrix at moderate temperatures) can effectively catalyze the *epoxidation of styrene* with *tert*-butyl hydroperoxide (TBHP) as the terminal oxidant [84]. Iron oxide nanoparticles supported on zirconia were tested in the gas-phase *conversion of cyclohexanol to cyclohexanone* in a fixed-bed flow type, Pyrex glass reactor, at 433–463 K [85]. Major detected products were cyclohexanone, cyclohexene and benzene, depending on the used catalyst. Experimental results showed that there was no leaching of metal, and that the catalyst was thus truly heterogeneous. In addition, core-shell Fe_2O_3/Pt nanoparticles with amorphous iron oxide cores exhibited superior catalytic activity with lower peak potential and enhanced CO_2 selectivity toward *methanol electrooxidation* in acidic medium [86]. This catalytic performance may be attributed to the uniform distribution of Pt particles on the amorphous Fe_2O_3 surface as well as interactions between the Pt particles and amorphous Fe_2O_3 cores. The catalytic activity of core-shell Fe_2O_3/Pt nanoparticles first increases and then decreases with decreasing Pt content. These nanomaterials also were found to have much higher structural stability and tolerance to the intermediates of methanol oxidation.

A magnetically separable core–shell iron oxide@nickel (IO@Ni) nanocatalyst, synthesized by reduction of Ni^{2+} ions in the presence of iron oxide (Fe^{2+}, Fe^{3+}) by a one-pot synthetic route using $NaBH_4$ as a reducing agent and starch as a capping agent, was found to have excellent activity for the *hydrogenation reactions* of aromatic nitro compounds under mild conditions using water as a green solvent (reaction 4). Excellent chemoselectivity and recyclability up to 30 cycles for the nitro group reduction was demonstrated. Nano propylsulphonated γ-Fe_2O_3 (NPS-γ-Fe_2O_3, reaction 5) was applied as a magnetically recyclable heterogeneous catalyst for the efficient one-pot synthesis of bis(pyrazolyl)methanes in water (reaction 6) [87]. The catalyst was easily isolated from the reaction mixture by a magnetic bar and reused at least five times without significant degradation in activity. Nanoporous α-Fe_2O_3 nanoparticles (about 100 nm in size and containing pores <10 nm) were synthesized *via* a hydrothermal method and applied in the catalytic *benzylation of benzene and benzyl chloride* (BC) in the fabrication of diphenylmethane (DPM) [88]. The BC conversion reached 100% in a reaction time of 3 min with 97.76% selectivity to DPM. The nanoporous α-Fe_2O_3 nanoparticles also have potential applications in other Friedel-Crafts alkylations, especially in large molecular reactions. Another example is related with chemical recycling of PET. Thus, easily recoverable superparamagnetic γ-Fe_2O_3 nanoparticles (10.5 nm in size and 147 $m^2 \cdot g^{-1}$ surface area, produced by calcining Fe_3O_4 nanoparticles prepared by the co-precipitation method) were used as a reusable catalyst for *PET glycolysis* [89]. At 300°C and a 0.05 catalyst/PET weight ratio, the maximum bis(2-hydroxyethlyl) terephthalate (BHET) monomer yielded reached more than 90% in 60 min. The catalyst was reused 10 times, giving almost the same BHET yield each time. In addition, heterogeneous photo-Fenton reaction which utilizes nanosized iron oxides as catalyst for maximizing the activity due to the enhanced physical or chemical properties brought about by the unique structures was described [90].

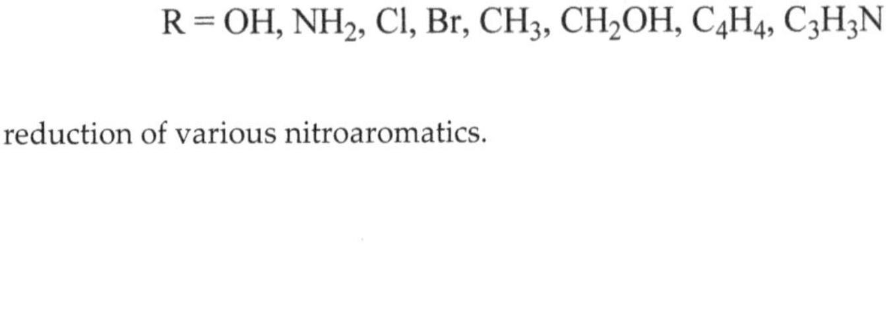

(4)

R = OH, NH_2, Cl, Br, CH_3, CH_2OH, C_4H_4, C_3H_3N

General scheme for the reduction of various nitroaromatics.

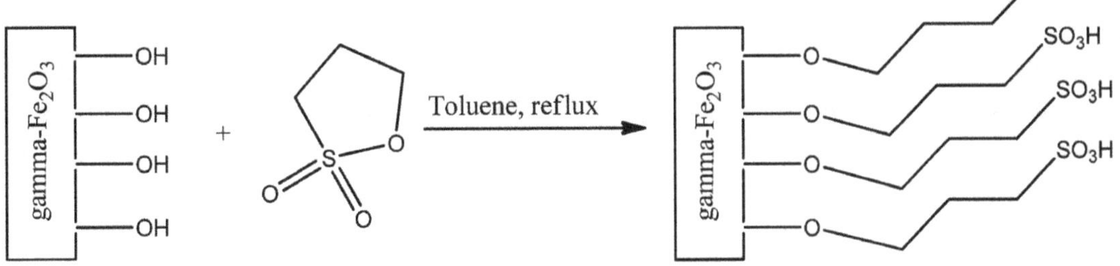

(5)

Synthesis of NPS-γ-Fe_2O_3.

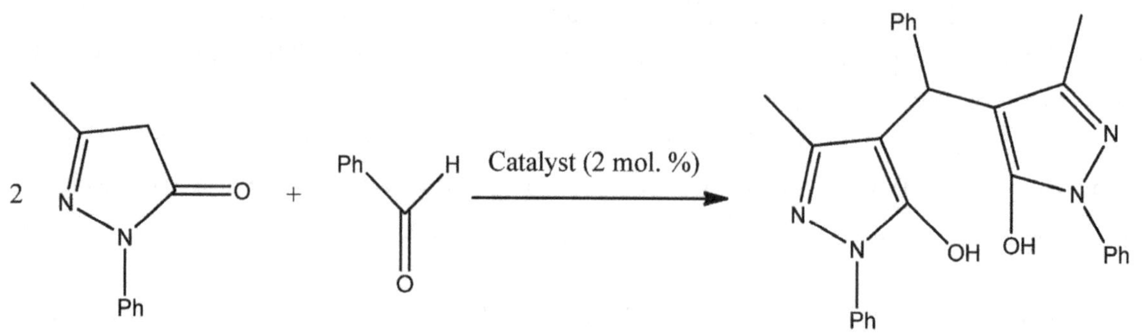 (6)

The reaction of benzaldehyde with 1-phenyl-3-methyl-5-pyrazolone.

Antibacterial activities. The doping effects of silver(I) and iron(III) on photocatalytic results using TiO_2 thin films were studied [91]. Ag and Fe doping and co-doping contents on nano-titania photocatalytic bactericidal films were obtained by sol-gel technique, thus uniting three classic active antibacterial species (Ag, Fe and TiO_2). The photocatalytic activity of TiO_2 films was confirmed by the sterilizing rate of the *E-coli* in each case. Applying fluorescent light irradiation, the optimal doping amounts of silver(I)/titania and iron(III)/titania were found to be 0.05% and 0.1%, respectively. In addition, a photocatalytic technique using visible light and carbon nanotubes and nano-sized Fe_2O_3 powder was used to *inhibit pathogenic bacterial growth* in water [92]. It was suggested that after careful design, this system can be used to disinfect drinking water, making it free of pathogenic bacteria.

3.4. Nano-Fe_3O_4 phases and their composites

Iron(II,III) oxide based nanostructures are slightly less explored in the organic synthesis. Thus, haemin-functionalized magnetic iron(II,III) oxide nanoparticles (Fe_3O_4/haemin) exhibited pronounced electrocatalytic activity towards *trichloroacetic acid* (TCA) like haemin itself (a linear detection range of 5–80 M and a detection limit of 0.3 M at 60 °C) [93]. This activity towards TCA was affected by detection temperature, which was controlled via electrically heated carbon paste electrodes. The maximal catalytic current was 5.8 times higher at 60 °C than at room temperature (25 °C). A process capable of synthesizing minor fractions of *aromatic hydrocarbons* (benzene, toluene, xylenes, and mesitylene) from CO_2 and H_2 at modest temperatures (T = 380 to 540°C) employing Fe/Fe_3O_4 nanoparticles as catalyst was designed [94]. The authors consider this technology as principally compatible with solar heat and hydrogen technology and having the potential to mitigate the impacts of global warming by making use of the existing distribution technology for gasoline. Also, *trisubstituted imidazoles* can be synthesized condensation reaction from 1,2-diketones, aromatic aldehydes, and ammonium acetate (reaction *7*) in high yield in the presence of sulphamic acid functionalized magnetic Fe_3O_4 nanoparticles (reaction *8*) as a solid acid catalyst under solvent-free classical heating conditions or using microwave irradiation [95].

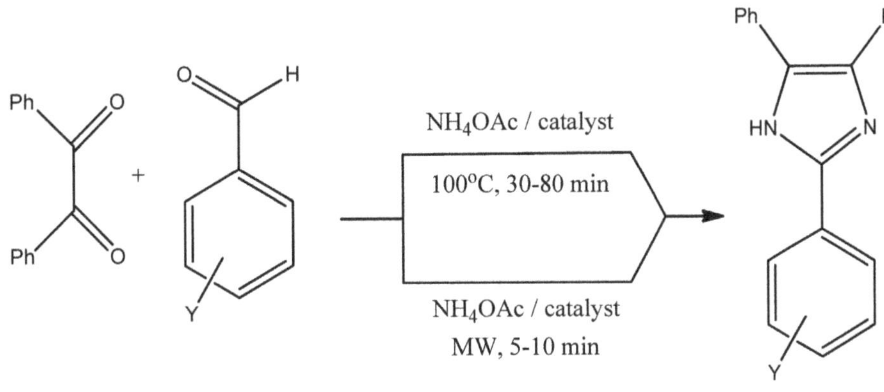

(7)

One-pot synthesis of 2,4,5-trisubstituted imidazoles catalyzed by sulphamic acid functionalized magnetic Fe_3O_4 nanoparticles under conventional heating conditions or using microwave irradiation.

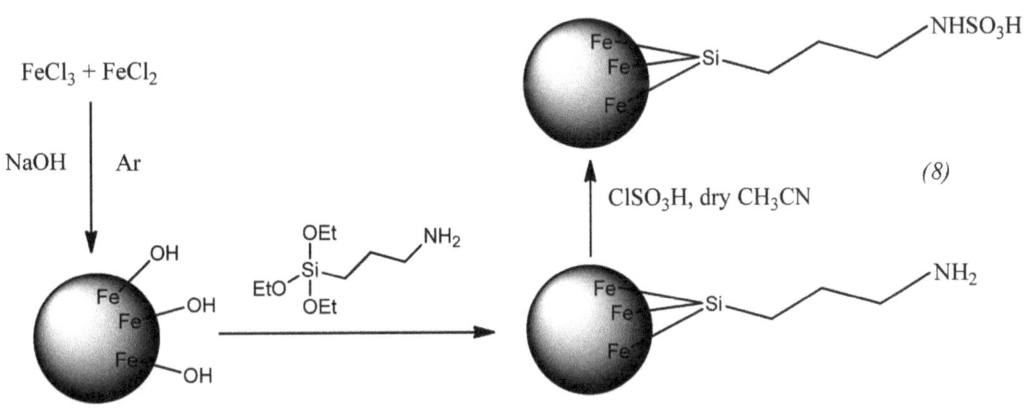

(8)

Preparation steps for fabricating sulphamic acid functionalized magnetic Fe_3O_4 nanoparticles.

Among other reactions, we note an efficient one-pot, three-component condensation reaction between 4-hydroxycoumarin, aryl glyoxals, and malononitrile catalyzed by Fe_3O_4 nanoparticles, which was carried out for the synthesis of several *dihydropyrano[c]chromenes* [96]. Also, an inexpensive and non-hazardous sulfuric acid functionalized magnetic Fe_3O_4 nanoparticles (efficiently catalyze one-pot multicomponent condensation of β-naphthol with aromatic and aliphatic aldehydes and amide derivatives (reaction *9*) at 100 °C under solvent-free conditions to afford the corresponding *amidoalkyl naphthols* in excellent yields and in very short reaction times [97]. Silver(0) nanoparticles supported on silica-coated Fe_3O_4 (synthesis see reactions *10*) serve as an efficient and recyclable heterogeneous catalyst for oxidant-free *dehydrogenation of alcohols* to the corresponding carbonyl compounds [98]. The catalyst can be easily recovered and reused for 8 reaction cycles without considerable loss of activity. At last, a nanocomposite of functionalized Ni(II) complex containing surface of pyridine, methoxysilanyl and amino groups with iron(II,III) oxide, Fe_3O_4@[-Ni(bpy)$_2$(py-tmos)] was found to be highly efficient green catalyst for the synthesis of a diverse range of 3,4-dihydropyrimidin-2(1H)-ones under solvent free conditions, and in addition it could be easily recovered by a simple magnetic separation and recycled at least 5 times without deterioration in catalytic activity [99].

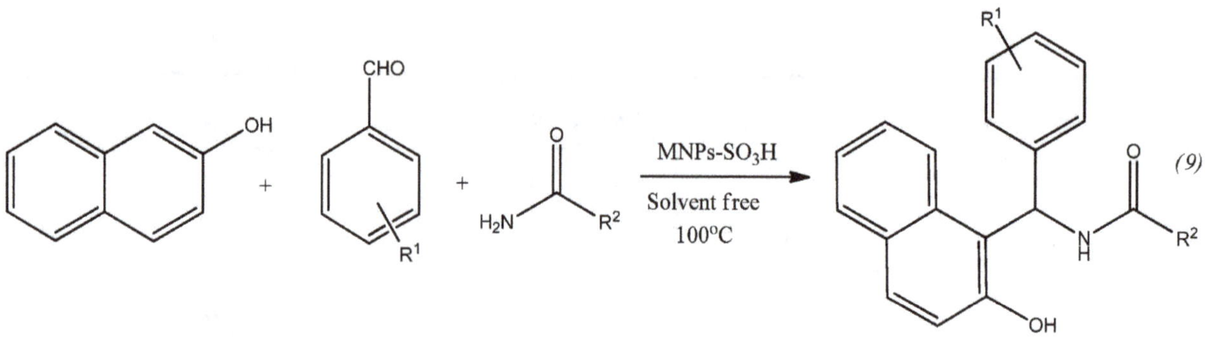

(9)

Synthesis of amidoalkyl naphthols

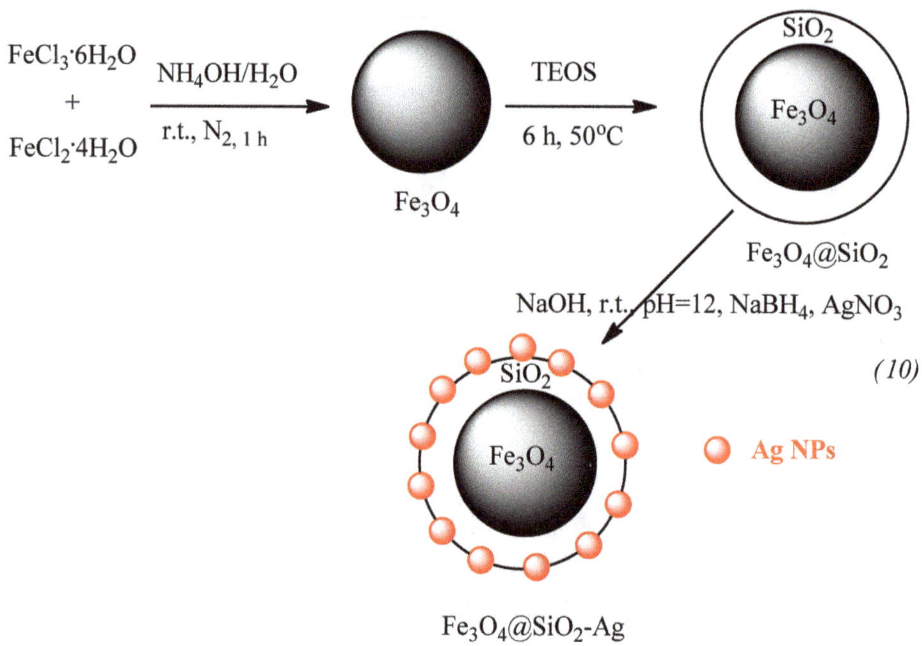

(10)

Preparation of magnetically recoverable heterogeneous nanocatalyst Fe_3O_4@SiO_2-Ag.

3.5. Ferrites

Cobalt ferrites having different sizes, from ultrasmall (2 nm) to 50 nm, can be fabricated by distinct techniques [100], mainly co-precipitation method (CPM), sometimes without using any capping agents/surfactants. Thus, the CPM was used to synthesize ultrasmall $CoFe_2O_4$ superparamagnetic nanoparticles (SPMNPs, 2–8 nm of an average size and high surface area of 140.9 m^2/g) without any surfactant [101]. Their catalytic activity was verified in the preparation of *arylidene barbituric acid derivatives* (reaction *11*) applying $CoFe_2O_4$ SPMNPs in aqueous ethanol as a reusable catalyst, which can be magnetically separated from the reaction system. Main advantages of this approach were found to be high yields, short reaction time and high turnover frequency, a clean reaction methodology, and chemoselectivity, among several others. More large-size $CoFe_2O_4$ magnetic nanoparticles (25 nm) were used as a catalyst for the *oxidation of various alkenes* in the presence of *tert*-butylhydroperoxide (*t*-BuOOH) with almost quantitative yields (reaction *12*) [102]. It seemed that this heterogeneously catalysis system proceeds by coordination of *t*-BuOOH to the metal (Fe^{3+} cations) on the surface of the catalyst. The separation of the catalyst from the reaction media was easily achieved with the aid of an external magnet, and the catalyst can be reused several times with no loss of activity. In addition, combination of synthesis techniques can also be used for cobalt ferrite preparation. Thus, synthesis of spinel $CoFe_2O_4$ MNPs (average sizes 40–50 nm) was achieved by a combined sonochemical and co-precipitation technique in aqueous medium, also without any surfactant or organic capping agent [103]. These uncapped NPs were utilized directly for *aldol reaction* in ethanol (reaction 13). After the reaction was over, the nanoparticules were compartmented by using an external magnet.

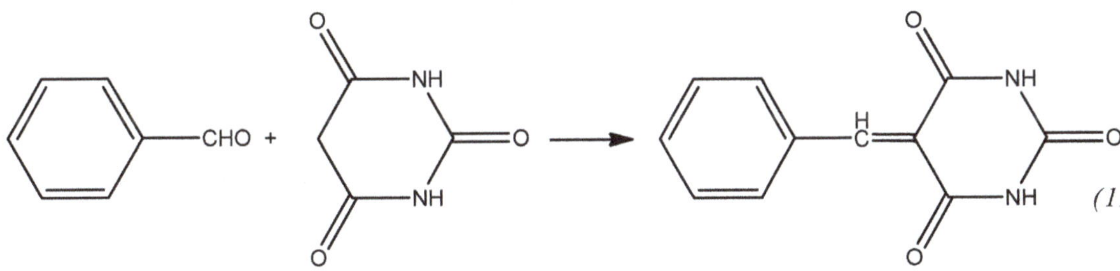

(11)

Optimization of reaction conditions using $CoFe_2O_4$ nanocatalyst. Yields 40–95%, best results in EtOH.

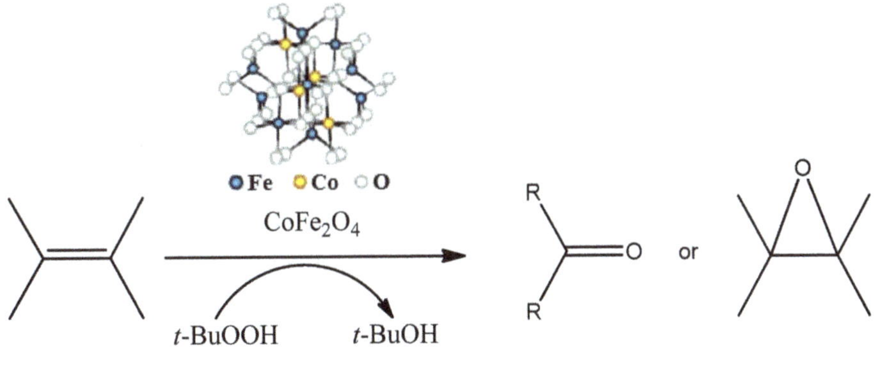

(12)

Oxidation of alkenes using $CoFe_2O_4$ catalyst.

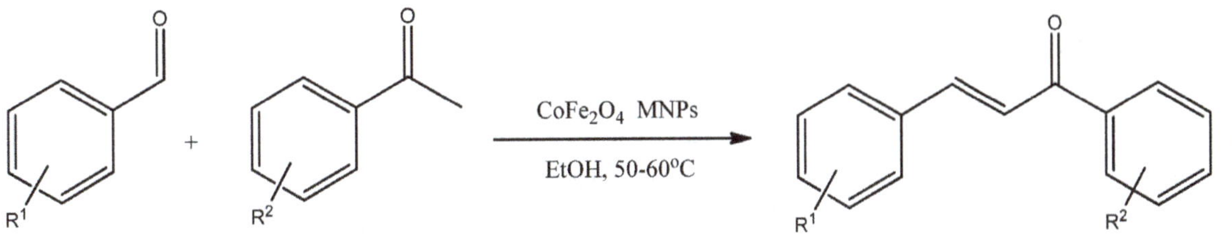

(13)

Figure 4. Aldol condensation reaction in presence of cobalt ferrite MNPs.

Nickel ferrites. Pure or doped nickel ferrites out of the nano-range size are common and frequently used in several catalytic processes. For instance, high reactivity of $NiFe_2O_4$ (111) surfaces (higher that in Fe_3O_4) is well-known; $NiFe_2O_4$ is an effective metal-doped ferrite catalyst in a typical industrial process such as the water-gas shift (WGS) reaction [104]. Similarly, $NiFe_2O_4$ was examined as catalyst in photocatalytic water oxidation using $[Ru(bpy)_3]^{2+}$ as a photosensitizer and $S_2O_8^{2-}$ as a sacrificial oxidant [105]. The catalytic activity of $NiFe_2O_4$ is comparable to that of a catalyst containing Ir, Ru, or Co in terms of O_2 yield and O_2 evolution rate under ambient reaction conditions. As an example of non-nano-sized doped ferrites, their catalysts (granules of ~1 mm diameter) of nickel, cobalt and copper, prepared by co-precipitation hydrothermal route and impregnated with palladium, cerium and lanthanum as promoters [106], were tested for carbon monoxide oxidation activities. The catalysis of $NiFe_2O_4$ nanoparticles on the *hydrogen storage performances* of magnesium hydride synthesized by high-energy ball milling was studied [107], showing that the initial dehydrogenation temperature of 7 mol.% $NiFe_2O_4$-doped MgH_2 is 191°C, which is 250°C lower than that of pristine MgH_2. The enhancement in the H_2 storage performances of MgH_2 by adding $NiFe_2O_4$ nanoparticles is primarily ascribed to intermetallic Fe_7Ni_3 and (Fe, Ni) phases during the desorption procedure, which act as the real catalyst in the 7 mol.% $NiFe_2O_4$-doped sample. As an example of another application, a magnetic acidic catalyst comprising Preyssler ($H_{14}[NaP_5W_{30}O_{110}]$) heteropoly acid supported on silica coated nickel ferrite nanoparticles ($NiFe_2O_4@SiO_2$) was investigated for the synthesis of *bis(dihydropyrimidinone)benzene and 3,4-dihydropyrimidin-2(1H)-ones derivatives* by the Biginelli reaction [108]. With the catalyst, the reactions occurred in less than 1 h with good to excellent yields.

Copper ferrites. Non-nanosized range copper ferrites have certain catalytic applications, such as, for example, for CO conversion to CO_2 [109]. In a difference of pure nickel ferrites, copper ferrite NPs are applied in organic catalysis in more uniform particle size (mainly about 20 nm). Thus, nano material on the basis of copper ferrite (20 nm) was applied as reusable heterogeneous initiator in the preparation of 1,4-dihydropyridines. The interaction of substituted *aromatic aldehydes, ethyl acetoacetate and ammonium acetate* (reaction *14*) was observed in presence of $CuFe_2O_4$ nano powders in ethanol at ambient conditions. The nano catalyst can be magnetically recovered and reused [110]. The same 20-nm size copper ferrite nano material also was reported as reusable heterogeneous initiator in the synthesis of β,γ-unsaturated ketones and allylation to acid chlorides in THF at r.t. without any additive/co-catalyst (reactions 15–16) [111]. The notable advantages are less expensive, heterogeneous reusable catalyst; mild reaction conditions, high yields of products, shorter reaction times, no isomerization during the reaction, and easy workup. In addition, 20-nm $CuFe_2O_4$ was applied as reusable heterogeneous initiator in the synthesis of *α-aminonitriles* by one-pot reaction of different aldehydes with amines and trimethylsilyl cyanides at r.t. in water as a solvent (reactions *17–18*) [112]. α-Aminonitriles are important in preparing a wide variety of amino acids, amides, diamines, and nitrogen containing heterocycles. In addition, a strategy for the synthesis of *benzoxazoles* from substituted N-(2-halophenyl)benzamides (reaction *19*) was developed [113], where inexpensive, readily available, air-stable, recyclable copper(II) ferrite serves as a nanocatalyst. Also, larger-size cubic copper ferrite $CuFe_2O_4$ nanopowders (24–51 nm in size) were synthesized *via* a hydrothermal route using industrial wastes (ferrous sulfate containing free sulfuric acid ≈10%, 0.01% Zn^{2+} and 2% silica; copper waste 12.5% Cu, 8.7% Cl^- with minor Ni 0.001%) [114]. Study of photocatalytic degradation of the *methylene blue* (MB, $C_{16}H_{18}ClN_3S$) dye using copper ferrite powders showed a good catalytic efficiency (95.9%) at hydrothermal temperature 200 °C for hydrothermal time 24 h at pH = 12 due to high surface area (118.4 m^2/g).

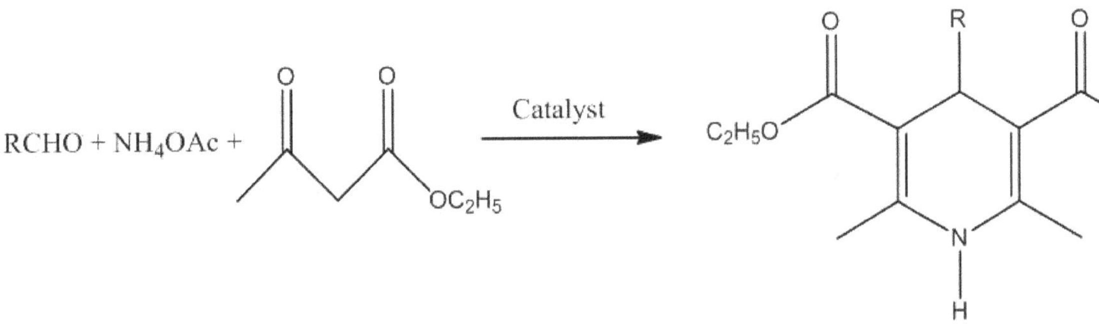

(14)

Catalyst: Copper ferrite (1 mol.%), R = a) Ph, b) 4-Me-*o*-C_6H_4, c) 4-ClC_6H_4, d) 4-NO_2-C_6H_4, e) 4-MeC_6H_4, (f) 3-NO_2-C_6H_4, (g) *n*-C_9H_9, (h) 2-NO_2-C_6H_4, (i) 2-furyl, (j) 2-Me-*o*-C_6H_4.

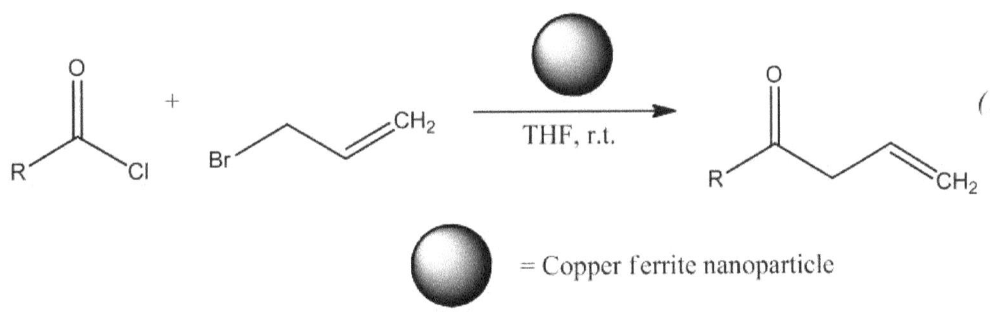

(15)

Synthesis of *β,γ*-unsaturated ketone using allyl bromide. R = (a) C_6H_5, (b) 2-ClC_6H_4, (c) 2-Br, 5-F, C_6H_3, (d) 2-Br, 5-F, C_6H_3, (e) furanyl, (f) 5-phenyl, 3-Methyl, 4-Isoxazolyl, (g) 5-(2,5-dichloro)phenyl, 3-methyl, 4-isoxazolyl, (h) $(CH_3)_3C$-, (i) $C_{11}H_{23}$-, (j) $C_{15}H_{31}$-.

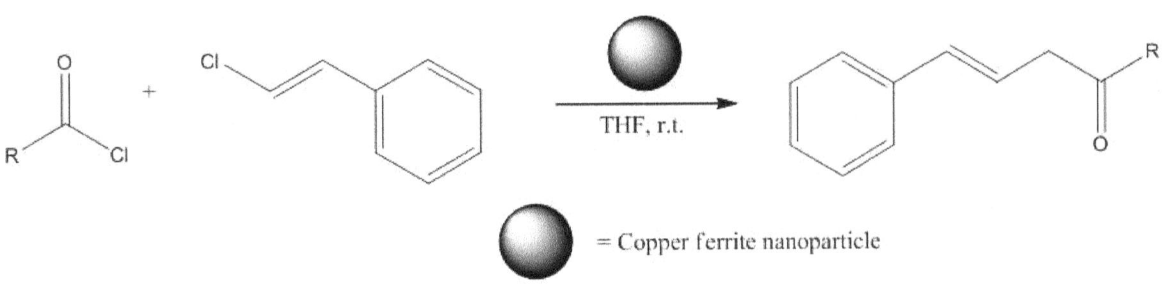

(16)

Synthesis of *β,γ*-unsaturated ketone using cinnamyl chlorides. Synthesis of *β,γ*-unsaturated ketone using cinnamyl chlorides. R = (a) C_6H_5, (b) 2-ClC_6H_4, (c) furanyl, (d)-$CH(CH_3)_2$.

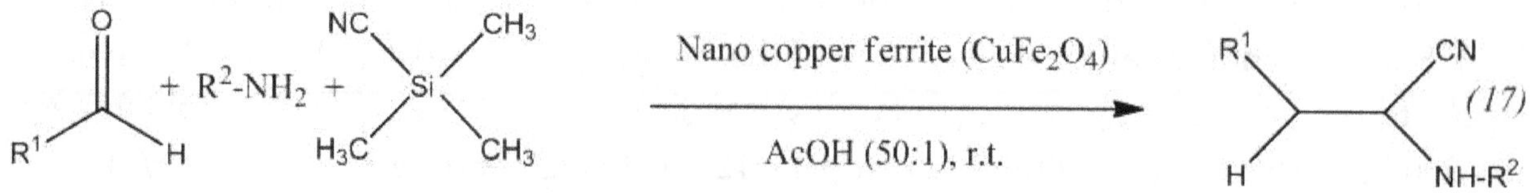

(17)

The synthesis of α-aminonitriles in the presence of nano $CuFe_2O_4$ in water as green solvent at r.t.

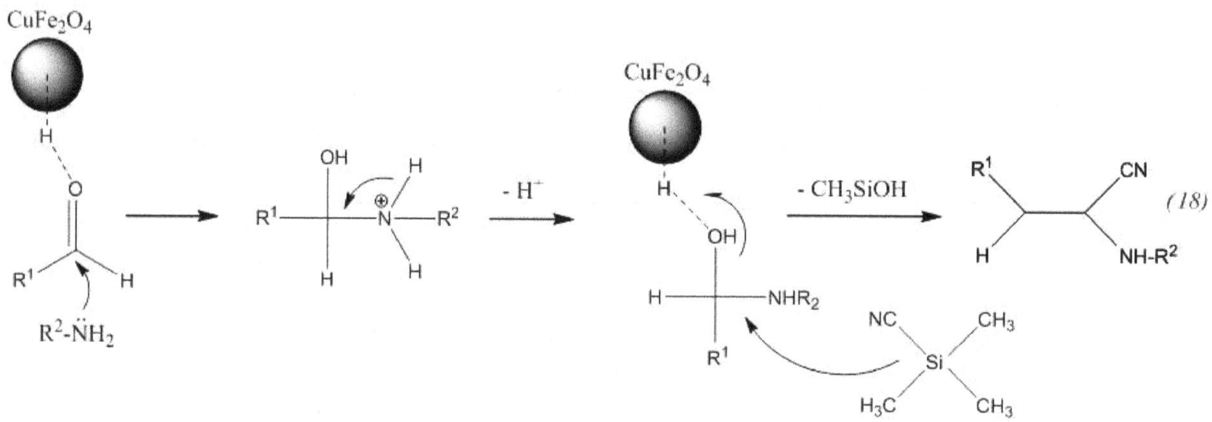

(18)

Suggested mechanism for the synthesis of α-aminonitriles derivatives in presence of acidic nano copper ferrite.

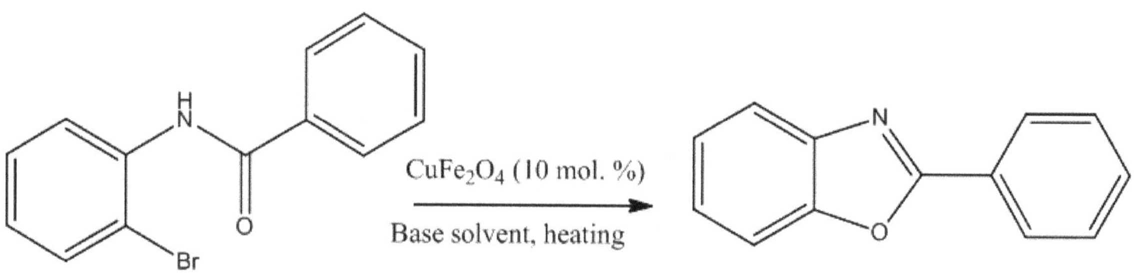

(19)

Catalyzed cyclization of N-(2-bromophenyl)benzamide to 2-phenyl-1,3-benzoxazole.

Zinc ferrite. Non-nano-sized zinc ferrites ($ZnFe_2O_4$) have been used in oxidative organic reactions. Thus, the catalytic behavior for *oxidative conversion of methane and oxidative coupling of methane* was investigated over pure and neodymium substituted zinc ferrites prepared by combustion method [115]. The catalytic activity proved to be strongly related to the oxide structure as well as to the specific defects created by substitution. The pure zinc ferrite ($ZnFe_2O_4$) and $ZnNd_2O_4$ exhibited high activity for coupling reaction whereas the neodymium substituted ferrites ($ZnFe_{1.75}Nd_{0.25}O_4$, $ZnFe_{1.5}Nd_{0.5}O_4$ and $ZnFeNdO_4$) was low active in this reaction. The order of the catalytic activities expressed as yields to C_2^+ were $ZnNd_2O_4$ > $ZnFe_2O_4$ > $ZnFe_{1.75}Nd_{0.25}O_4$ > $ZnFeNdO_4$ > $ZnFe_{1.5}Nd_{0.5}O_4$.

Analyzing pure zinc ferrite nanocatalysts, we note that mainly ultrasmall particles are currently applied in catalytic purposes. Thus, a nanosized highly ordered mesoporous zinc ferrite (ZF, 7–10 nm in size) was synthesized *via* co-precipitation method, further sulfated with ammonium sulfate solution to obtain sulfated ZF (SZF) and was used for the synthesis of nopol by Prins condensation of β-pinene and paraformaldehyde (reaction *20*) [116]. 70% *conversion of β-pinene* with 88% selectivity to nopol was observed; the spent catalyst was regenerated and reused successfully up to four cycles with slight loss in catalytic activity. The influence of various reaction parameters such as solvent, reaction temperature, effect of substrate stoichiometry and catalyst loading was investigated. In particular, very low conversion (9%) of β-pinene was observed in protic solvents such as methanol (solvent effect); in case of apolar–aprotic solvents such as hexane, ethyl acetate and toluene, increase in the β-pinene conversion as well as nopol selectivity was observed. In case of reaction temperature, as the temperature increased to 110 °C, β-pinene conversion increased to 72% with slight drop in selectivity for nopol (57%). 95 °C was identified as optimal reaction temperature for further studies. At molar ratio of β-pinene to paraformaldehyde 1:3 molar ratio, maximum conversion was observed. The β-pinene conversion increased with an increase the catalyst loading (0.12 to 0.16 g) without affecting the nopol selectivity.

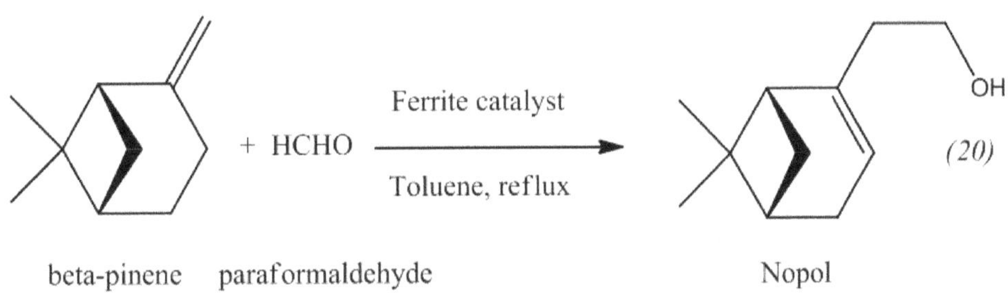

(20)

Prins condensation reaction of β-pinene and paraformaldehyde.

Other simple ferrites. The catalytic behavior of supported Au NPs for the process of *oxidation of benzyl alcohol* was elucidated in presence of gold nanoparticles [117]. Mg^{2+} ions, being present in the ferrite structure, led to an improvement of the catalytic activity of supported Au NPs to *ca.* 35% conversion, when an additional base was absent. Modifying the support with addition of MgO, the catalytic activity of supported Au nanoparticles was further improved to *ca.* 50% conversion; however, the catalyst was found to be deactivated in successive recycling tests. As well as nano-Fe_2O_3 mentioned above in the corresponding section, nano-$MnFe_2O_4$ particles (20–30 nm in size), synthesized by co-precipitation phase inversion method and low-temperature combustion method, using $MnCl_2$, $FeCl_3$, $Mn(NO_3)_2$, $Fe(NO_3)_3$, NaOH and $C_6H_8O_7$, were applied for thermal *decomposition of ammonium perchlorate* [118]. The catalytic mechanism was explained by the favorable electron transfer space provided by outer *d* orbit of transition metal ions and the high specific surface absorption effect of $MnFe_2O_4$ particles. Manganese ferrite nanoparticles were also applied in the synthesis of *spirooxindoles* (compounds **I-III**) *via* a one-pot and three-component reaction of isatins, malononitrile, and anilinolactones in the presence of a catalytic amount of $MnFe_2O_4$ NPs in PEG-400, as a nontoxic, green, and reusable solvent [119].

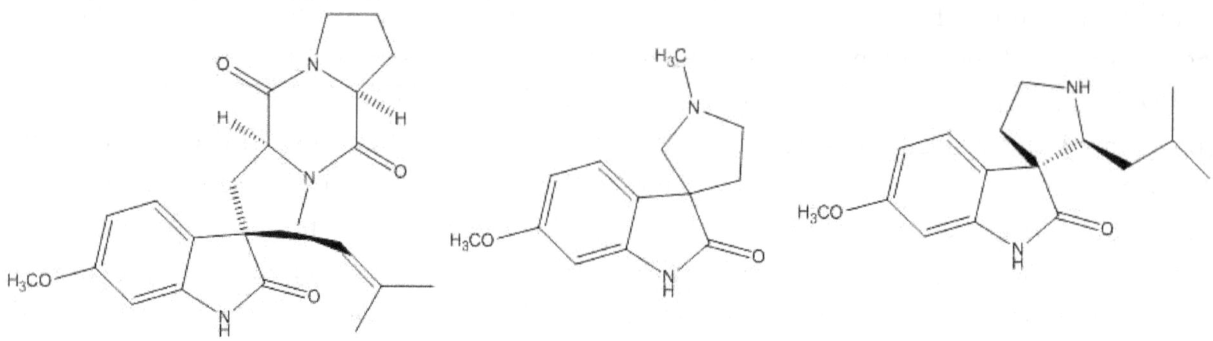

I Spirotryprostatin A **II** Horsfiline **III** Elacomine

Selected spirooxindole natural products.

Mixed-metal or core-shell ferrites. Ferrites containing 2 metal ions, additionally to iron, are much more widespread in the nano-catalysis; their nanoparticle size can vary in a broad range, from ultrasmall particles (5–8 nm) up to 100 nm or more (in case of supported NPs). Both nano-sized and out-of-nano-sized mixed-metal ferrite NPs can be synthesized by a variety of methods, in particular classic sol-gel and co-precipitation methods or microwave heating ($MnZnFe_2O_4$ [120]).

***Cobalt**-based ferrite nanoparticles.* For cobalt-containing ferrite NPs, as well as for zinc ferrite above, one of important applications is the methanol decomposition to CO and hydrogen. Thus, $Cu_{1-x}Co_xFe_2O_4$ ($0<x<1$, 8–40 nm in size) was applied as a nanodimensional powder for this purpose [121]. The stabilization of the cubic structure with the substitution of copper ions by cobalt in mixed Cu-Co ferrites was observed. Cobalt containing ferrites exhibited higher and more stable catalytic activity and selectivity in *methanol decomposition* to CO and hydrogen in comparison with the $CuFe_2O_4$ one. Photocatalytic properties of the cobalt zinc ferrite $Co_{1-x}Zn_xFe_2O_4$ ($0<x<1$) nanoparticles (10.5–14.8 nm in size), prepared by a hydrothermal method, were studied on the example of *degradation of methyl blue* in aqueous solution [122]. It was elucidated that the oxidation-reduction potential of methyl blue aqueous solution in presence of the ferrite nano-particles at pH=7 under natural sunlight irradiation was negative and increased with increase in Zn content. The degradation rate of methyl blue also decreases as increase in Zn content in sunlight.

***Nickel**-based ferrite nanoparticles.* Similar to cobalt ferrites, several nickel-containing mixed or core-shell ferrites have been reported as nanocatalysts but in more narrow size range (18–50 nm). Thus, a magnetically separable catalyst consisting of ferric hydrogen sulfate (FHS) supported on silica-coated nickel ferrite nanoparticles (50 nm) was prepared [123]. This catalyst was shown to be an efficient heterogeneous catalyst for the *synthesis of 1,8-dioxodecahydroacridines* (reaction *21*) under solvent-free conditions. The catalyst can be recycled several times with no significant loss of catalytic activity.

Other mixed-metal ferrite nanoparticles. Ferrite nanoparticles, containing other metals and applied in the catalysis, are represented more chaotically in the available literature. Thus, the spinel ferrites $Cu_{1-x}Cd_x[Fe_{1-x}Al_xCr_{1-x}Mn_x]O_4$, where $0<x<1$, having unknown particle size, were

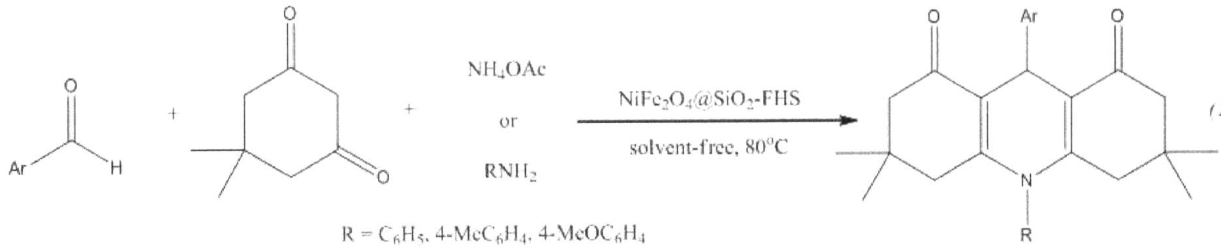

(21)

Synthesis of 1,8-dioxodecahydroacridines in the presence of $NiFe_2O_4$@SiO_2-FHS.

prepared by the coprecipitation technique [124]. Catalytic studies using *decomposition of H_2O_2* as a model reaction between 303 and 343 K for 1–5 h using first order rate law suggested higher catalytic power for the composition $x = 0$ and then it decreases gradually. For the mixed spinel ferrite system $Mn_{1-x}Cu_xFe_2O_4$ ($x = 0, 0.25, 0.5, 0.75, 1.0$), the formation of phase pure spinels with FCC cubic structure with particle size ranging from 5.21 nm to 20 nm was observed at 333 K applying co-precipitation method with $MnCl_2$, $Fe(NO_3)_3 \cdot 9H_2O$ and $Cu(NO)_3 \cdot 3H_2O$ as precursors [125]. These ferrites were used as catalysts in the *alkylation of aniline*, showing a maximum conversion of 80.5% of aniline with selectivity of 98.6% towards N-methylaniline at 673 K, methanol/aniline molar ratio of 5:1 and weight hour space velocity of 0.2 h^{-1}. It was found that the yield is maximum for $CuFe_2O_4$. In addition, the catalytic performance of the ferrites was found to be proportional to surface area as well as acidity.

4. Conclusions

Iron-based nanoparticles, utilized in catalytic reactions described in this chapter, possess different sizes, from ultrasmall (2 nm) to 100 nm. They are obtained mainly by wet-chemical sol-gel or co-precipitation methods, sometimes combined with simple calcination at high temperatures, sonochemical technique, mechanical high-energy ball milling, or spark plasma sintering, among other methods. Microwave heating or hydrothermal route are also frequently used. Due to magnetic properties, these nano catalysts can be easily recovered from reaction systems and reused up to several runs almost without loss of catalytic activity.

Catalytic processes with application of iron-based nanocomposites are in a wide range. Notable attention is paid to methanol decomposition to CO and methane or to CO and hydrogen. Other catalyzed organic reactions consist of oxidation of various alkenes, aldol, alkylation and dehydrogenation reactions, synthesis of various organic compounds such as, for example, quinoxaline derivatives [126], β,γ-unsaturated ketones, arylidene barbituric acid derivatives, α-aminonitriles, nopol, 1,4-dihydropyridines, and 1,8-dioxodecahydro-acridines. Degradation/decomposition processes are also reported, for instance decomposition of H_2O_2 or photocatalytic degradation of methylene blue. Some of catalyzed reactions might have great practical applications, for instance transesterification of soybean oil to biodiesel. In addition, small iron-based particles could also be considered [127] as substituents of noble metals in a variety of catalytic transformations. Several iron nanomaterials could have biological applications, such as peroxidase-like catalytic activity of Fe_3O_4 ultrasmall NPs [128].

We note that the total number of nano-iron composites applications for catalytic purposes is still not high, so it could be a perfect research niche for further applications of these nanomaterials in a variety of organic processes.

Author details

Boris I. Kharisov[1], Oxana V. Kharissova[1*], H.V. Rasika Dias[2], Ubaldo Ortiz Méndez[1], Idalia Gómez de la Fuente[1], Yolanda Peña[1] and Alejandro Vázquez Dimas[1]

*Address all correspondence to: okhariss@mail.ru

1 Universidad Autónoma de Nuevo León, Monterrey, México

2 The University of Texas at Arlington, Arlington, TX, USA

References

[1] Li, X.Q.; Elliott, D.W.; Zhang, W. Zero-Valent Iron Nanoparticles for Abatement of Environmental Pollutants: materials and engineering aspects. Crit. Rev. Solid State Mater. Sci. 2006, 31, 111–122.

[2] Hofmann-Amtenbrink, M.; von Rechenberg, B.; Hofmann, H. Superparamagnetic nanoparticles for biomedical applications. In: Nanostructured Materials for Biomedical Applications, 2009, Transworld Research Network, Kerala, India.

[3] Mahmoudi, M.; Sahraian, M. A.; Shokrgozar, M. A.; Laurent, S. Superparamagnetic Iron Oxide Nanoparticles: Promises for Diagnosis and Treatment of Multiple Sclerosis. ACS Chem. Neurosci. 2011, 2, 118–140.

[4] Huber, D. L. Iron nanoparticles, in: Dekker encyclopedia of nanoscience and nanotechnology, Schwarz, J. A.; Contescu, C. I.; Putyera, K., Eds., 2008, vol. 3, pp. 1681–1687, CRC Press, Taylor and Francis Group, Boca Raton, FL.

[5] C.S. Rajan.Nanotechnology in Groundwater Remediation. International Journal of Environmental Science and Development, 2011, 2(3), 182-187.

[6] Cheong, S.; Ferguson, P.; Hermans, I.F.; Jameson, G.N.L.; Prabakar, S.; Herman, D.A.J.; Tilley, R.D. Synthesis and Stability of Highly Crystalline and Stable Iron/Iron Oxide Core/Shell Nanoparticles for Biomedical Applications. ChemPlusChem, 2012, 77, 135–140.

[7] Harm, U.; Schuster, J.; Mangold, K.-M. Modification of iron nanoparticles for ground water remediation. DECHEMA, Karl-Winnacker Institut, 2010, http://kwi.dechema.de/kwi_media/Downloads/ec/F564_Nanoeisen_Harm.pdf.

[8] Bai J, Wang J-P. High-magnetic-moment core-shell-type FeCo-Au/Ag nanoparticles. Appl. Phys. Lett., 2005, 87(15), 152502/1-152502/3.

[9] Figuerola, A.; Di Corato, R.; Manna, L.; Pellegrino, T. From iron oxide nanoparticles towards advanced iron-based inorganic materials designed for biomedical applications. Pharmaceutical Research, 2010, 62(2), 126–143.

[10] Qing Wei Ding, Tian Wei Qian, Hong Fang Liu, Xue Wang. Preparation of Zero-Valent Iron Nanoparticles and Study of Dispersion. 2011, Applied Mechanics and Materials, 55–57, 1748–1752.

[11] Jae-min Lee; Ji-hun Kim; Jin-wook Lee; Jae-hwan Kim; Ho-seok Lee; Yoon-seok Chang. Synthesis of Fe-nano Particles Obtained by Borohydride Reduction with Solvent. Paper A-068, in: Bruce M. Sass (Conference Chair), Remediation of Chlorinated and Recalcitrant Compounds—2008. Proceedings of the Sixth International Conference on Remediation of Chlorinated and Recalcitrant Compounds (Monterey, CA; May 2008). ISBN 1-57477-163-9, published by Battelle, Columbus, OH, www.battelle.org/chlorcon.

[12] Khodabakhshi, A.; Amin, M.M.; Mozaffari. M. Synthesis of magnetite nanoparticles and evaluation of its efficiency for arsenic removal from simulated industrial wastewater. Iran. J. Environ. Health. Sci. Eng., 2011, 8 (3), 189–200.

[13] Park, H.; Ayala, P.; Deshusses, M.A.; Mulchandani, A.; Choi, H.; Myung, N.V. Electrodeposition of maghemite (γ-Fe_2O_3) nanoparticles. Chem. Eng. J., 2008, 139, 208–212.

[14] Grabis, J.; Heidemane, G.; RAŠMANE, D. Preparation of Fe_3O_4 and γ-Fe_2O_3 Nanoparticles by Liquid and Gas Phase Processes. Mat. Sci. (MEDŽIAGOTYRA). 2008, 14 (4), 292–295.

[15] Ballard, S.G. Apparatus and methods for the production of powders. US20056972115 (2005).

[16] Kim, W.-B., Park, J.-S., Suh, C.-Y., Kil, D.-S., Lee, J.-C. 2007, US20070209477.

[17] Lei, J.P.; Dong, X.L.; Zhu, X.G. et al. Formation and characterization of intermetallic Fe–Sn nanoparticles synthesized by an arc discharge method. Intermetallics, 2007, 15(12), 1589–1594.

[18] Gupta, M.; Wong, E.; Leong, W. Microwaves and metals. Wiley-Interscience, 2007, 256 pp.

[19] Komarneni, S.; Katsuki, H.; Li, D.; Bhalla, A.S. Microwave–polyol process for metal nanophases. J. Phys.: Condens. Matter., 2004, 16, S1305-S1312.

[20] Becker, M.F.; Brock, J.R.; Cai, H. et al. Metal nanoparticles generated by laser ablation. Nanostruct. Mater. 1998, 10(5), 853–863.

[21] Rodrigues, A.R.; Soares, J.M.; Machado, F.L.A.; de Azevedo, W.M.; de Carvalho, D.D. Synthesis of α-Fe particles using a modified metalmembrane incorporation technique. J. Magnetism Magnetic. Mat., 2007, 310(2), Part 3: 2497–2499.

[22] Klaine, S.J.; Alvarez, P.J.J.; Batley, G.E.; Fernandes, T.F.; Handy, R.D.; Lyon, D.J.; Mahendra, S.; Mclaughlin, M.J.; Lead, J.R. Nanomaterials in the environment: behavior, fate, bioavailability, and effects. Environmental Toxicology and Chemistry, 2008, 27(9), 1825–1851.

[23] Yuvakkumar, R.; Elango, V.; Rajendran, V.; Kannan, N. Preparation and characterization of zero-valent iron nanoparticles. Digest Journal of Nanomaterials and Biostructures, 2011, 6(4), 1771–1776.

[24] Bönnemann, H.; Brijoux, W.; Brinkmann, R.; Dinjus, E.; Joussen, T.; Korall, B. Erzeugung von kolloiden Übergangsmetallen in organischer Phase und ihre Anwendung in der Katalyse. Angew. Chem., 1991, 103, 1344–1346; Angew. Chem. Int. Ed. Engl., 1991, 30, 1312.

[25] Harutyunyan, A.; Grigorian, L.; Tokune, T. Method for synthesis of metal nanoparticles. 2005, US20056974493.

[26] Klinke, C.; Kern, K. Iron nanoparticle formation in a metal-organic matrix: from ripening to gluttony. Nanotechnology, 2007, 18, 215601.

[27] Lu, Y.; Zhu, Z.; Liu, Z. Carbon-encapsulated Fe nanoparticles from detonation-induced pyrolysis of ferrocene. Carbon, 2005, 43(2), 369–374.

[28] Anastas, P. T.; Horvath, I. T. Green Chemistry for a Sustainable Future. Wiley; 1st edition, 2012, 350 pp.

[29] Chao-Jun Li, Paul T. Anastas. Handbook of Green Chemistry — Green Processes. Wiley-VCH; 3 Volume Set edition, 2012, 1326 pp.

[30] Ahluwalia, V. K. Green Chemistry: Environmentally Benign Reactions, Second Edition. CRC Press; 2nd edition, 2012, 326 pp.

[31] Patel, J. T.; Patel, O. B.; Raval, B. P. Green Chemistry: New Avenues in Chemical Research: Focus in Healthcare. LAP LAMBERT Academic Publishing, 2012, 60 pp.

[32] Luque, R. Green Chemistry. Nova Science Publishers, 2011.

[33] Hoag, G. E.; Collins, J. B.; Varma, R. S.; Nadagouda, M. Green synthesis of metal nanoparticles using plant extracts. PCT Int. Appl. 2009, WO 2009140694 A2 20091119.

[34] Kipkurgat Erick Tanui. Green Synthesis and Characterization of Iron Nanoparticles. http://chemistry.uonbi.ac.ke.

[35] Yao, C.; Ma, H.; Tong, Y. Electrochemical preparation and magnetic study of amorphous nanostructured Nd-Fe-Co-Ni-Mn high entropy alloy film. Yingyong Huaxue, 2011, 28(10), 1189–1194.

[36] Glebov, A.V.; Glebov, V.A.; Popova, O.I. Development of nanofilm Fe-Pt magnets for superdense-recording heads. Tsvetnye Metally, 2009, (12), 67–70.

[37] Wang, J.; Cheng, R. Preparation and application of activated carbon supported iron nanomaterial, with application to pentachlorophenol degradation. 2010, CN 101708457.

[38] Bystrzejewski, M. Synthesis of carbon-encapsulated iron nanoparticles via solid state reduction of iron oxide nanoparticles. Journal of Solid State Chemistry, 2011, 184(6), 1492–1498.

[39] Wei, Z.; Wang, X.; Yang, H. Preparation of carbon-encapsulated Fe core-shell nanostructures by confined arc plasma. Materials Science Forum, 2011, 688(Nano-Scale and Amorphous Materials), 245–249.

[40] Rao, C.N.R.; Thomas, P.J.; Kulkarni, G.U. Nanocrystals: Synthesis, Properties and Applications (Springer Series in Materials Science). Springer, 2007, 182.

[41] Bingshe, X.; Junjie, G.; Xiaomin, W.; Xuguang, L.; Hideki, I. Synthesis of carbon nanocapsules containing Fe, Ni or Co by arc discharge in aqueous solution. Carbon, 2006, 14(13), 2631–2634.

[42] Nadagouda, M.N.; Lytle, D.A. Microwave-assisted combustion synthesis of nano iron oxide/iron coated activated carbon, anthracite, cellulose fiber, and silica, with arsenic adsorption studies. Journal of Nanotechnology, 2011, 972486, 8 pp.

[43] Yuliang An; Jieshan Qiu. Synthesis of carbon encapsulated iron nanoparticles by carbonization of starch with iron as catalyst. 2007, The American Carbon Society. http://acs.omnibooksonline.com/data/papers/2004_C085.pdf

[44] An Yuilang; Wu Xiaojuan; Sui Zhiming; yuan Xia, Liu yanqiu. A novel method for synthesis of homogeneous carbon encapsulated Fe nanoparticles based on natural biopolymer. J. Rare Earths, 2007, 25, 452.

[45] Salah El-Din, T.A.; Elzatahry, A.A.; Aldhayan, D.M.; Al-Enizi, A.M.; Al-Deyab, S.S. Synthesis and Characterization of Magnetite Zeolite Nano Composite. Int. J. Electrochem. Sci.,2011, 6, 6177–6183.

[46] Liu, H.; Lan, G.; Yan, Y.; Tang, H.; Liu, H.; Li, Y. Direct hydrothermal synthesis of novel ordered magnetic mesoporous nanocomposites with high content of iron. Gongye Cuihua, 2011, 19(8), 11–15.

[47] Shi, Y.-f.; Zhou, X.; Zhong, L.-b.; Xu, W.-l.; Wang, Y.; Zhang, Q.-q. Synthesis and formation mechanism of core-shell Fe_3O_4 coated gold nanomaterials. Dongnan Daxue Xuebao, Yixueban, 2011, 30(1), 6–10.

[48] Liu, X.; Xu, M.; Zhao, R.; Zhong, J. Iron phthalocyanine prepolymer/Fe_3O_4 nano hybrid magnetic material and its preparation method. Edited by: Casciano, D.A.; Sahu, S.C. Faming Zhuanli Shenqing, 2011, CN 102086304.

[49] Goodarz Naseri, M.; Saion, E.B.; Abbastabar Ahangar, H.; Halim Shaari, A.; Hashim, M. Simple Synthesis and Characterization of Cobalt Ferrite Nanoparticles by a Thermal Treatment Method. J. Nanomater. Volume 2010, Article ID 907686, 8 pp.

[50] Li, X.Q.; Elliot, D.W.; Zhang, W.X. Zero-valent iron nanoparticles for abatement ofenvironmental pollutants: materials and engineering aspects. Crit. Rev. Solid State, 2006, 31, 111–22.

[51] Pan, J.R.; Huang, C.; Hsieh, W.-P.; Wu, B.J. Reductive catalysis of novel TiO_2/Fe^0 composite under UV irradiation for nitrate removal from aqueous solution. Separation and Purification Technology, 2012, 84, 52–55.

[52] Kustov, L.M.; Al-Abed, S.R.; Virkutyte, J.; Kirichenko, O.A.; Shuvalova, E.V.; Kapustin, G.I.; Mishin, I.V.; Nissenbaum, V.D.; Tkachenko, O.P.; Finashina, E.D. Novel Fe-Pd/SiO_2 catalytic materials for degradation of chlorinated organic compounds in water. Pure Appl. Chem. 2014, 86(7), 1141–1158.

[53] Amir, A.; Lee, W.Enhanced reductive dechlorination of tetrachloroethene by nanosized zero valent iron with vitamin B12. Chem. Engin. J., 2011, 170, 492–497.

[54] Dror, I.; Jacov, O.M.; Cortis, A.; Berkowitz, B. Catalytic Transformation of Persistent Contaminants Using a New Composite Material Based on Nanosized Zero-Valent Iron. ACS Appl. Mater. Interfaces, 2012, 4, 3416–3423.

[55] Varanasi, P.; Fullana, A.; Sidhu, S. Remediation of PCB contaminated soils using iron nano-particles. Chemosphere, 2007, 66(6), 1031–1038.

[56] Dehghani, M., Shahsavani, E.; Farzadkia, M.; Reza Samaei, M. Optimizing photo-Fenton like process for the removal of diesel fuel from the aqueous phase. J. Environ. Health Sci. Eng., 2014, 12, 87.

[57] Messele, S.A.; Bengoa, C.; Stuber, F.; Fortuny, A.; Fabregat, A.; Font, J. Catalytic wet peroxide oxidation of phenol using nanoscale zero-valent iron supported on activated carbon. Desalination and Water Treatment, 2015, 55, 1–10. http://www.tandfonline.com/doi/full/10.1080/19443994.2014.1002011#preview

[58] Torres Galvis, H.M.; Bitter, J.H.; Khare, C.B.; Ruitenbeek, M.; Dugulan, A.I.; de Jong, K.P. Supported Iron Nanoparticles as Catalysts for Sustainable Production of Lower Olefins. Science, 2012, 335, 835.

[59] Zhang, C.; Zhou, L.; Yang, J.; Yu, X.; Jiang, Y.; Zhou, M.Nanoscale zero-valent iron/AC as heterogeneous Fenton catalysts in three-dimensional electrode system. Environ. Sci. Pollut. Res. 2014, 21, 8398–8405.

[60] Zhang, S.; Wang, D.; Zhang, X.; Fan, P. Zero-Valent Iron Immobilized on Multi-Walled Carbon Nanotubes for Heterogeneous Catalytic Ozonation of Methylene Blue as Model Compound. Clean – Soil, Air, Water, 2014, 42 (5), 609–616.

[61] Datta, K.K.R.; Petala, E.; Datta, K.J.; Perman, J.A.; Tucek, J.; Bartak, P., Otyepka, M.; Zoppellaro, G.; Zboril, R. NZVI modified magnetic filter paper with high redox and catalytic activities for advanced water treatment technologies. Chem. Commun., 2014, 50, 15673–15676.

[62] Varma, R.S. Nano-catalysts with magnetic core: sustainable options for greener synthesis. Sustainable Chemical Processes, 2014, 2, 11.

[63] Wei, S.; Wang, Q.; Zhu, J., Sun, L.; Line, H.; Guo, Z. Multifunctional composite core–shell nanoparticles. Nanoscale, 2011, 3, 4474–4502.

[64] Govan, J.; Gun'ko, Yu.K. Recent Advances in the Application of Magnetic Nanoparticles as a Support for Homogeneous Catalysts. Nanomaterials, 2014, 4, 222–241.

[65] Liu, W.-J.; Qian, T.T.; Jiang, H. Bimetallic Fe nanoparticles: Recent advances in synthesis and application in catalytic elimination of environmental pollutants. Chemical Engineering Journal, 2014, 236, 448–463.

[66] Hsieh, S.; Lin, P.-Y. FePt nanoparticles as heterogeneous Fenton-like catalysts for hydrogen peroxide decomposition and the decolorization of methylene blue. J. Nanopart. Res. 2012, 14, 956.

[67] Mao, Z.; Wu, Q.; Wang, M.; Yang, Y.; Long, J.; Chen, X. Tunable synthesis of SiO_2-encapsulated zero-valent iron nanoparticles for degradation of organic dyes. Nanoscale Research Letters, 2014, 9, 501.

[68] Li, Q.-x.; Sun, Z.-y.; Wang, T. Edited by: Casciano, D.A.; Sahu, S.C. A study on preparation, characterization and photocatalytic activity of iron-doped nano-TiO_2 thin films based on self-assembled monolayers. Kuangwu Xuebao, 2011, 31(1), 102–107.

[69] Katsenovich, Y.P.; Miralles-Wilhelm, F.R. Evaluation of nanoscale zerovalent iron particles for trichloroethene degradation in clayey soils. Science of the Total Environment, 2009, 407, 4986–4993.

[70] Ji, Y.; Wu, Y.; Zhao, G.; Wang, D.; Liu, L.; He, W.; Li, Y. Porous bimetallic Pt-Fe nanocatalysts for highly efficient hydrogenation of acetone. Nano Research, 2015, 8(8), 2706-2713. http://link.springer.com/article/10.1007%2Fs12274–015-0777-z

[71] Parimala, L.; Santhanalakshmi, J. Studies on the Iron Nanoparticles Catalyzed Reduction of Substituted Aromatic Ketones to Alcohols. Journal of Nanoparticles. Volume 2014, Article ID 156868, 10 pp.

[72] Hudson, R.; Chazelle, V.; Bateman, M.; Roy, R.; Li, C.; Moores, A. Sustainable synthesis of magnetic ruthenium-coated iron nanoparticles and application in the catalytic transfer hydrogenation of ketones. ACS Sustainable Chem. Eng., 2015, 3 (5), 814–820.

[73] Matsumoto, T.; Sadakiyo, M.; Lee Ooi, M.; Kitano, S.; Yamamoto, T.; Matsumura, S.; Kato, K.; Takeguchi, T.; Yamauchi, M. CO_2-Free Power Generation on an Iron Group Nanoalloy Catalyst via Selective Oxidation of Ethylene Glycol to Oxalic Acid in Alkaline Media. Sci. Reports, 2014, 4, 5620, 6 pp.

[74] Mazumder, V.; Chi, M.; More, K.L.; Sun, S. Core/Shell Pd/FePt Nanoparticles as an Active and Durable Catalyst for the Oxygen Reduction Reaction. J. Am. Chem. Soc. 2010, 132, 7848–7849.

[75] Hudson, R.; Riviere, A.; Cirtiu, C.M.; Luska, K.L.; Moores, A. Iron-iron oxide core–shell nanoparticles are active and magnetically recyclable olefin and alkyne hydrogenation catalysts in protic and aqueous media. Chem. Commun., 2012, 48, 3360–3362.

[76] Zhang, Z.; Zhou, Y.; Zhang, Y.; Zhou, S.; Xiang, S.; Sheng, X.; Jiang, P. A highly reactive and magnetic recyclable catalytic system based on AuPt nanoalloys supported on ellipsoidal Fe@SiO_2. J. Mater. Chem. A, 2015, 3, 4642–4651.

[77] Mahinroosta, M. Catalytic effect of commercial nano-CuO and nano-Fe_2O_3 on thermal decomposition of ammonium perchlorate. Journal of Nanostructure in Chemistry, 2013, 3, 47.

[78] Lesin, V.I.; Pisarenko, L.M.; Kasaikina, O.T. Colloidal Catalysts Based on Iron (III) Oxides. 1. Decomposition of Hydrogen Peroxide. Colloid Journal, 2012, 74 (1), 85–90.

[79] Sharma, G.; Jeevanandam, P. A Facile Synthesis of Multifunctional Iron Oxide@Ag Core–Shell Nanoparticles and Their Catalytic Applications. Eur. J. Inorg. Chem., 2013, 6126–6136.

[80] Zaleska, A. Doped-TiO_2: A Review. Recent Patents on Engineering, 2008, 2, 157–164.

[81] Aufan, M. Nanoparticules d'oxydes mrtalliques: relations entre la reactivite de surface et des reponses biologiques. http://www.tel.archives-ouvertes.fr/docs/00/30/85/03/PDF/auffan_PhD.pdfSimilares.

[82] Jing, C.; Xiaoguang Meng; Suqin Liu; Salem Baidas; Ravi Patraju; Christos Christodoulatos; George P. Korfiatis. Surface complexation of organic arsenic on nanocrystalline titanium oxide. Journal of Colloid and Interface Science, 2005, 290(1), 14–21.

[83] Tada, H.; Soejima, T.; Ito, S.; Kobayashi, H. Photoinduced desorption of sulfur from gold nanoparticles loaded on metal surfaces. J. Amer. Chem. Soc. 2004, 126(49), 15952–15953

[84] Bepari, R.A.; Bharali, P.; Das, B.K. Controlled synthesis of α-and γ-Fe_2O_3 nanoparticles via thermolysis of PVA gels and studies on α-Fe_2O_3 catalyzed styrene epoxidation. Journal of Saudi Chemical Society, 2014, In press.

[85] Sadiq, M.; Zamin, G.; Razia; Ilyas, M. Synthesis and Characterization of Iron Oxide Nanoparticles Supported on Ziconia and Its Application in the Gas-Phase Oxidation of Cyclohexanol to Cyclohexanone. Modern Research in Catalysis, 2014, 3, 12–17

[86] Liu, Y.-T.; Yuan, Q.-B.; Duan, D.-H.; Zhang, Z.-L.; Hao, X.-G.; Wei, G.-Q.; Liu, S.-B. Electrochemical activity and stability of core-shell Fe_2O_3/Pt nanoparticles for methanol oxidation. Journal of Power Sources, 2013, 243, 622–629.

[87] Sobhani, S.; Pakdin-Parizi, Z.; Nasseri, R. Nano n-propylsulphonated γ-Fe_2O_3: A novel magnetically recyclable heterogeneous catalyst for the efficient synthesis of bis(pyrazolyl)methanes in water. J. Chem. Sci. 2013, 125 (5), 975–979.

[88] Cuong, N.D.; Hoa, N.D.; Hoac T.T.; Khieu, D.Q.; Quang, D.T.; Quang, V.V.; Hieu, N.V. Nanoporous hematite nanoparticles: Synthesis and applications for benzylation of benzene and aromatic compounds. J. Alloys Comp. 2014, 582, 83–87.

[89] Bartolome, L.; Imran, M.; Lee, K.G.; Sangalang, A.; Keun Ahnd, J.; Hyun Kim, D. Superparamagnetic γ-Fe_2O_3 nanoparticles as an easily recoverable catalyst for the chemical recycling of PET. Green Chem., 2014, 16, 279–286.

[90] Wang, C.; Liu, H.; Sun, Z. Heterogeneous Photo-Fenton Reaction Catalyzed by Nanosized Iron Oxides for Water Treatment. International Journal of Photoenergy, 2012, Article ID 801694, 10 pp.

[91] Han, J.; Zhang, S.-j.; Lu, Z.-g.; Wang, Y.-j. Doping Fe^{3+} to nano-Ag TiO_2 for photocatalytic performance improvement. Yingyong Guangxue, 2010, 31(5), 718–723.

[92] Sharon, M.; Pal, B.; Kamat, D.V. Photocatalytic killing of pathogenic bacterial cells using nanosize Fe_2O_3 and carbon nanotubes. Journal of Biomedical Nanotechnology, 2005, 1(3), 365–368.

[93] Yin, Z.-Z.; Li, Y.; Jiang, L.-P.; Kumar Rana, R.; Jun-Jie Zhu. Synthesis and electrocatalytic activity of haemin-functionalised iron(II,III) oxide nanoparticles. Analytica Chimica Acta, 2013, 781, 48–53.

[94] Wang, H.; Hodgson, J.; Shrestha, T.B.; Thapa, P.S.; Moore, D.; Wu, X.; Ikenberry, M.; Troyer, D.L.; Wang, D.; Hohn, K.L.; Bossmann, S.H. Carbon dioxide hydrogenation to aromatic hydrocarbons. Beilstein J. Nanotechnol. 2014, 5, 760–769.

[95] Safari, J.; Zarnegar, Z. Sulphamic acid-functionalized magnetic Fe_3O_4 nanoparticles as recyclable catalyst for synthesis of imidazoles under microwave Irradiation. J. Chem. Sci. 2013, 125 (4), 835–841.

[96] Khodabakhshi, S.; Karami, B. Mojtaba Baghernejad Iron(II,III) oxide nanoparticle-catalyzed selective synthesis of unknown dihydropyrano[c]chromenes under green conditions. Monatsh Chem., 2014, 145, 1839–1843.

[97] Safari, J.; Zarnegar, Z. A magnetic nanoparticle-supported sulfuric acid as a highly efficient and reusable catalyst for rapid synthesis of amidoalkyl naphthols. Journal of Molecular Catalysis A: Chemical, 2013, 379, 269–276.

[98] Bayat, A.; Shakourian-Fard, M.; Ehyaei, N.; Mahmoodi Hashemi, M. Silver nanoparticles supported on silica-coated ferrite as magnetic and reusable catalysts for oxidant-free alcohol dehydrogenation. RSC Adv., 2015, 5, 22503–22509.

[99] Girija, G.; Bhojya Naik, H.S.; Vinay Kumar, B.; Sudhamani, C.N.; Harish, K.N. Fe_3O_4 nanoparticle supported Ni(II) complexes: A magnetically recoverable catalyst for Biginelli reaction. Arabian Journal of Chemistry, 2014, Available from: http://www.sciencedirect.com/science/article/pii/S1878535214001701, In press.

[100] Goodarz Naseri, M.; Saion, E.B.; Abbastabar Ahangar, H.; Halim Shaari, A.; Hashim, M. Simple Synthesis and Characterization of Cobalt Ferrite Nanoparticles by a Thermal Treatment Method. J. Nanomater. Volume 2010, Article ID 907686, 8 pp.

[101] Kaur Rajput, J.; Kaur, G. $CoFe_2O_4$ nanoparticles: An efficient heterogeneous magnetically separable catalyst for "click" synthesis of arylidene barbituric acid derivatives at room temperature. Chinese Journal of Catalysis, 2013, 34, 1697–1704.

[102] Kooti, M.; Afshari, M. Magnetic cobalt ferrite nanoparticles as an efficient catalyst for oxidation of alkenes. Scientia Iranica F, 2012, 19 (6), 1991–1995.

[103] Kamal Senapati, K.; Phukan, P. Magnetically separable cobalt ferrite nanocatalyst for aldol condensations of aldehydes and ketones. Bulletin of the Catalysis Society of India, 2011, 9, 1–8.

[104] Kumar, P.V.; Short, M.P.; Yip, S.; Yildiz, B.; Grossman, J.C. High Surface Reactivity and Water Adsorption on $NiFe_2O_4$ (111) Surfaces. J. Phys. Chem. C., 2013, 117 (11), 5678–5683.

[105] Hong, D.; Yamada, Y.; Nagatomi, T.; Takai, Y.; Fukuzumi, S.Catalysis of Nickel Ferrite for Photocatalytic Water Oxidation Using $[Ru(bpy)_3]^{2+}$ and $S_2O_8^{2-}$. J. Am. Chem. Soc. 2012, 134, 19572–19575.

[106] Radhakrishnan Nair, T.D.; Aniz, C.U. Effect of Redox Nature of Impregnated Ferrite Catalysts on Their Carbon Monoxide Oxidation Activity. RRJMS, 2013, 1 (2), 45–52.

[107] Wan, Q.; Li, P.; Shan, J.; Zhai, F.; Li, Z.; Qu, Z. Superior Catalytic Effect of Nickel Ferrite Nanoparticles in Improving Hydrogen Storage Properties of MgH_2. J. Phys. Chem. C, 2015, 119, 2925–2934.

[108] Eshghi, H.; Javid, A.; Khojastehnezhad, A.; Moeinpour, F.; Bamoharram, F.F.; Bakavoli, M.; Mirzaei, M. Preyssler heteropolyacid supported on silica coated $NiFe_2O_4$ nanoparticles for the catalytic synthesis of bis(dihydropyrimidinone)benzene and 3,4-dihydropyrimidin-2(1H)-ones. Chinese Journal of Catalysis, 2015, 36, 299–307.

[109] Lou, J.-C.; Chang, C.-K. Catalytic Oxidation of CO Over a Catalyst Produced in the Ferrite Process. Env. Eng. Sci., 2006, 23 (6), 1024–1032.

[110] Kasi Viswanath, I.V.; Murthy, Y. L. N. One-Pot, Three-Component Synthesis of 1, 4-Dihydropyridines by Using Nano Crystalline Copper Ferrite. Chem. Sci. Trans., 2013, 2 (1), 227–233.

[111] Murthy, Y. L. N.; Diwakar, B.S.; Govindh, B.; Nagalakshmi, K.; Kasi Viswanath, I.V.; Singh, R. Nano copper ferrite: A reusable catalyst for the synthesis of β,γ-unsaturated ketones. J. Chem. Sci., 2012,124 (3), 639–645.

[112] Gharib, A.; Noroozi Pesyan, N.; Vojdani Fard, L.; Roshani, M. Catalytic Synthesis of α-Aminonitriles Using Nano Copper Ferrite ($CuFe_2O_4$) under Green Conditions. Organic Chemistry International, Volume 2014, Article ID 169803, 8 pp.

[113] Yang, D.; Zhu, X.; Wei, W.; Jiang, M.; Zhang, N.; Ren, D.; You, J.; Wang, H. Magnetic Copper Ferrite Nanoparticles: An Inexpensive, Efficient, Recyclable Catalyst for the Synthesis of Substituted Benzoxazoles via Ullmann-Type Coupling under Ligand-Free Conditions. Synlett, 2014, 25, 729–735.

[114] Rashad, M.M.; Mohamed, R.M.; Ibrahim, M.A.; Ismail, L.F.M.; Abdel-Aal, E.A. Magnetic and catalytic properties of cubic copper ferrite nanopowders synthesized from secondary resources. Advanced Powder Technology, 2012, 23, 315–323.

[115] Papa, F.; Patron, L.; Carp, O.; Paraschiv, C.; Balint, I. Catalytic behavior of neodymium substituted zinc ferrites in oxidative coupling of methane. Rev. Roum. Chim. 2010, 55 (1), 33–38.

[116] Jadhav S.V.; Mohan Jinka, K.; Bajaj, H.C. Nanosized sulfated zinc ferrite as catalyst for the synthesis of nopol and other fine chemicals. Catalysis Today, 2012, 198, 98–105.

[117] de Moura, E.M.; Garcia, M.A.S.; Gonçalves, R.V.; Kiyohara, P.K.; Jardimc, R.F.; Rossi, L.M. Gold nanoparticles supported on magnesium ferrite and magnesium oxide for the selective oxidation of benzyl alcohol. RSC Adv., 2015, 5, 15035–15041.

[118] Han, A.; Liao, J.; Ye, M.; Yan, L.; Peng, X. Preparation of Nano-$MnFe_2O_4$ and Its Catalytic Performance of Thermal Decomposition of Ammonium Perchlorate. Chinese Journal of Chemical Engineering, 2011, 19(6), 1047–1051.

[119] Ghahremanzadeh, R.; Rashid, Z.; Zarnanic, A.-H.; Naeimi, H. Manganese ferrite nanoparticle catalyzed tandem and green synthesis of spirooxindoles. RSC Adv., 2014, 4, 43661–43670.

[120] Zhenyu, L.; Guangliang, X.; Yalin, Z. Microwave assisted low temperature synthesis of MnZn ferrite nanoparticles. Nanoscale Res. Lett. 2007, 2, 40–43.

[121] Velinov, N.; Koleva, K.; Tsoncheva, T.; Paneva, D.; Manova, E.; Kunev, B.; Mitov, I. Copper-cobalt ferrites as catalysts for methanol decomposition. 11th European Congress on Catalysis – EuropaCat-XI, Lyon, France, September 1st-6th, 2013.

[122] Wang, L.L.; He, H.Y. Surface Alkaline-Acidic Characteristics and Photo catalytic Properties of $Co_{1-x}Zn_xFe_2O_4$ Nano-particles Synthesized by Hydrothermal Method. J. Sci. Res. & Reports, 2014, 3 (2), 263–274; Article no. JSRR.2014.001.

[123] Khojastehnezhad, A.; Rahimizadeh, M.; Eshghi, H.; Moeinpour, F.; Bakavoli, M. Ferric hydrogen sulfate supported on silica-coated nickel ferrite nanoparticles as new and green magnetically separable catalyst for 1,8-dioxodecahydroacridine synthesis. Chinese J. Cat., 2014, 35, 376–382.

[124] Kumar B. Gupta, S.; Venkatachalam, A. Synthesis and characterization of spinel ferrites $Cu_{1-x}Cd_x[Fe_{1-x}Al_xCr_{1-x}Mn_x]O_4$. Rasayan J. Chem., 2010, 3 (4), 745–750.

[125] Dixit, R.; Gupta, P.; Saxena, R.; Dwivedi, R. Methylation of Aniline over Mn-Cu Ferrites Catalysts. Global Journal of Science Frontier Research Chemistry, 2013, 13 (7), 10 pp.

[126] Dandia, A.; Singh, R.; Joshi, J.; Maheshwari, S. Magnetically separable $CuFe_2O_4$ nanoparticles: an efficient catalyst for the synthesis of quinoxaline derivatives in tap-water under sonication. Eur. Chem. Bull., 2013, 2 (10), 825–829.

[127] Manova, E.; Estournès, C.; Paneva, D.; Reh, J.-L.; Tsoncheva, T.; Kunev, B.; Mitov, I. Mössbauer study of nanodimensional nickel ferrite – mechanochemical synthesis and catalytic properties. ICAME 2005 (Proceedings of the 28th International Conference on the Applications of the Mössbauer Effect (ICAME 2005), Montpellier, France, 4–9 September 2005, Volume I (Part I–II/V). Editors: Lippens, P.-E.; Jumas, J.-C.; Génin, J.-M. R.), 2007, 215–220.

[128] Wu, Y.; Song, M.; Xin, Z.; Zhang, X.; Zhang, Y.; Wang, C.; Li, S.; Gu, N. Ultra-small particles of iron oxide as peroxidase for immunohistochemical detection. Nanotech., 2011, 22, 225703, 8 pp.

5

Preparation of Functionalized Graphene and Gold Nanocomposites – Self-assembly and Catalytic Properties

Jingxin Zhou, Tifeng Jiao, Qingrui Zhang and Jie Hu

Abstract

Nanocomposites and nanomaterials have been attracting more attention in various fields. Nanocomposites can be prepared with a variety of special physical, thermal, and other unique properties. They have better properties than conventional microscale composites and can be synthesized using simple and inexpensive techniques. A composite material consists of an assemblage of two materials of different natures completing and allowing us to obtain a material of which the set of performance characteristics is greater than that of the components taken separately. In our recent research, some functionalized nanocomposites and nanomaterials have been prepared and investigated. In addition, some of the analytical methods, theoretical treatments, and synthetic tools, which are being applied in the area of self-assembly and supramolecular chemistry, will be highlighted. In this chapter, we summarize our main research contributions in recent years in two sections: (1) preparation and catalytic properties of some functionalized graphene nanocomposites; (2) preparation and catalytic properties of some functionalized gold nanocomposites. These works not only provided important inspirations for developing graphene-hybridized materials but also opened new possibilities to improve the photocatalytic activity of photocatalyst.

Keywords: nanocomposites, nanomaterials, catalytic property, self-assembly

1. Introduction

Nanocomposites and nanomaterials have been attracting increasing attention in various fields [1–6]. Nanocomposites can be prepared with a variety of special physical, thermal, and other unique properties. They have better properties than conventional microscale composites and can be synthesized using simple and inexpensive techniques. In addition, as the assemblage of various separate components, the designed composite materials would demonstrate greater characteristics and performances. At the same time, nanomaterials and nanocomposites often have unique properties that could enable composite materials with multiple unique properties simultaneously; however, it is often challenging to achieve these properties in large-scale nanocomposites and functionalized nanomaterials. In order to obtain the stated objective composites, it is important to design different models of the interactions in nanocomposites and next controllable properties. Moreover, it is well known that various self-assembly techniques are important to fabricate well-defined hierarchical and organized nanostructures with special properties and controlled capacity [7–10]. For different research systems, such as inorganic–organic hybrids, various nanoparticles, colloidal microspheres, and supramolecular nanostructures, the self-assembly process can demonstrate obvious advantages, especially in preparing special nanostructures. The incorporation of nanocomposites with self-assembly shows new performances for designed composite nanomaterial with different organized style and processing routes.

In addition, as a special sp^2-bonded single layer carbon material, graphene has been attracting more attentions in various application fields, especially as photocatalytic material due to its natural characteristics, such as high chemical stability, large surface area, and excellent adsorption capacity [11]. And some new nanocomposites from graphene materials as photocatalysts have been developed in recent years. For example, the research group of Ullah prepared ternary Pt-graphene/TiO_2 nanocomposites by microwave-assisted method and investigated the photocatalytic degradation for dye rhodamine B [12]. Bai and coworkers reported the preparation of new $ZnWO_4$/graphene hybrid materials by in situ reduction of graphene oxide and characterized the degradation capacity of methylene blue as photocatalysts [13]. Sun's group reported the synthesis of $ZnFe_2O_4$/ZnO nanocomposites anchored on graphene substrates via an ultrasound-aided method [14]. Xu and coworkers prepared reduced graphene oxide/Bi_2WO_6 composite materials as photocatalysts, and compared the photocatalytic activities of composites with pure Bi_2WO_6 [15]. In addition, more research works have been reported to enhance the photocatalytic activity of graphene-based composites as photocatalysts [16–19]. Moreover, in the last several decades, the preparations of gold nanoparticles have been attracting much attention [20–22]. Up to now, various processes [23–25] and designed capping agents [26–29] have demonstrated the synthesis of functional gold nanoparticles with designed nanostructures and tailored behaviors. It is well known that many factors, such as the particle size, shape, colloid stability, and surface functionalized groups, seem to be critical for the applications fields [30–32]. In addition, because the self-assembled nanostructures are important and closely related to molecular skeletons in designed template compounds, the design and synthetic routes are the key step in the whole process.

In recent research work reported by our group, various research systems about graphene and gold nanocomposites/nanomaterials have been designed and characterized. And the involved analytical methods and self-assembly techniques have been also highlighted. So in the present chapter, we would like to demonstrate our recent research contributions in two parts: (1) preparation and catalytic properties of some functionalized graphene nanocomposites; (2) preparation and catalytic properties of some functionalized gold nanocomposites. These works not only provided important inspirations for developing graphene-hybridized materials but also opened new possibilities to improve the photocatalytic activity of photocatalyst.

2. Preparation and catalytic properties of some functionalized graphene nanocomposites

Recently, our group has reported the design and fabrication of hybrid organogels by self-assembly of composites containing cationic compounds and GO [33]. It is interesting to note that the obtained gelation performances can be regulated via different functional substituted headgroups in used compounds. The experimental data indicated that ammonium headgroup in molecular skeletons seemed more favorable for the composite gelation than pyridinium segment. The obtained results suggested that the self-assembly modes in present GO-based composites could be manipulated by controlling efficient headgroup effect. In addition, various weak forces between present building blocks seemed also responsible for the formation of different nanostructures. Based on the obtained results data in present composite gel system, a reasonable mechanism about self-assembly modes in gels is shown in Figure 1. For the prepared CTAB-GO gel, various organized building blocks are obtained in different solvents because of the van der Waals force of substituent chains and the strong electrostatic interaction of ammonium headgroups with oxygen-containing functional groups at GO surface. In addition, for the cases of C16Py-GO and BPy-GO composite gels, the strong π–π stacking between carbon net in GO plane and additional pyridine headgroups showed as more competitive with other forces, such as electrostatic interaction and van der Waals force. Thus, the present research work demonstrates new exploration for the design of GO-based composite gels and self-assembled soft matters.

In addition, we have also demonstrated the formation of organogels by self-assembly of cationic gemini amphiphile–GO composites [34]. Their gelation behaviors in various organic solvents can be controlled by regulating molecular symmetry. The obtained data indicated that the designed functional groups and molecular symmetry could change the self-assembly modes and produce different self-assembled nanostructures. It seemed that longer alkyl chains in molecular skeletons could be helpful to enhance the intermolecular hydrophobic force in the process of self-assembly. So the changes of building blocks and stacking modes between present GO-based composites and different solvents are responsible for the formation of various nanostructures, as seen in Figure 2. It clearly indicated that the formed nanostructures in present composite materials were obviously different, such as nano-wrinkle, nano-lamella, and nano-belt. The obtained different morphologies in self-assembled gels can be mainly due to various formation mechanisms upon special self-assembly modes via weak interactive forces between building blocks. Finally, a reasonable self-assembled mechanism for symmetry effects in formation of present organized nanostructures is demonstrated. Thus, the present GO-based composite gel materials will give a helpful clue for the design and preparation of functional GO composite nanomaterials.

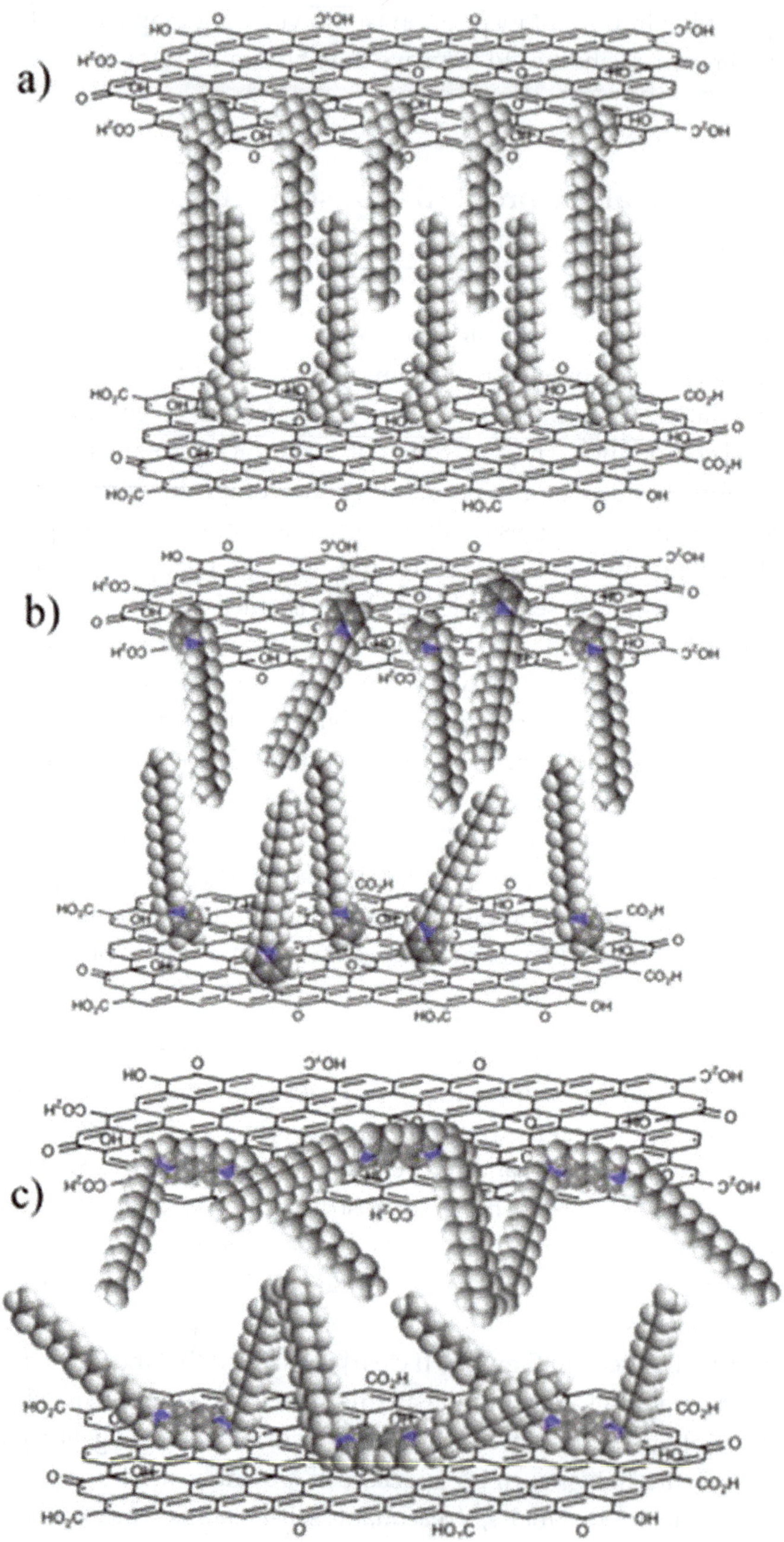

Figure 1. Scheme of different assembly modes in cationic amphiphiles–graphene oxide gels. CTAB-GO (a), C16Py-GO (b), and BPy-GO (c).

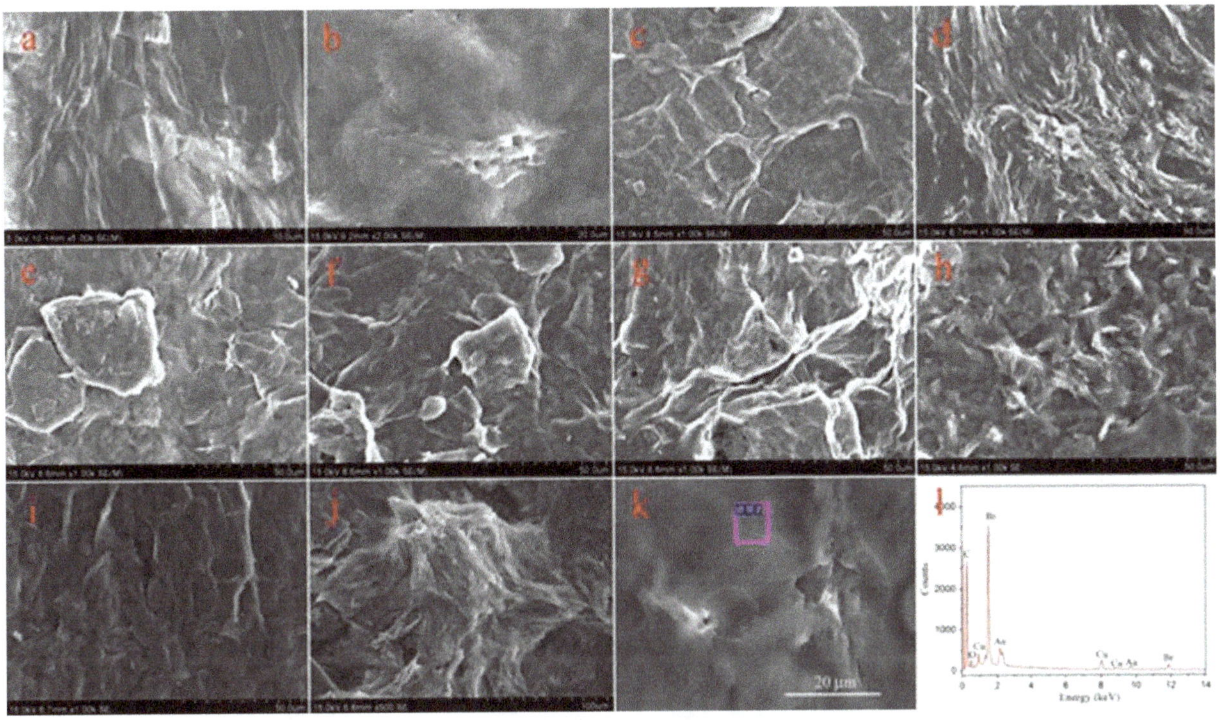

Figure 2. SEM images of xerogels for GO sheet, C18-6-6/GO gels, C18-6-12/GO gels, and C18-6-18/GO gels in different solvents.

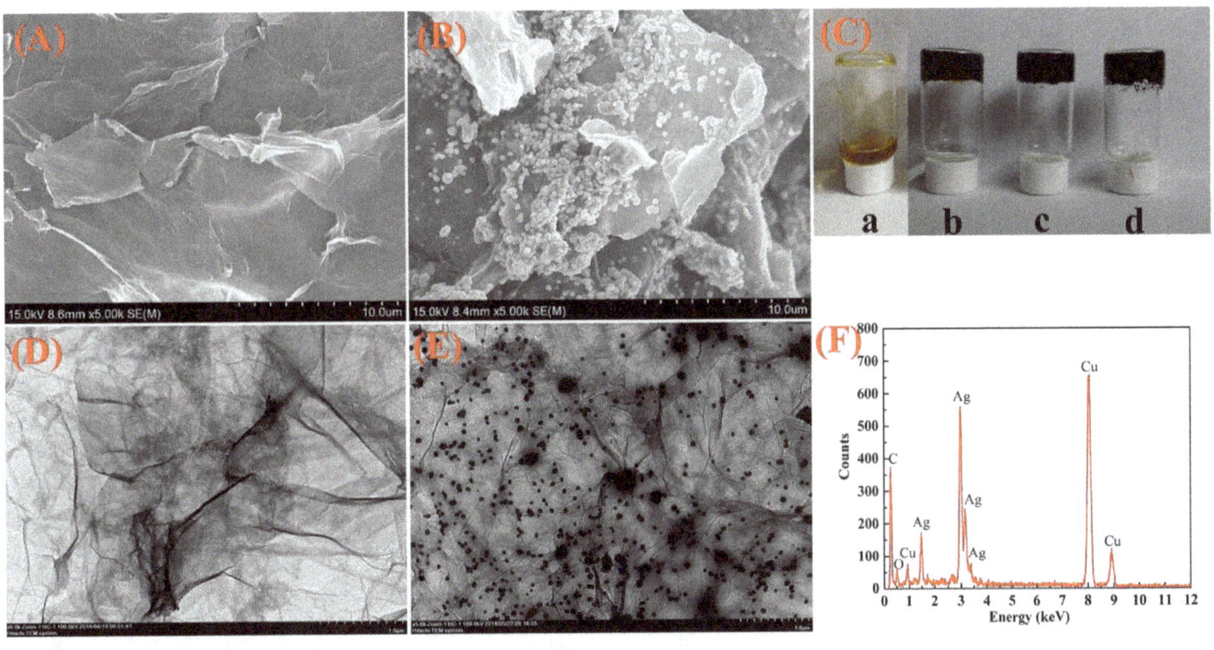

Figure 3. SEM and TEM images for the RGO/PEI and RGO/PEI/Ag hydrogels.

In another composite system, we have reported the design and preparation of silver nanoparticle-containing RGO-based composite hydrogel materials via an in situ reduction process [35]. The obtained experimental data indicated that the prepared composite gels were composed of 3D net-like nanostructures, as seen in Figure 3. In addition, the used preparation method included the in situ reduction of GO and silver acetate in hydrogel structures to fabricate present RGO-based composite hydrogel. So the formed silver nanoparticles were uniformly anchored on RGO surface in composite gel. Moreover, the photocatalytic behaviors for removing dye pollutants are also characterized for the silver nanoparticle-containing RGO-based composite hydrogel, as seen in Figure 4. It is interesting to note that the obtained photocatalytic composite materials can be reused from an aqueous degradation system, indicating the important and potential applications for dye removal and wastewater treatment.

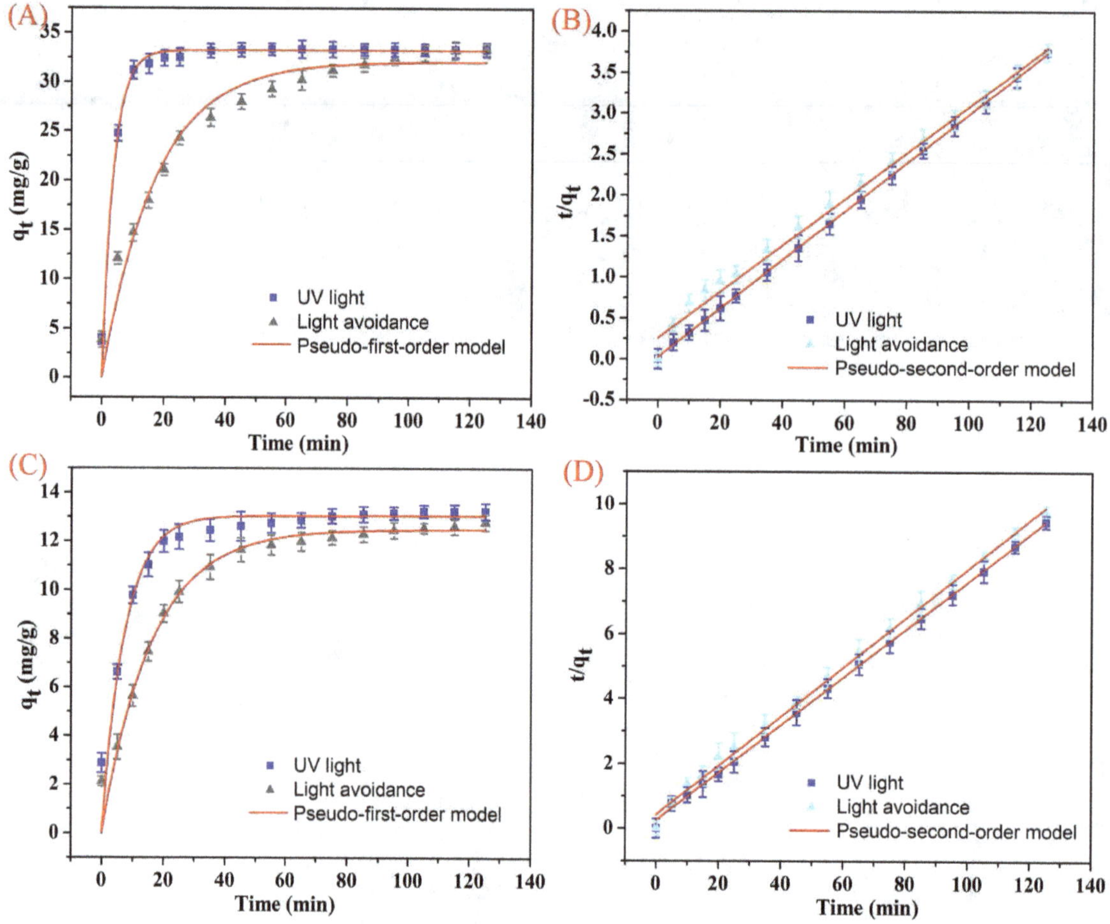

Figure 4. Degradation kinetics curves of as-prepared RGO/PEI/Ag nanocomposites on MB (a, b) and RhB (c, d) at 298 K.

In order to investigate the mechanism of hybrid graphene composites, we have also synthesized some $LaMnO_3$–graphene composites as photocatalysts by a sol–gel method [36]. It is found that $LaMnO_3$ perovskite phase was successfully fixed on graphene surface with well-dispersion capacity, as seen in Figure 5. The data indicated that the photocatalytic capacity of as-formed $LaMnO_3$–graphene composite materials were better than pristine $LaMnO_3$ material, with the detailed results in Figure 6. The enhancement of photocatalytic properties can be mainly due to the high separation efficiency of photo-induced electron–hole pairs originated from the excellent conductivity of graphene in composite and the large interfacial contact between components, which is helpful to increase the dyes adsorption and improve the transfer efficiency in photocatalytic process. This research demonstrated new inspiration for designing photocatalytic graphene-based hybrid materials.

In another work, organized $La_{1-x}Sr_xMnO_3$/graphene thin films were prepared by both sol–gel and spin-coating methods [37]. In experimental process, the formed sol nanoparticles were adsorbed on graphene surface via electrostatic force in aging time. Then, the formed $LaMnO_3$ nanoparticles increase sizes and form crystal domains on graphene surface in the next calcination step. Present obtained nanostructures and morphology were investigated during various characterization techniques. Figure 7 shows the XRD patterns of graphene, $LaMnO_3$, $LaMnO_3$/graphene, and $La_{0.9}Sr_{0.1}MnO_3$/graphene thin film. The diffraction peak of graphene appeared in the vicinity of 23°, which was similar to the diffraction peak position of graphite. However, this peak broadened and weakened because the size of the graphite sheet decreased, the integrity of the crystal structure declined, and the degree of disorder increased. The pattern of $LaMnO_3$ is in agreement with PDF33-0713, indicating its perovskite structure with complete crystal shape. In the process of acid red 3GN photodegradation, $LaMnO_3$/graphene thin film had sound stability and better photocatalytic ability than $LaMnO_3$ thin film. As shown in Figure 8, a mild photodecomposition effect was observed in the degradation of dye. The absorption peak of acid red 3GN dye solution at 509 nm was from the initial 0.5446 to 0.5138 after irradiating for 48 h without photocatalysts. The experimental results indicated that graphene enhanced the dye adsorption, inhibiting the reunion of light-induced e^-–h^+ and improving photocatalytic capacity. In addition, it should be noted that a red shift of absorption edge was found by doping Sr, which seemed helpful to increase the photocatalytic performance of the obtained composite film.

In addition, another new $LaMn_{1-x}Co_xO_3$/graphene composite material as photocatalyst had been designed and prepared by sol–gel method [38]. The experimental data indicated that $LaMnO_3$ perovskite phase was anchored on graphene surface with the special perovskite structure. In addition, the photocatalytic capacity was characterized by the degradation of diamine green B. In the photodegradation process, the graphene component in composite can accelerate the dye adsorption, while doping Co component improves the photocatalytic performance. Thus, the reasonable charge transfer mechanism that occurred in the obtained $LaMnO_3$/graphene composite during photocatalytic process is demonstrated in Figure 9. Firstly, diamine green B molecules could shift to the active surface of prepared composites from solution and organized in self-assembled face-to-face mode via π–π stacking with aromatic graphene net. Due to the effect of these holes and electron transfers, charge recombination is pushed in obtained $LaMnO_3$/graphene composite and improves the efficiency of photocatalytic capacity.

Moreover, we have also reported the preparation of some graphene-based $LaNiO_3$ composite films by both sol–gel method and spin-coating technique [39]. The obtained experimental results indicated that the size of formed $LaNiO_3$ nanoparticles was about 20 nm, well dispersed on graphene surface. The photocatalytic capacity of present hybrid films had been demonstrated by degradation of acid red A. In comparison with pure $LaNiO_3$ films, the designed $LaNiO_3$/graphene composite films showed better photocatalytic behavior. It is interesting to

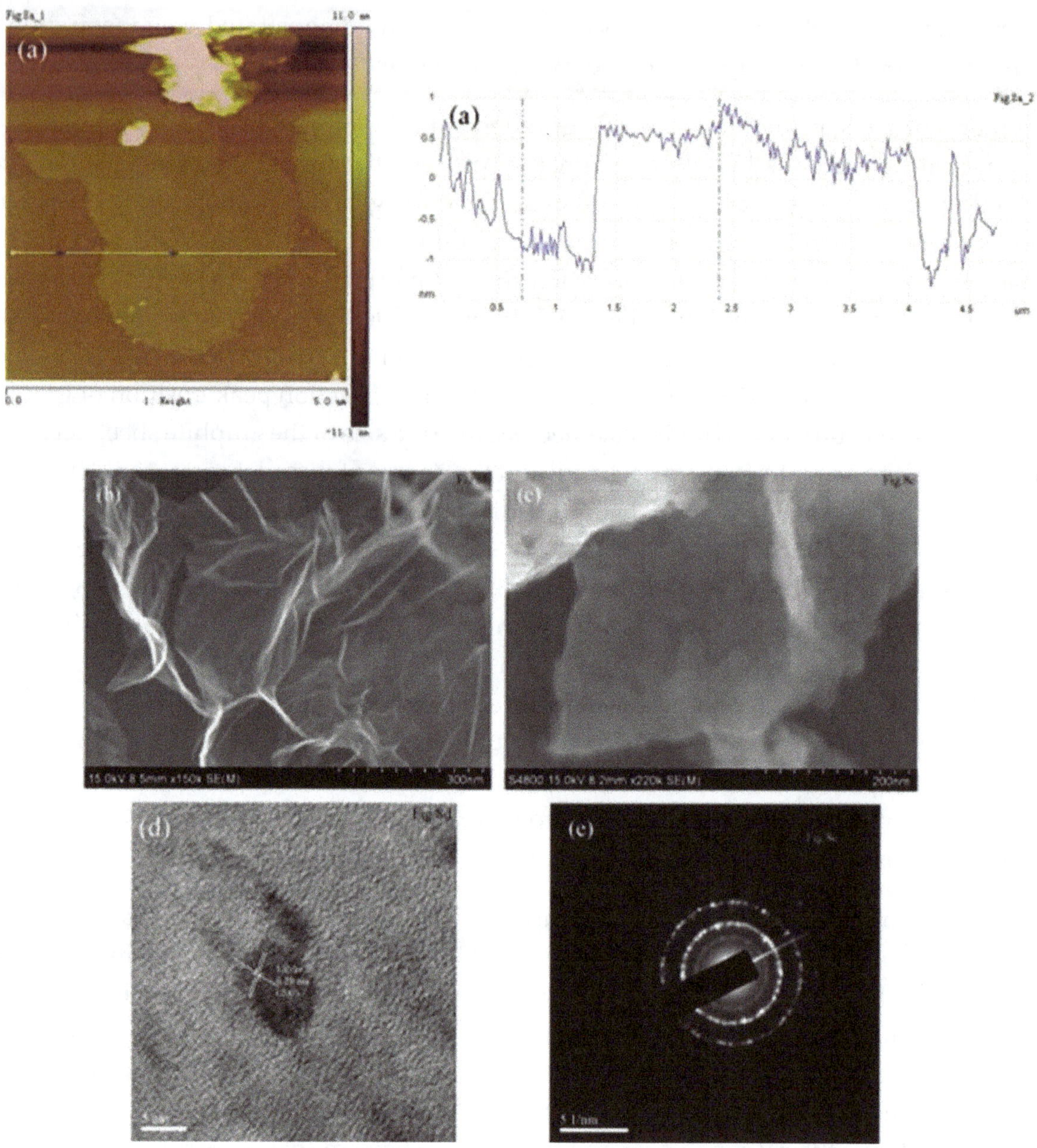

Figure 5. (a) AFM image of the as-synthesized graphene; (b) SEM image of graphene; (c, d, e) SEM, HRTEM, and SA-ED images of $LaMnO_3$–graphene composites.

note that when the content of graphene shifted to the value of about 4%, the photocatalytic efficiency of the obtained composite films was double that of pure $LaNiO_3$ films.

In another system, $La_{1-x}Ca_xMnO_3$ perovskite–graphene composites are synthesized as catalysts for Zn–air cell cathodes [40]. The results indicated that perovskite phase adhered on the surface of graphene sheets, and adding graphene significantly improved the electrochemical performance of $LaMnO_3$. The XPS spectrum of $La_{0.6}Ca_{0.4}MnO_3$–graphene composite is shown in Figure 10. The peak contained all the elements of $La_{0.6}Ca_{0.4}MnO_3$. In addition, the obtained graphene showed gauze-like fold nanostructures, mainly originated from the oxygenic functional groups and the surface defects during preparation process. So the formed porous

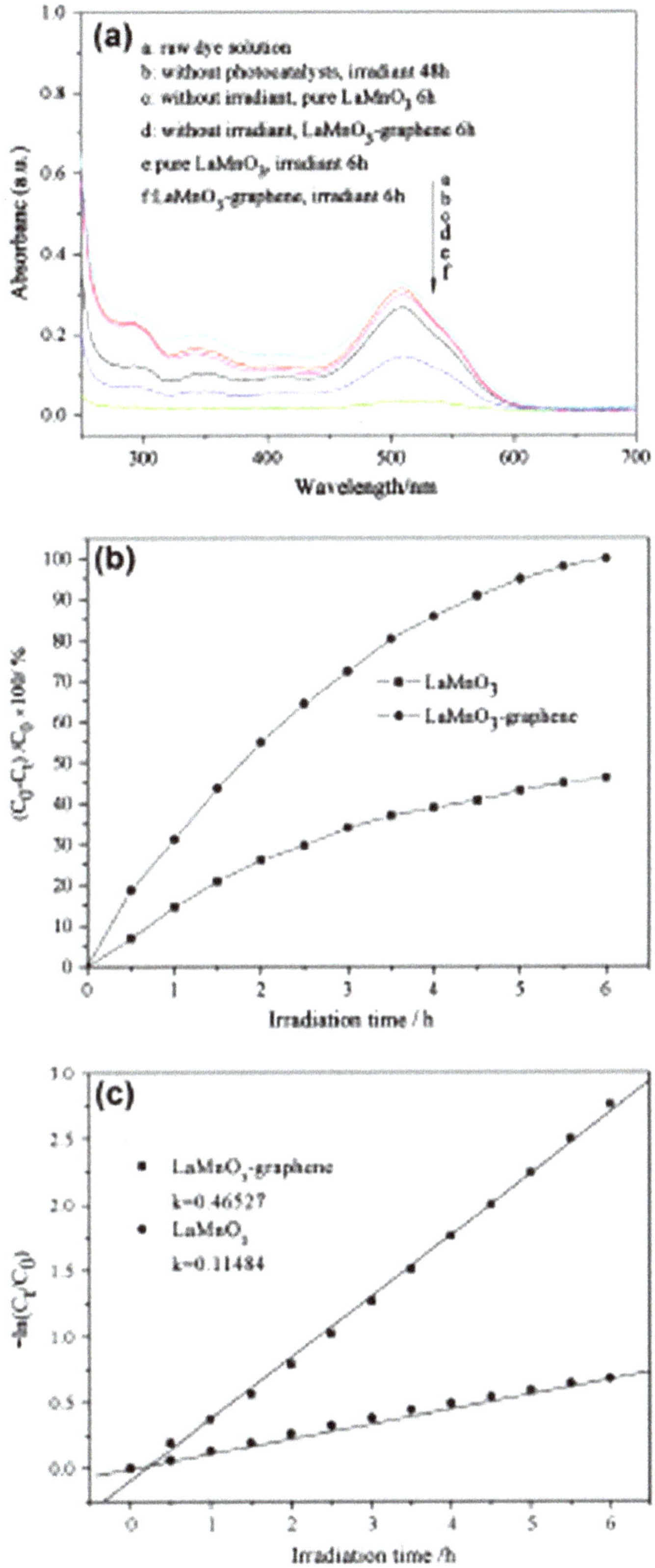

Figure 6. (a) UV–Vis spectral changes of the degradation of acid red A by $LaMnO_3$–graphene and $LaMnO_3$; (b) photocatalytic activities of $LaMnO_3$–graphene composite and $LaMnO_3$; (c) kinetics of photocatalytic degradation by the $LaMnO_3$ and $LaMnO_3$–graphene.

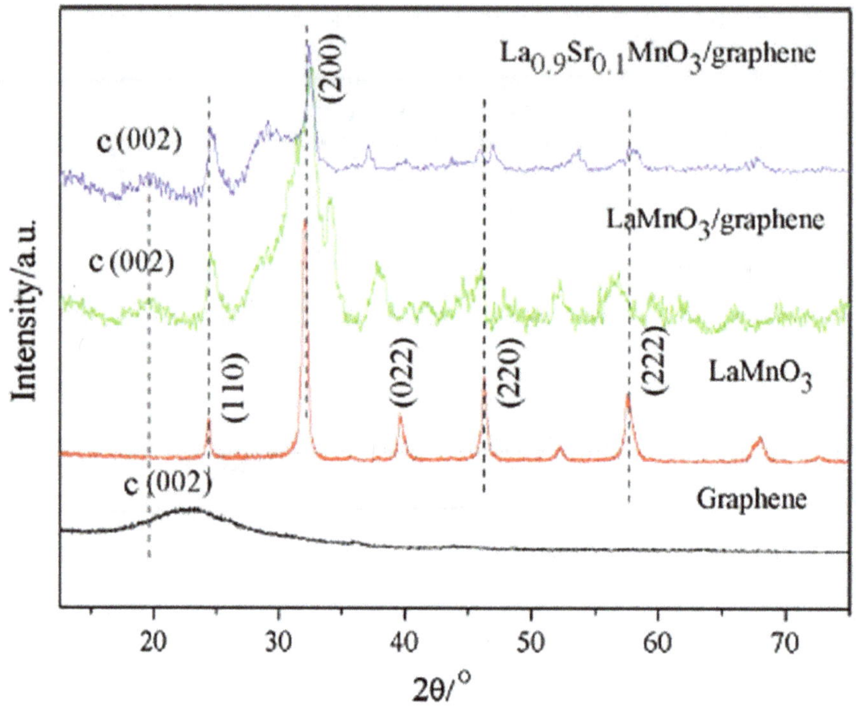

Figure 7. XRD pattern of graphene, $LaMnO_3$, $LaMnO_3$/graphene, and $La_{0.9}Sr_{0.1}MnO_3$/graphene.

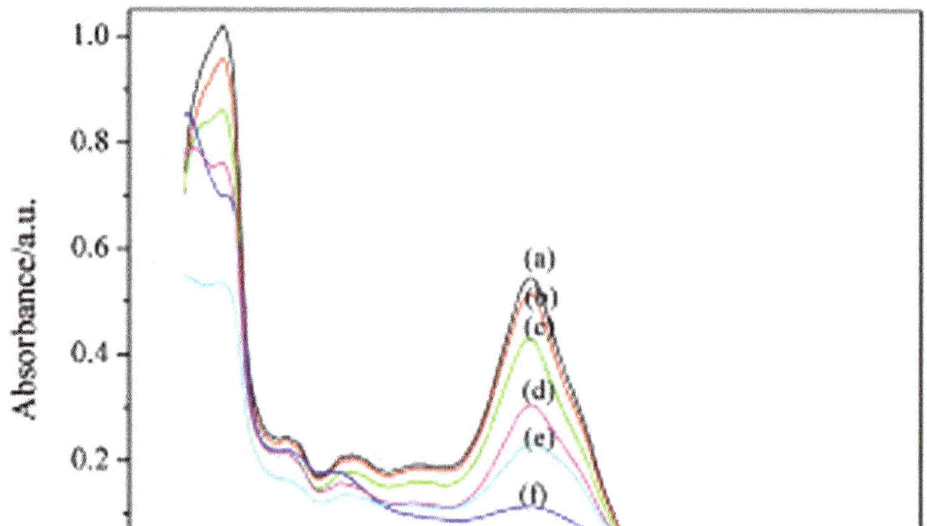

Figure 8. UV–Vis absorb spectrums of acid red-3GN under different conditions: (a) raw dye solution, (b) irradiation 48 h without catalyst, (c) $LaMnO_3$ thin film adsorb dye 4 h without irradiation, (d) $LaMnO_3$/graphene or $La_{0.9}Sr_{0.1}MnO_3$/graphene thin film adsorb dye 4 h without irradiation, (e) irradiation 4 h with $LaMnO_3$ thin film as catalyst, (f) irradiation 4 h with $LaMnO_3$/graphene thin film as catalyst.

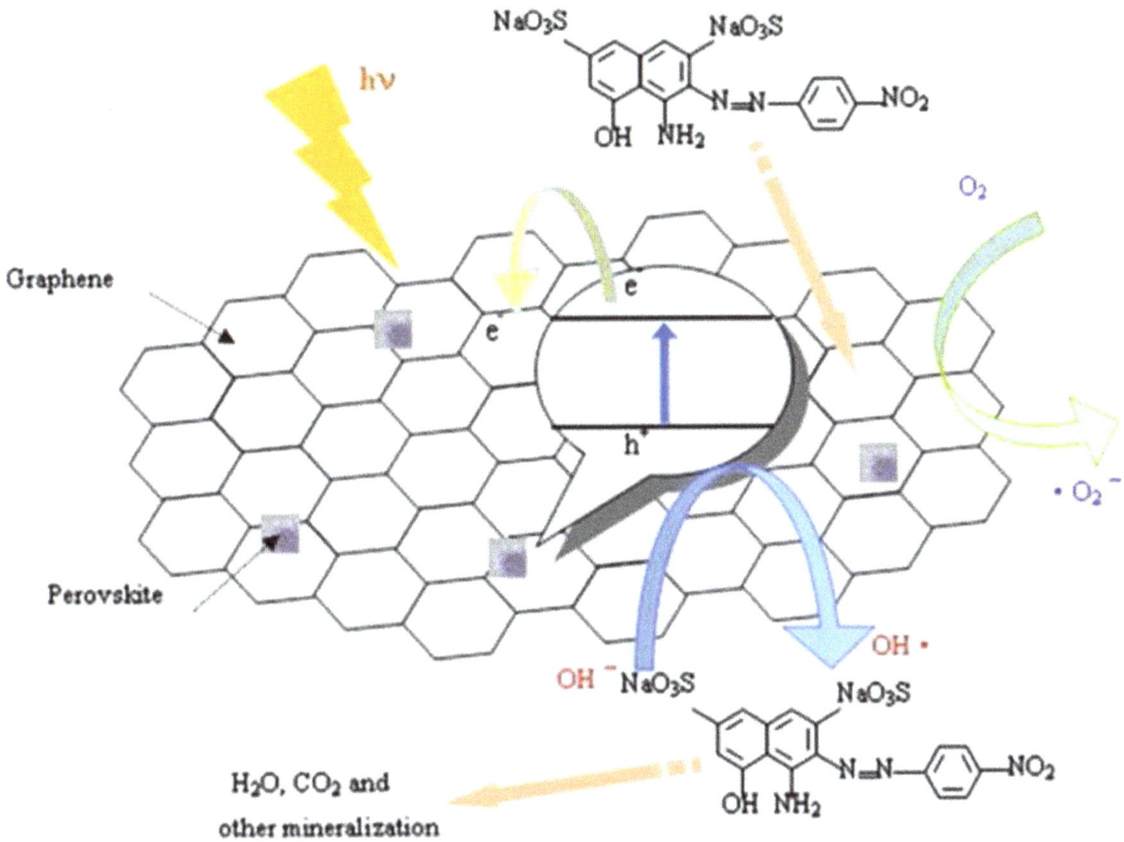

Figure 9. Proposed mechanism for photocatalytic degradation of diamine green B over graphene-based perovskite photocatalysts under light irradiation.

3D nanostructure can obviously enhance the three-phase domains and improve the mass transfer process as catalyst materials of air electrode. The experimental data indicated that the voltage plateau was superior with 10 wt% ratio value of graphene. Moreover, Ca doping maintained the perovskite structure and obviously enhanced the electrocatalytic activity for ORR, and $La_{0.6}Ca_{0.4}MnO_3$–graphene composite demonstrated the best catalytic capacity. Thus, the obtained research work indicates that the prepared graphene-based $La_{1-}xCaxMnO_3$ composites are important material for design of air electrodes catalysts.

Recently, we have also reported the preparation of $LaMnO_3$/graphene thin films with the perovskite-type as new photocatalyst via sol–gel process and spin-coating method [41]. The obtained results indicated that the addition of graphene did not change the perovskite structure, with formed $LaMnO_3$ particles at about 22 nm well dispersed on graphene surface. Figure 11 displays the nitrogen adsorption–desorption isotherms and pore size distribution curves calculated by BJH method for $LaMnO_3$ and $LaMnO_3$/graphene powders. The larger surface area can effectively absorb the dye, thus increasing the contact probability of pollutant molecular and catalyst. Determination of contact angle indicated that the contact angle of glass substrate decreased and the hydrophilicity improved after treating with H_2SO_4 and APTES. The UV–Vis photocatalytic activity of the photocatalysts was evaluated by the degradation of diamine green B. $LaMnO_3$/graphene thin films had better photocatalytic ability than $LaMnO_3$ and TiO_2 films.

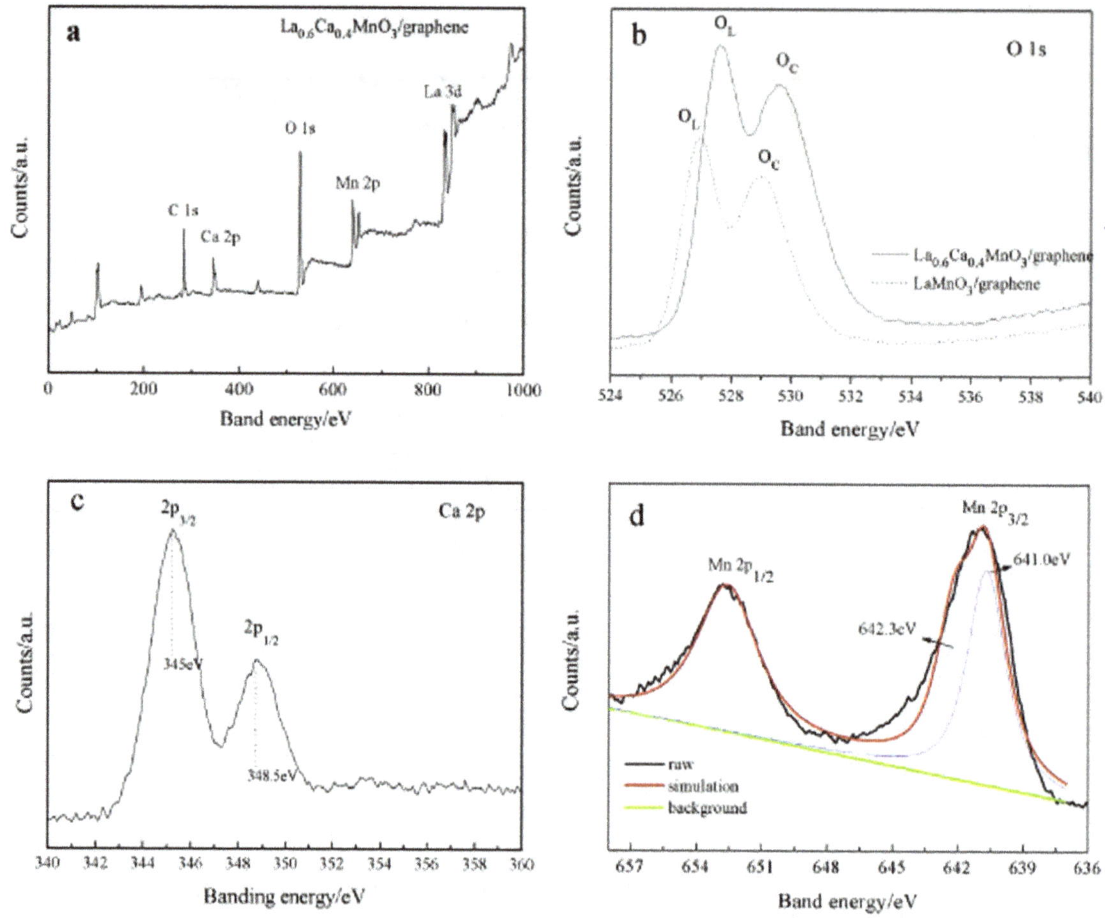

Figure 10. XPS spectrum of $La_{0.6}Ca_{0.4}MnO_3$–graphene composite. (a) Overall spectrum, high-resolution curves of (b) O 1s region, (c) Ca 2p region, and (d) Mn 2p region.

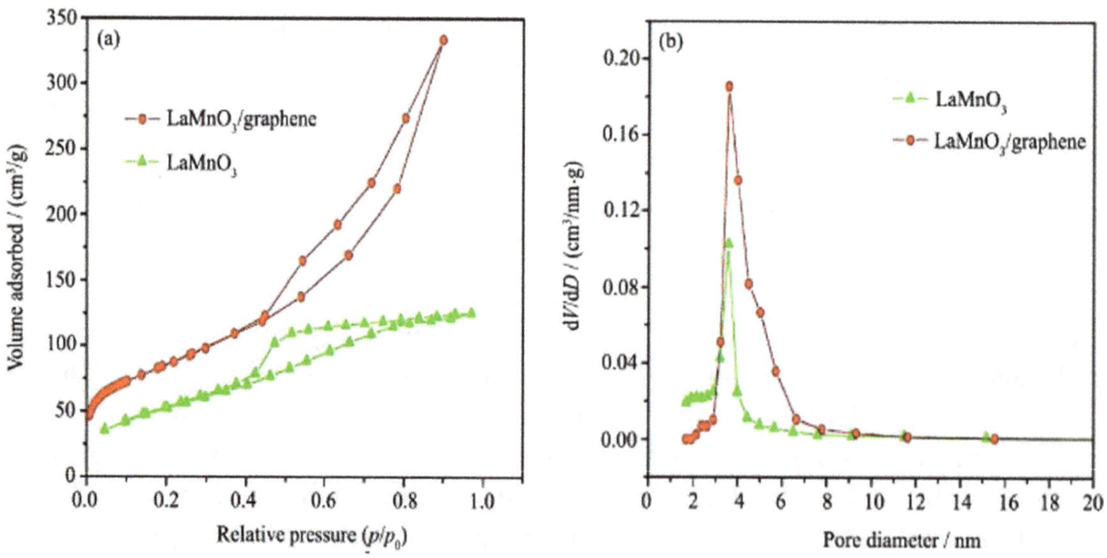

Figure 11. N_2 adsorption–desorption isotherms (a) and pore size distributions of samples (b).

3. Preparation and catalytic properties of some functionalized gold nanocomposites

In this section, many examples have been investigated, including the synthesis of gold nanoparticles by some compounds with hydrophilic spacers and aromatic headgroups at different interface. The obtained experimental data indicated that various gold nanostructures could be prepared by controlling different substituted headgroups in template compounds. In addition, the photocatalytic capacities of as-prepared gold nanoparticles on the degradation of organic dyes were also demonstrated.

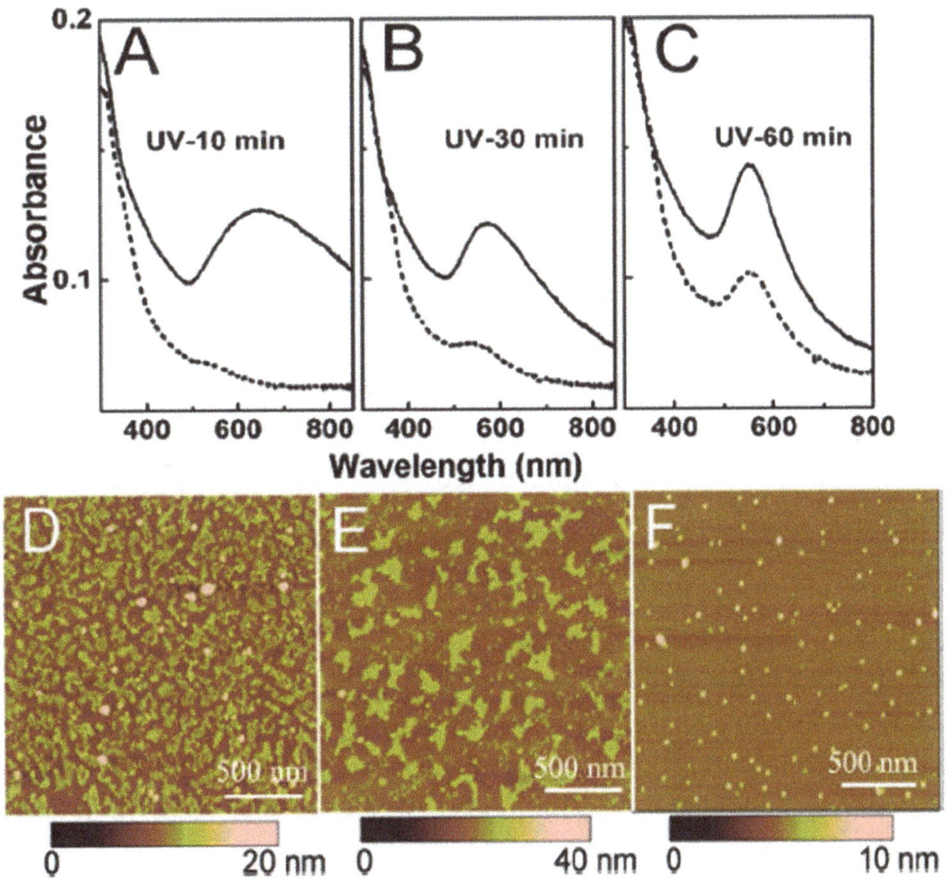

Figure 12. UV–Vis spectra and AFM images of organized LB films containing gold nanoparticles.

Firstly, the effect of gemini compounds on in situ self-assembly and fabrication of gold nanoparticles in organized molecular films have been demonstrated [42]. In comparison with traditional compounds, gemini molecules demonstrated well the capacity to fix chloroaurate ions onto various solid substrates, suggesting an important route to prepare metallic nanoparticles in organized films by chemical reduction method. In addition, gemini compounds could produce a 2D confined environment in LB films to accommodate the formed gold nanoparticles. Thus, the size, shape, and interparticle distances can be regulated by changing various reducing methods. Various nanostructures could be fabricated, such as nanoparticles, nanowires, and the tree-branched domains. Based on the obtained experimental results, different gold nanostructures could be synthesized by regulating various reductants or reducing processes. Moreover, we tried to control gold nanostructures by optical method. The obtained composite film was irradiated with UV light and then reduced by $AuCl_4^-$ ions with hydroquinone, as seen in Figure 12. The present research work provided the new clue for the synthesis of metallic particles in films from special compounds, showing important exploration in designing various gold nanostructures.

In addition, we presented a facile synthetic method to the gold nanostructures using a series of gemini amphiphiles through liquid–liquid biphase method [43]. The gemini amphiphiles themselves could serve as both capping and reducing agents. The spacer and concentration of the gemini amphiphiles played an important role in the formation of gold nanoprisms. It is predicted that gold and other novel metal nanostructures may be produced by gemini amphiphiles whose properties can be well controlled by designing different headgroups, spacers, or alkyl chains. In order to make clear the gold nanostructures, the chloroform solution was cast onto copper grid for TEM measurement, as shown in Figure 13. For gold nanostructures generated by GN1, a few polygon gold nanoparticles were found. In contrast, uniform gold nanoparticles with a size of 11.4 ± 1.2 nm were observed in GN2 chloroform solution. Interestingly, in GN3 chloroform solution, triangular nanoplates and nanoparticles were both observed. The obvious difference suggests that the spacer has an influence on the shapes of gold nanostructures. On the other hand, no gold nanoparticle can be observed in aqueous phase no matter how long the reaction proceeded.

Considering the above research background, we have prepared new kinds of gold nanoparticles via some bolaform Schiff base compounds with hydrophilic spacers and aromatic headgroups in molecular skeletons [44]. By stirring the mixed solution of aqueous $AuCl_4^-$ ions with chloroform solution of used Schiff base molecules, the metal ions shifted to the chloroform phase and reduced to the formation of different gold nanoparticles. The data indicated that different gold nanostructures could be obtained by regulating the molecular skeletons of used bolaform compounds, including spacers and headgroups, as well as the relative ratios of compounds to metal ions. In addition, the characterization of morphologies and spectra indicated that the present designed bolaform amphiphilic compounds could act as both capping agents and reducing agents. So the UV–Vis spectra in different conditions are demonstrated, as seen in Figure 14. From the obtained UV–Vis data, it clearly indicated that the spacer and azomethine segments in bolaform molecules could be positively charged at interface in transferred process. In addition, the photocatalytic capacity of the obtained gold nanoparticles on dye degradation was demonstrated in Figure 15, showing the influence of molecular skeletons in the used compounds on the regulation of prepared gold nanoparticles and next catalytic behaviors. The present obtained data suggested that various gold nanostructures could be designed and synthesized by changing substituted skeletons in used template compounds

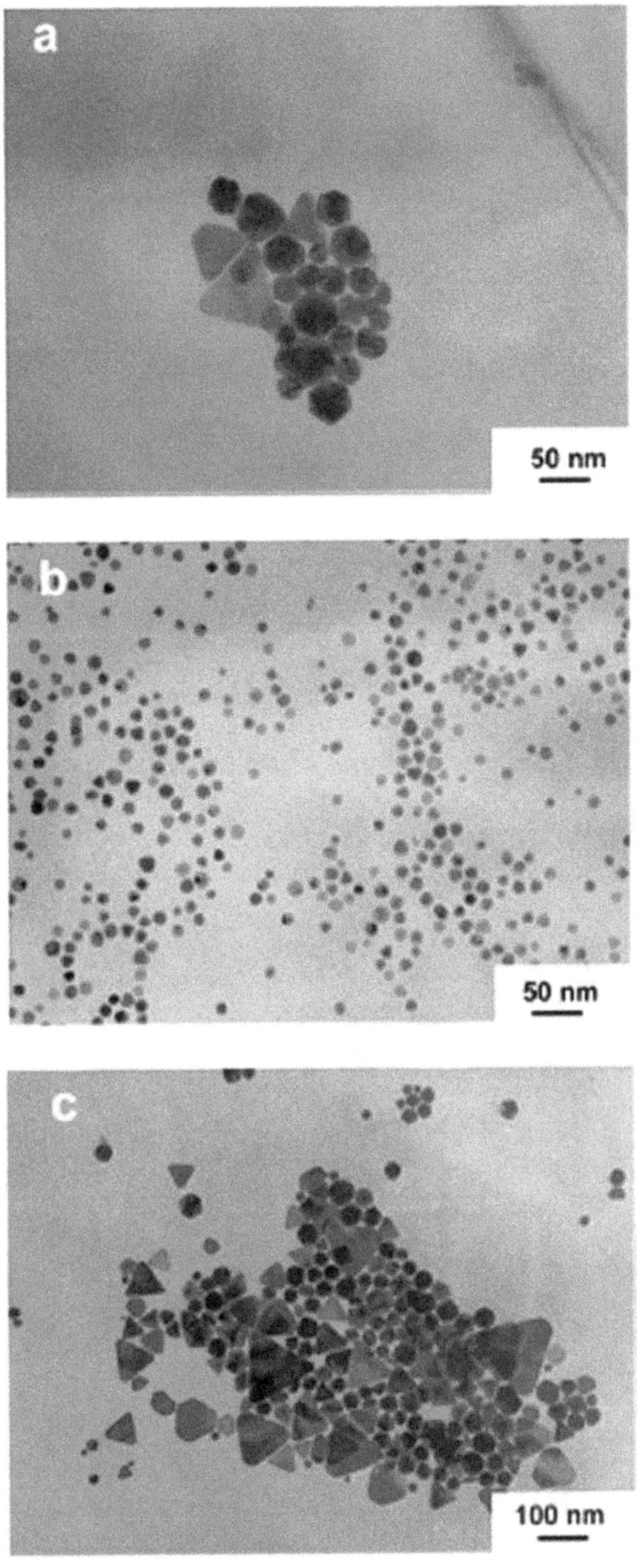

Figure 13. TEM images of reduced gold in GN1 (a), GN2 (b), and GN3 (c) chloroform solution after 36 h of reaction respectively.

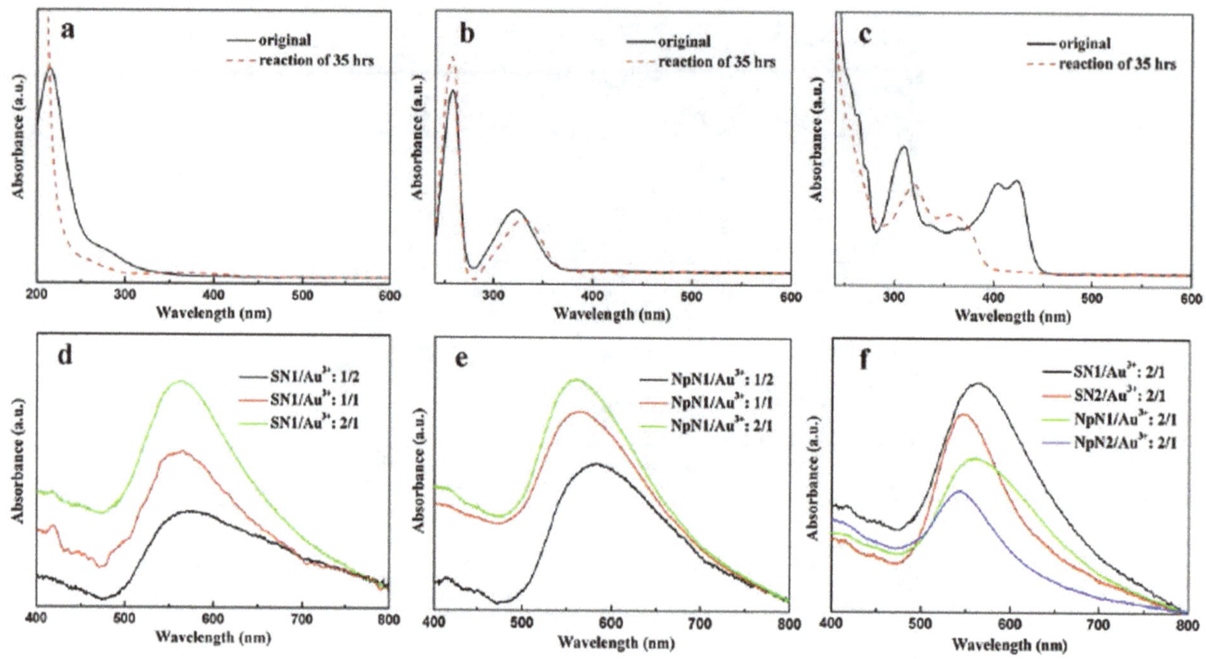

Figure 14. UV–Vis spectra of 0.1 mM $AuCl_4^-$ ions and prepared gold nanostructures in aqueous solution.

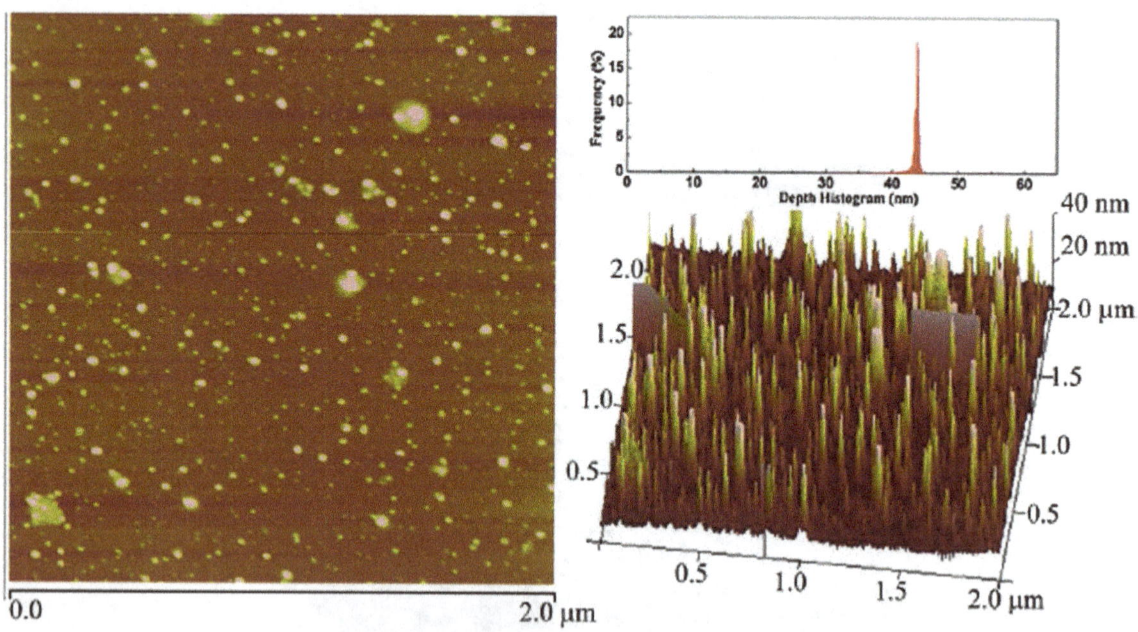

Figure 15. AFM images of height and 3D surface plot view with depth histogram of gold nanoparticles on mica prepared from instant ultrasonic ethanol solution with SN2 to chloroaurate ion ratio of 2:1 after 35 h of reaction.

In addition, some other gold nanoparticles were synthesized by two bolaform cholesteryl imide derivatives with different lengths of ethyleneamine spacers at a liquid–liquid interface [45]. Spectral and morphological measurements indicated that both bolaform amphiphiles could act as both capping agents and reducing agents. To further characterize the prepared gold nanostructures, TEM measurements have been demonstrated, as seen Figure 16. The images indicated that the size distribution of obtained gold nanostructures could be regulated by changing various spacers in used molecular skeletons. In addition, the effect of molar ratio of the template compound to $AuCl_4^-$ ions was also investigated in details. The experimental data indicated that various nanostructures, such as hexagonal, polygon nanoparticles, and nanoplates, could be synthesized. Moreover, the photocatalytic capacity of prepared gold nanostructures on dye degradation was also characterized, suggesting the importance of compounds' skeletons in regulating the formation of gold nanoparticles and changing relative catalytic properties, as demonstrated in Figure 17. The obtained research data would give new clue for the preparation of gold nanostructures by designing special template compounds.

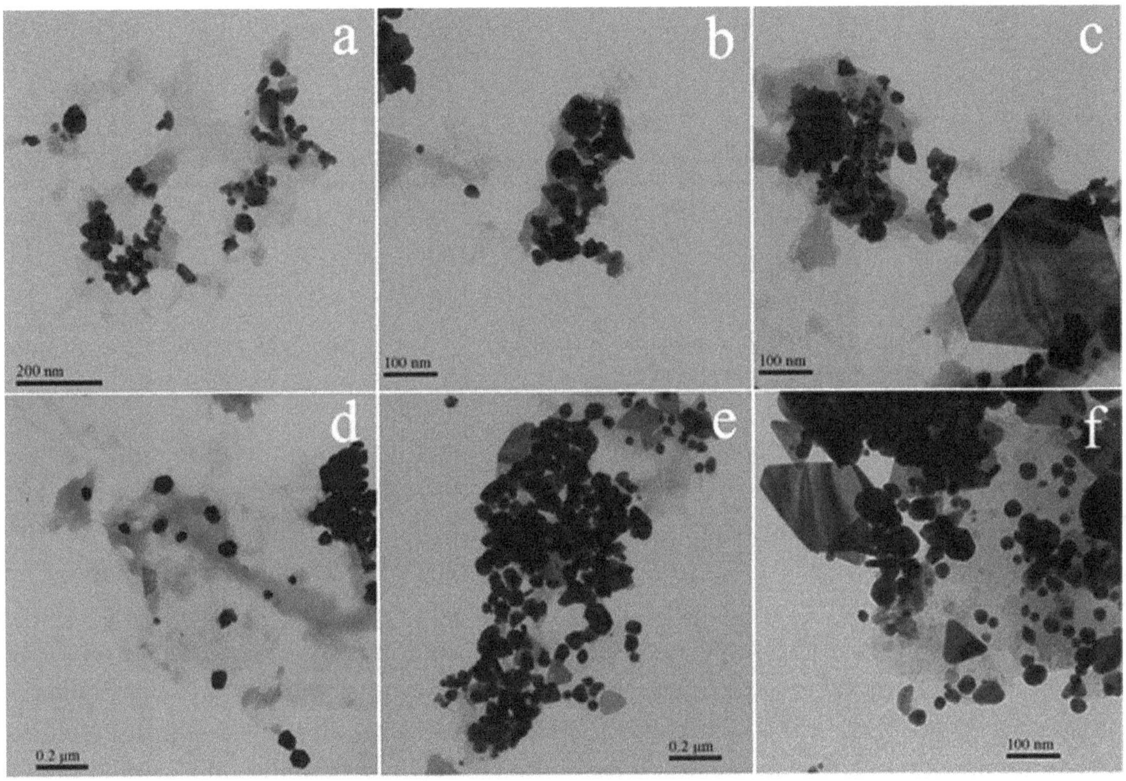

Figure 16. TEM images of gold nanoparticles by using different compounds to chloroaurate ion ratio.

4. Conclusion and perspectives

We are working on the preparation, self-assembly, and application of functionalized nanocomposites and nanomaterials. In this chapter, various kinds of nanocomposites including graphene nanocomposites and gold nanoparticles have been designed and characterized. For preparation of graphene composites, hybridization with functional little organic molecules, polymers, and various metal oxides/salts have been achieved, and the self-assembly process and photocatalytic capacities have also been demonstrated. As for the synthesis of gold

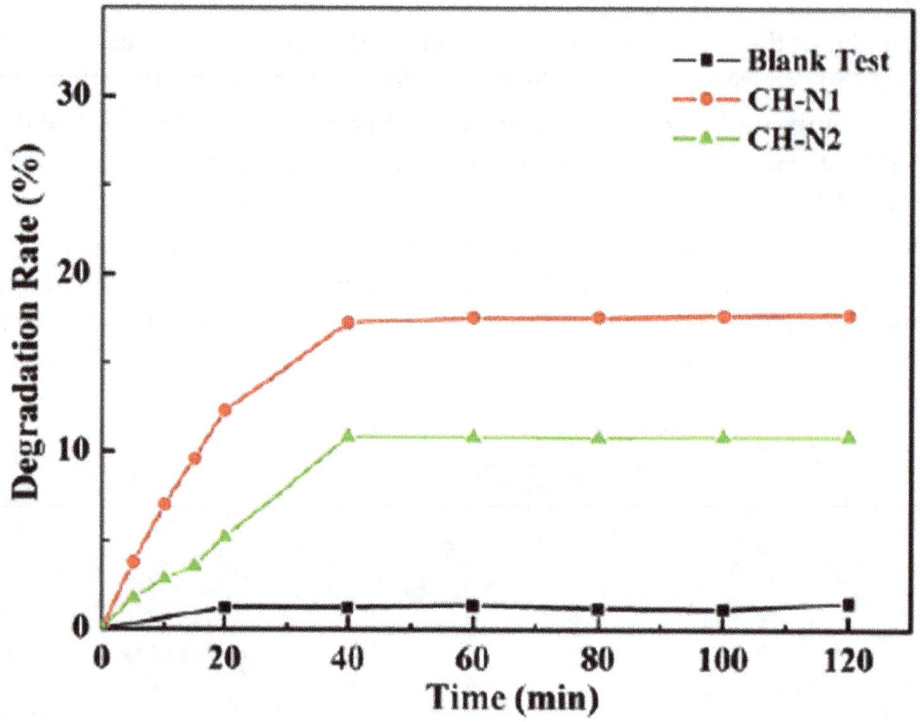

Figure 17. Photocatalytic activities of gold nanoparticles by using two amphiphiles with amphiphile to chloroaurate ion ratio of 2:1 on the degradation of methyl orange.

nanoparticles, the different prepared conditions, such as LB films/bulk, molecular skeletons/substituted headgroups, and kinds of template compounds, have been investigated to control and regulate the designed gold nanostructures. Thus, the photocatalytic behaviors of as-obtained gold nanoparticles on dyes degradation were also characterized. The above-mentioned research work may provide new and potential perspective for the preparation and analysis of nanocomposite and nanomaterials. In closing, in recent several decades, preparation of novel nanocomposites are promising and emerging as attentive research platforms based on special properties and application fields. In addition, the self-assembly techniques belonging to supramolecular chemistry seem to be key subject in physical chemistry, related to various fundamental and scientific fields such as electron and energy transfer, organic electronics, catalysis engineering, and so on. Thus, the present preliminary research works only give a cursory browse of recent progress in this field. Future research orientation on synthesis of nanocomposites and nanomaterials would mainly focus on the special nanostructures with less-expensive fabrication process to gain high-performance and low-cost nanocomposite and nano-devices.

Acknowledgements

The authors would like to extend their thanks to the National Natural Science Foundation of China (Nos. 21473153, 21207112, and 51402253); the Natural Science Foundation of Hebei

Province (No. B2013203108); the Science Foundation for Excellent Youth Scholars from Universities and Colleges of Hebei Province (No. YQ2013026); the Support Program for the Top Young Talents of Hebei Province; the Scientific and Technological Research and Development Program of Qinhuangdao City (No. 201502A006); the China Postdoctoral Science Foundation (No. 2015M580214); and Open Foundation of National Key Laboratory of Biochemical Engineering (Institute of Process Engineering, Chinese Academy of Sciences) for providing funds for this research.

Author details

Jingxin Zhou[1], Tifeng Jiao[1,2*], Qingrui Zhang[1] and Jie Hu[1]

*Address all correspondence to: tfjiao@ysu.edu.cn

1 Hebei Key Laboratory of Applied Chemistry, School of Environmental and Chemical Engineering, Yanshan University, Qinhuangdao, P. R. China

2 National Key Laboratory of Biochemical Engineering, Institute of Process Engineering, Chinese Academy of Sciences, Beijing, P. R. China

References

[1] Mutiso RM, Winey KI. Electrical properties of polymer nanocomposites containing rod-like nanofillers. *Prog Polymer Sci* 2015;40:63–84.

[2] Campbell TA, Ivanova OS. 3D printing of multifunctional nanocomposites. *Nano Today* 2013;8(2):119–20.

[3] Vashist SK, Luong JHT. Recent advances in electrochemical biosensing schemes using graphene and graphene-based nanocomposites. *Carbon* 2015;84:519–50.

[4] Hu K, Kulkarni DD, Choi I, Tsukruk VV. Graphene-polymer nanocomposites for structural and functional applications. *Prog Polymer Sci* 2014;39(11):1934–72.

[5] Jiao T, Hu J, Zhang Q, Xiao Y. Preparation and self-assembly of functionalized nanocomposites and nanomaterials-relationship between structures and properties. In: Larramendy M. (ed.) *Nanomaterials–Toxicity and Risk Assessment*. Rijeka: InTech; 2015, pp. 177–210.

[6] Jiao T, Wang S, Zhou J. Molecular design and supramolecular assemblies of novel amphiphiles with special molecular structures in organized molecular films. In: Rahman M. (ed.) *Nanomaterials*. Rijeka: InTech; 2011, pp. 315–46.

[7] Lee YS. *Self-Assembly and Nanotechnology Systems: Design, Characterization, and Applications*. New Jersey: John Wiley & Sons, Inc.; 2011, pp. 111–40.

[8] Jiao T, Wang Y, Zhang Q, Zhou J, Gao F. Regulation of substituent groups on morphologies and self-assembly of organogels based on some azobenzene imide derivatives. *Nanoscale Res Lett* 2013;8:160.

[9] Jiao T, Huang Q, Zhang Q, Xiao D, Zhou J, Gao F. Self-assembly of organogels via new luminol imide derivatives: diverse nanostructures and substituent chain effect. *Nanoscale Res Lett* 2013;8:278.

[10] Jiao T, Gao F, Zhang Q, Zhou J, Gao F. Spacer effect on nanostructures and self-assembly in organogels via some bolaform cholesteryl imide derivatives with different spacers. *Nanoscale Res Lett* 2013;8:406.

[11] Liang D, Cui C, Hu H, Wang Y, Xu S, Ying B, Li P, Lu B, Shen H. One-step hydrothermal synthesis of anatase TiO_2/reduced graphene oxide nanocomposites with enhanced photocatalytic activity. *J Alloys Compd* 2014;582:236–40.

[12] Ullah K, Zhu L, Meng ZD, Ye S, Sun Q, Oh WC. A facile and fast synthesis of novel composite Pt-graphene/TiO_2 with enhanced photocatalytic activity under UV/Visible light. *Chem Eng J* 2013;231:76–83.

[13] Bai XJ, Wang L, Zhu YF. Visible photocatalytic activity enhancement of $ZnWO_4$ by graphene hybridization. *ACS Catal* 2012;2:2769–78.

[14] Sun L, Shao R, Tang LQ, Chen ZD. Synthesis of $ZnFe_2O_4$/ZnO nanocomposites immobilized on graphene with enhanced photocatalytic activity under solar light irradiation. *J Alloys Compd* 2013;564:55–62.

[15] Xu JJ, Ao YH, Chen MD. A simple method for the preparation of Bi_2WO_6-reduced graphene oxide with enhanced photocatalytic activity under visible light irradiation. *Mater Lett* 2013;92:126–8.

[16] Li T, Shen JF, Li N, Ye MX. Hydrothermal preparation, characterization and enhanced properties of reduced graphene-$BiFeO_3$ nanocomposite. *Mater Lett* 2013;91:42–4.

[17] Lv T, Pan LK, Liu XJ, Lu T, Zhu G, Sun Z. Enhanced photocatalytic degradation of methylene blue by ZnO-reduced graphene oxide composite synthesized via microwave-assisted reaction. *J Alloys Compd* 2011;509:10086–91.

[18] Jiang LX, Li KX, Yan LS, Dai YH, Huang ZM. Preparation of Ag(Au)/graphene-TiO_2 composite photocatalysts and their catalytic performance under simulated sunlight irradiation. *Chin J Catal* 2012;33(12):1974–81.

[19] Zhang DF, Pu XP, Ding GQ, Shao X, Gao YY, Liu JX, Gao MC, Li Y. Two-phase hydrothermal synthesis of TiO_2-graphene hybrids with improved photocatalytic activity. *J Alloys Compd* 2013;572:199–204.

[20] Ghosh SK, Pal T. Interparticle coupling effect on the surface plasmon resonance of gold nanoparticles: from theory to applications. *Chem Rev* 2007;107:4797–862.

[21] Katz E, Willner I. Integrated nanoparticle-biomolecule hybrid systems: synthesis, properties, and applications. *Angewandte Chemie International Edition* 2004;43(45): 6042–108.

[22] Daniel MC, Astruc D. Gold nanoparticles: assembly, supramolecular chemistry, quantum-size-related properties, and applications toward biology, catalysis, and nanotechnology. *Chem Rev* 2004;104:293–346.

[23] Liao J, Bernard L, Langer M, Schoenberger C, Calame M. Reversible formation of molecular junctions in 2D nanoparticle arrays. *Adv Mater* 2006;18(18):2444–7.

[24] Klajn R, Bishop KJM, Grzybowski BA. Light-controlled self-assembly of reversible and irreversible nanoparticle suprastructures. *Proc Nat Acad Sci USA* 2007;104:10305–9.

[25] Koplin E, Niemeyer CM, Simon U. Formation of electrically conducting DNA-assembled gold nanoparticle monolayers. *J Mater Chem* 2006;16:1338–44.

[26] Meister A, Drescher S, Mey I, Wahab M, Graf G, Garamus VM, Hause G, Mo1gel HJ, Janshoff A, Dobner B, Blume A. Helical nanofibers of self-assembled bipolar phospholipids as template for gold nanoparticles. *J Phys Chem B* 2008;112:4506–11.

[27] Jadzinsky PD, Calero G, Ackerson CJ, Bushnell DA, Kornberg RD. Structure of a thiol monolayer-protected gold nanoparticle at 1.1 Å resolution. *Science* 2007;318:430–3.

[28] Wen Y, Jiang X, Yin G, Yin J. Multi-responsive amphiphilic gold nanoparticles (AuNPs) protected by poly(ether amine) (PEA). *Chem Commun* 2009;43:6595–7.

[29] Song WJ, Du JZ, Sun TM, Zhang PZ, Wang J. Gold nanoparticles capped with polyethyleneimine for enhanced siRNA delivery. *Small* 2010;6:239–46.

[30] Ko S, Park TJ, Kim HS, Kim JH, Cho YJ. Directed self-assembly of gold binding polypeptide-protein: a fusion proteins for development of gold nanoparticle-based SPR immunosensors. *Biosens Bioelectron* 2009;24:2592–7.

[31] George J, Thomas KG. Surface plasmon coupled circular dichroism of Au nanoparticles on peptide nanotubes. *J Am Chem Soc* 2010;132:2502–3.

[32] Cho EC, Au L, Zhang Q, Xia Y. The effects of size, shape, and surface functional group of gold nanostructures on their adsorption and internalization by cells. *Small* 2010;6:517–22.

[33] Jiao T, Wang Y, Zhang Q, Yan X, Zhao X, Zhou J, Gao F. Self-assembly and headgroup effect in nanostructured organogels via cationic amphiphile-graphene oxide composites. *PLoS ONE* 2014;9(7):e101620.

[34] Jiao T, Wang Y, Zhang Q, Yan X, Zhao X, Huo Q, Zhou J, Gao F. Organogels via gemini amphiphile-graphene oxide composites: self-assembly and symmetry effect. *Sci Adv Mater* 2015;7(9):1677–85.

[35] Jiao T, Guo H, Zhang Q, Peng Q, Tang Y, Yan X, Li B. Reduced graphene oxide-based silver nanoparticle-containing composite hydrogel as highly efficient dye catalysts for wastewater treatment. *Sci Rep-UK* 2015;5:11873.

[36] Hu J, Ma J, Wang L, Huang H. Synthesis and photocatalytic properties of $LaMnO_3$-graphene nanocomposites. *J Alloys Compd* 2014;583:539–45.

[37] Hu J, Ma J, Wang L, Huang H. Preparation of $La_{1-x}Sr_xMnO_3$ /graphene thin films and their photocatalytic activity. *Mater Sci Eng B* 2014;180:46–53.

[38] Hu J, Ma J, Wang L, Huang H, Ma L. Preparation, characterization and photocatalytic activity of Co-doped $LaMnO_3$/graphene composites, *Powder Technol* 2014;254:556–62.

[39] Hu J, Wang L, Ma J, Huang H. Study on the preparation and photocatalytic performance of the $LaNiO_3$/graphene composite film. *Rare Metal Mater Eng* 2014;43(7):1736–41.

[40] Hu J, Wang L, Shi L, Huang H. Preparation of $La_{1-x}Ca_xMnO_3$ perovskite-graphene composites as oxygen reduction reaction electrocatalyst in alkaline medium. *J Power Sources* 2014;269:144–51.

[41] Hu J, Men J, Ma J, Huang H. Preparation of $LaMnO_3$/graphene thin films and their photocatalytic activity. *J Rare Earths* 2014;32(12):1127–35.

[42] Zhong L, Jiao T, Liu M. Synthesis and assembly of gold nanoparticles in organized molecular films of gemini amphiphiles. *Langmuir* 2008;24:11677–83.

[43] Zhong L, Jiao T, Liu M. A facile method to the synthesis of gold nanoprisms using a series of gemini amphiphiles. *J Nanosci Nanotechnol* 2009;9:2726–30.

[44] Jiao T, Wang Y, Guo W, Zhang Q, Yan X, Chen J, Wang L, Xie D, Gao F. Synthesis and photocatalytic property of gold nanoparticles by using a series of bolaform Schiff base amphiphiles. *Mater Res Bull* 2012;47(12):4203–9.

[45] Jiao T, Wang Y, Zhang Q, Yan X, Chen J, Zhou J, Gao F. Preparation and photocatalytic property of gold nanoparticles by using two bolaform cholesteryl imide derivatives. *J Dispers Sci Technol* 2013;34(12):1675–82.

6

Advanced Nanomatericals for Solar Photocatalysis

Le Li and Minqiang Wang

Abstract

Heterogeneous photocatalysis using semiconductors and renewable solar energy has been regarded as one of the most promising processes to alleviate and even solve both the world crises of energy supply and environmental pollution. Recently, numerous semiconducting materials and its composites have been studied for their photocatalytic applications. In this chapter, we briefly summarize recent progress in the binary oxide system (including TiO_2 and α-Fe_2O_3), ternary oxide (Bi system), and the semiconducting materials and their composites which have remarkable applications in photocatalytic degradation of toxic pollutants, hydrogen production and as an adsorbent for wastewater treatment. In addition, we highlight the challenges and opportunities when we implement photocatalytic materials to help on the development of energy research and find ways to approach major problems.

Keywords: Photocatalysis, TiO_2, α-Fe_2O_3, $BiVO_4$, Bi_2WO_4, BiOX, Binary oxide, Ternary oxide

1. Introduction

Ever-increasing environmental issues and consumption of fossil fuels have stimulated extensive research on the utilization of sustainable solar energy [1]. The extensive use of fossil fuels has led to a serious energy crisis and environmental pollution, which are the two major challenges facing the world in the 21st century. Among the many advanced technologies available today, heterogeneous photocatalysis in view of semiconductors taking advantage of regeneration solar energy has been identified as one of the most prospective strategies for resolving both the environmental and energy problems and has thus caused much attention during the recent decades [2–4]. In the past decades, numerous results have been reported for photocatalysis and their applications to produce hydrogen from water (see Fig.1 a) [5–8], convert solar energy into electric energy (see Fig.1 b) [9,10], degrade organic pollutants (see

Fig.1 c) [11–13] and reduce CO_2 into organic fuels (see Fig.1 d) [13–15]. Besides the naturally abundant in nonrenewable energy sources such as solar energy can be renewed into chemical or electrical and thermal energies by using semiconductors having persisting materials in the process of photocatalysis [17-20]. Generally speaking, the mechanism of a typical power-driven photocatalysis process is mainly owing to three critical related synergistic steps: (i) light absorption and charge excitation; (ii) charge separation and transport from the semiconductor particle to its surface active sites; (iii) surface photocatalytic chemical reactions, and this process is similar to the fundamental mechanism of photocatalysis in power systems.

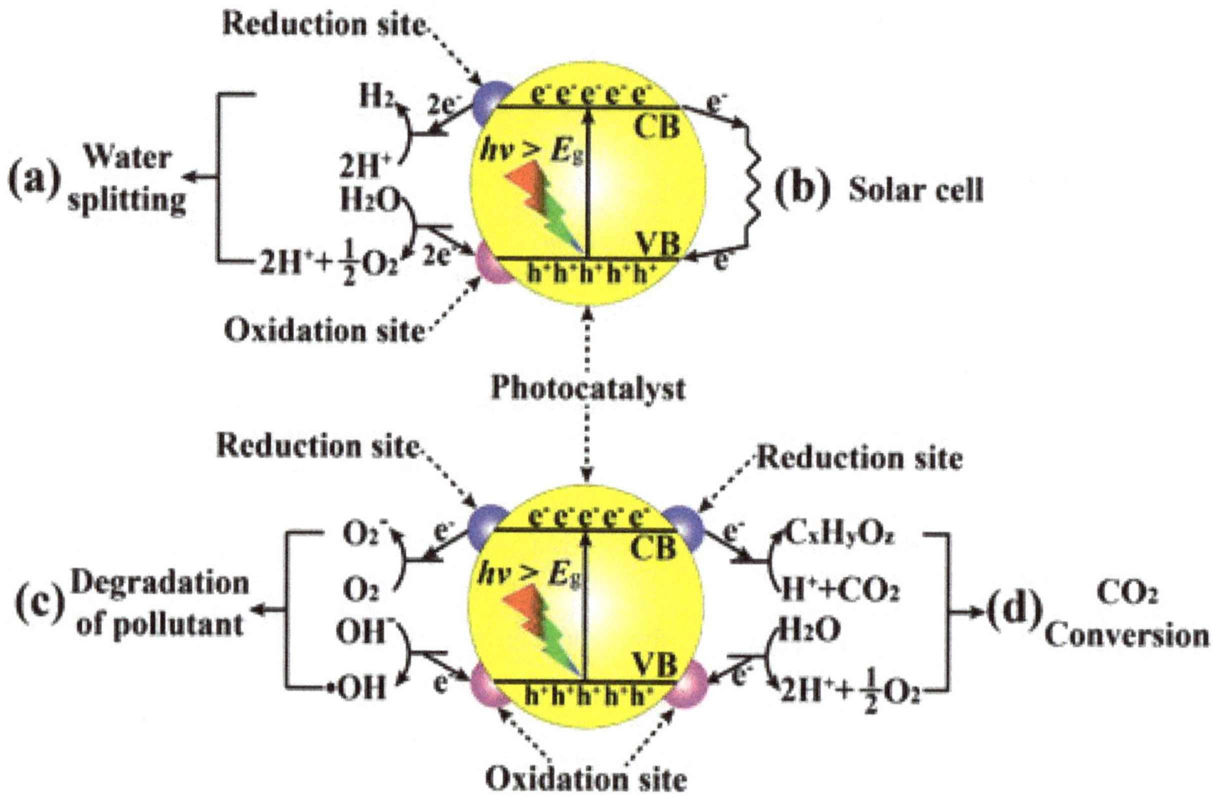

Figure 1. Photocatalytic mechanisms of water splitting, solar cell, degradation of pollutants, CO_2 reduction via one-step photoexcitation. CB and VB represent the conduction and valence bands, respectively [16].

Typically, the electron-hole pair with specific reduction and oxidation potential will be created on its conduction band (CB) and valence band (VB) under the irradiation of incident light with energy greater than the band gap of a given semiconductor. Here, the band gap of the semiconductor determines the utilization rate of the energy of the incident light, and the CB and VB values are the origin of the reduction and oxidation abilities of the photoexcited electrona and holes [21]. However, in practical process, the performance of photocatalysts is mainly related to two conditions: (i) the energy (*hv*) of the incident photon should be larger than the energy gap (*Eg*) of the photocatalyst; (ii) the redox potential of reactants should be located between the CB and VB of the semiconductor photocatalyst. One the one hand, the former condition indicates a narrow band gap, which can facilitate the efficient utilization of incident solar-light. On the other hand, the latter condition demonstrate that a more higher CB potential and a more lower VB potential, which are thermodynamically beneficial for the reduction and oxidation reactions of the reactants, respectively. But a high CB and low VB potential means a broad band gap of a photocatalyst, which leads to the poor solar–light utilization as discussed in condition (i). It is obvious that these conditions above are mutually contradiction, and it is important to find the balance point to design the photocatalyst. However, for a single component photocatalyst, it is difficult to possess both wide light –absorption range and strong redox ability concurrently. Besides, in the single-component structure, the photogenerated electrons in the CB can easily return to the VB or trap in the defect state and recombine with the holes, which seriously reduces the utilization efficiency of solar energy [22-24]. Hence, designing appropriate heterogeneous photocatalytic systems should be an effective way to overcome this problem.

The chapter is divided into four main sections. In the first part, we describe the importance of binary oxide system photocatalytic matericals in the case of two prominent and widely studied matel oxides: Titania (TiO_2), hematite (α-Fe_2O_3). In the second part, we do focus on materials with a specific ternary oxide photocatalytic matericals, such as Bi systems photocatalytic materials. In the third part, we discuss the semiconducting materials and its composites which have promising applications in the area of energy and environment especially in photocatalysis. In the end, we highlight the challenges and opportunities on the way to implement photocatalytic materials to help on the development of energy research and finding ways to approach for the major problems. Hence, we believe that a comprehensive chapter on advanced nanomatericals for solar photocatalytic is desirable for the further development of the novel photocatalytic materials and deeper understanding of photocatalytic mechanisms will be achieved in the near future, through more fundamental interdisciplinary research.

2. Binary oxide system

2.1. TiO_2

TiO_2 has turned out to be one of the most commonly investigated semiconductors due to its low cost, long–term thermodynamic stability in aqueous solution, low toxicity, and high efficiency in the removal of pollutants in water and air as well as hydrogen generation [25-28]. There following are the four commonly known polymorphs of TiO_2 found in nature: anatase (tetragonal), rutile (tetragonal), brookite (orthorhombic), and TiO_2(B) (monoclinic) [29,30]. Rutile TiO_2 has a tetragonal crystal structure and contains six atoms per unit cell as shown in Fig. 2 [31]. Rutile is the most thermodynamically stable polymorph of TiO_2 at all temperatures, exhibiting lower total free energy than metastable phases of anatase and brookite. Anatase TiO_2 has a crystalline structure that corresponds to the tetragonal system but the distortion of the TiO_6 octahedron is slightly larger for the anatase phase. Anatase is the most commonly used in photocatalytic applications due to its inherent superior photocatalytic properties [32-34].

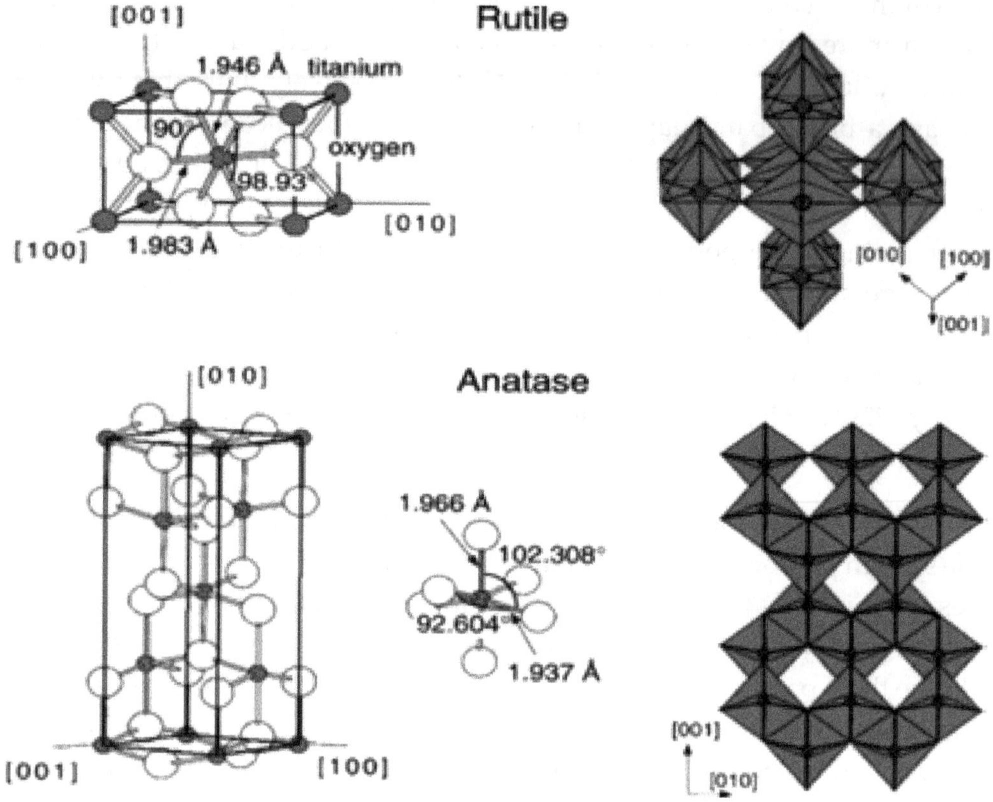

Figure 2. Representations of the TiO_2 anatase and rutile forms [31].

Anatase is the least thermodynamically stable TiO_2 polymorph as a bulk phase, although, from energy calculations, it appears as the most stable phase when the grain size is below 10–20 nm [35,36]. The crystalline structure of the TiO_2 oxides can be described in terms of TiO_6 octahedral chains. These differ by the distortion of each octahedron and the assembly pattern of the resulting octahedral chains. The Ti–Ti distances in the anatase structure are greater than in rutile, while the Ti–O distances are shorter [37]. These structural differences lead to different mass densities as well as different electronic band structures. As a result, the anatase phase is 9% less dense than rutile and it presents more pronounced localization of the Ti 3d states and further a narrower 3d band. Also, the O 2p–Ti 3d hybridization is different in the two structures. Anatase exhibits a valence and conduction band with more pronounced O 2p–Ti 3d character and less nonbonding self-interaction between similar ions (e.g., anion-anion and cation-cation interactions) [38]. The importance of the covalent vs ionic contributions to the metal-oxygen bond has already been discussed in a more general context for Ti oxides [39,40]. Therefore, it could be claimed that differential structural characteristics between anatase and rutile of TiO_2 are possibly attributed to the difference in the mobility of the charge carriers upon light excitation.

In 1972, K. Honda and A. Fujishima discovered the photosensitization effect of a TiO_2 electrode on the electrolysis of H_2O into H_2 and O_2 using a Pt metal electrode as cathode and a TiO_2 photoanode irradiated with UV light. They found that, under UV light irradiation of the TiO_2 electrode, the electrolysis of H_2O proceeded at a much lower bias voltage as compared with normal electrolysis [41]. From then on, the TiO_2-based photocatalyst have been extensively studied in the past few decades due to its proper energy bandgap that matches the UV –visible light irradiation, which favors many light-driven applications [42-47]. Moreover, TiO_2 has got many advantages and the nature of this material is naturally abundant, commercially available, economically viable, chemically stable, non-toxic and environmental eco –friendly [44]. However, TiO_2 has also faced few problems as photocatalysts in applying solar energy processes due to its low sunlight spectrum matching, limited activity and reduced sensitivity [48]. To overcome this shortage, recently many researchers have developed many different modification methods to TiO_2 material to make it as a potential challenging material for highly active photocatalyst [48-55]. Among these works, crystal growth, doping and heterostructuring of semiconductor photocatalysts are commonly used and can substantial tune the light-response range, redox potentials of photoinduced charge carriers, and electron-hole pair separation probability within the photocatalysts. Specifically, crystal growth can be critical in controlling the phase, shape, and size of photocatalysts, as well as their crystallinity and specific surface area. By rationally controlling crystal growth, the intrinsic surface atomic structure and resultant surface states of the derived photocatalysts can be adjusted. For materials design, doping effect can exert a substantial influence on modifying the electronic structure and the construction of heteroatomic surface structures of the aiming material. In particular, nonmetal doping (N [56,57], C [58-60], S [61,62], B [63-65], F [66-68], Br [69], I [70-73], P [74]) in photocatalyst has attracted increasing attention due to its effectiveness in realizing visible–light photocatalytic activity of wide bandgap semiconductor photocatalysts. The chemical states and locations of dopants are considered to be key factors in adjusting the spectral distribution of the induced electronic states of those dopants and reconstructing favorable surface structure for photocatalysis. The hybrids of two or more semiconductor systems, that is, heterostructures, seem to be possess advantageous in more efficiently utilizing solar light by combining different electronic structures when compared with sing-phase semiconductor photocatalysts. Furthermore, an efficient photo-excited electron or hole transfer from one component to another with proper band edge matching can greatly decrease the electron-hole recombination probability and increase the lifetime of charge carriers, which further promoting the photocatalytic efficiency. In addition to the basic requirements of electronic structure for each unit in the integrated photocatalytic systems, a favorable interface contact between the two materials is essential in promoting interface charge carrier transfer through different pathways. Fig. 3 demonstrates the connection between crystal growth, doping and hetero-structure of semiconductors for heterogeneous photocatalysis, (CB: conduction band; VB: valence band) [75].

According to the Wulff construction and calculated surface energy, the shape of anatase under equilibrium conditions is a slightly truncated tetragonal bipyramid enclosed with eight isosceles trapezoidal surfaces of {101} and two top squares of {001}, as shown in Fig. 4 [76] It is predicted that the percentage of {101} is as high as 94%. Although the surface energy of {010} (0.53 J m $^{-2}$) was calculated to be between {001} (0.90 J m^{-2}) and {101} (0.44 J m $^{-2}$) [77], it is surprising that no {010} will appears in the equilibrium shape of anatase. Anatase TiO_2 is usually exposed with low-index facets. Theoretical calculations indicate that the (101) surface (0.44 J m^{-2}) is the thermodynamically the most stable surface, the (001) surface (0.90 J m^{-2}) is

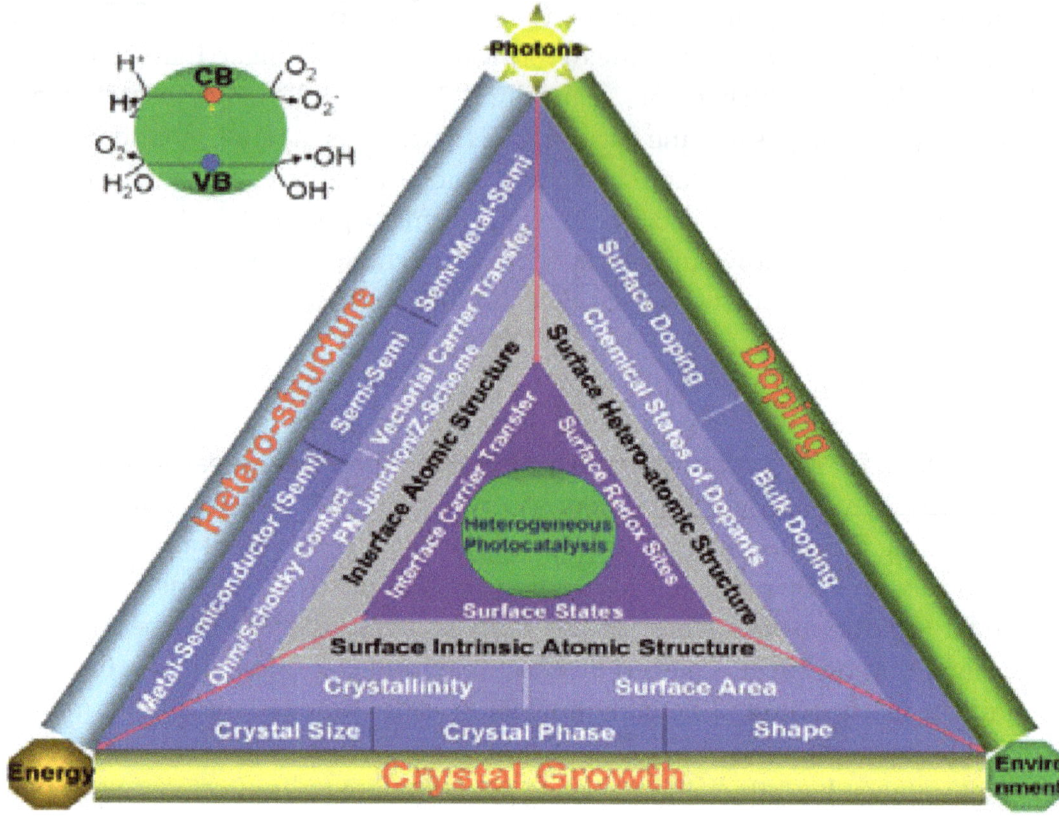

Figure 3. Correlation of key factors in crystal growth, doping and heterostructuring of semiconductors for photocatalysis. (CB: conduction band; VB:valence band) [75].

the most active and the (100) and (110) surfaces are between the (101) and (001) surfaces. As a consequence, facets that have a high surface energy diminish quickly in the minimization of surface energy during the crystal-growth process. Therefore, a large percentage of high active facets has become a popular target in the synthesis of anatase TiO_2 crystals. In the case of rutile, the predicted equilibrium shape of a macroscopic crystal was constructed with (110), (100), (001) and (011) faces (see Fig. 4) [78]. It is found that in the equilibrium shape, the most stable (110) face with the lowest surface energy of 15.6 meV au^{-2} dominates the shape, whereas (001) with the highest surface energy of 28.9 meV au^{-2} does not exist at all. Gong et al. demonstrated the systematic results of the structures and energetics of 10 stoichiometric 1×1 low-index surfaces with different possible terminations of brookite [79]. The determining factors of the relative stabilities of different faces are found to be negatively related to the concentration of exposed coordinatively unsaturated Ti atoms. The equilibrium shape of brookite crystal is shown in Fig. 4, we can observe that the most of it is composed of (111), (210), (010) and reconstructed (001) facets. It is worth noting that brookite (210) is one of the most stable facets, which has a very similar atomic structure to the most stable facet (101) of anatase. However, their electronic states are different, which may result in different chemical reactivities [80].

Usually, different facets of a single–crystalline material possess distinctive adsorption, catalytic reactivity and selectivity, which are caused by its different geometric and electronic structures [81]. Since Lu and his coworkers first reported that the uniform anatase single

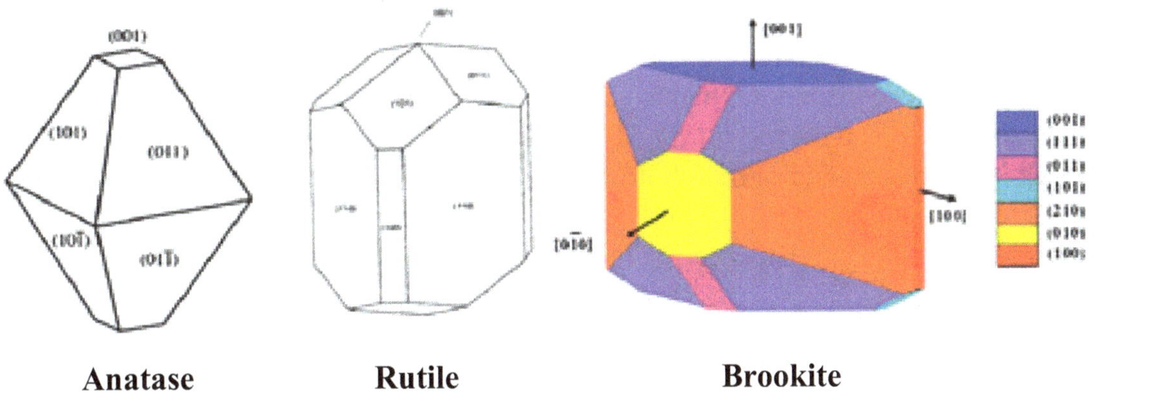

Figure 4. The equilibrium shape of a TiO_2 crystal in the anatase, rutile and brookite, according to the Wulff construction and the calculated surface energies [76, 78,79].

crystals with 47% {001} facets displayed superior photoactivity [82], crystal facet engineering has proven to be an effective strategy to finely tune the efficiency and selectivity of heterogeneous photocatalysts for different applications. Besides, different crystal facets can also facilitate the separation of electrons and holes [83]. To date, many improved synthesis procedures have been successfully developed and lots of exciting advances have been achieved [84-99]. Lu and coworkers demonstrated that, under UV light irradiation, the sheet-like anatase TiO_2 crystal dominated by {001} facets is capable of producing OH that is more than five times higher than that of Degussa P25 TiO_2 [85]. They concluded that the high density unsaturated five-fold Ti and their unique electronic structures of the {001} facets should be responsible for the improved photoactivity. Similar results also have been reported by other groups. For example, Han et al. [84] reported that the photocatalytic ability of TiO_2 nanosheets with {001} facets was higher than that of P25 in the degradation of methyl orange(MO) molecules. Zhang et al. [86] successfully synthesized the a remarkable 80% level of reactive {001} facets microsheet anatase TiO_2 single–crystal photocatalyst, which exhibited much better photocatalytic performance in the oxidative decomposition of organic pollutant. By tuning the percentage of the {001} facets, the photoreactivity was enhanced from 40.0% to 84.5%. The reactive {001} facets played an important role in the photocatalytic reaction owing to their strong ability to dissociatively adsorb water to form hydrogen peroxide and peroxide radicals. Although high-energy {001} facets have been widely studied, anatase TiO_2 crystals with higher-energy {100} facets have been less well developed. Recently, Li and Xu [100] reported a facile hydrothermal route for the synthesis of tetragonal–faceted nanorods (NRs) of anatase TiO_2 with highly exposed higher-energy {100} facets, which exhibited higher reactivity owing to the large percentage of {100} facets compared with crystals that have normal majority {101} facets.

These results discussed above demonstrated that a higher density of surface–unsaturated atoms will lead to a high surface energy of the crystal facets, which generally exhibit better photocatalytic performance. However, recent studies have shown that a high surface energy does not always make the crystal facets highly reactive in photocatalytic reactions. For example, Liu et al. [97] demonstrated a raised conduction band of nanosized single crystals of anatase TiO_2 with 82% {101} facets compared to the crystals with 72% {001} facets, which is determined by UV/Vis adsorption spectroscopy. This different electronic–band difference will further lead to a difference in atomic coordination, and this decrease will result in an enhanced photoactivity in the splitting of water into hydrogen. This example shows that the band-gap of crystal facets or crystal plates will change as the change in the arrangement of surface atoms. As a result, the redox power of the photoexcited electrons and holes will be correspondingly changed.

Generally speaking, the {101} facets are more reductive than {001} facets, which could act as possible tanks of photogenerated electrons, while {001} facets act as oxidation sites, which play a major role in the photooxidative processes [101,102]. For example, Pan et al. reported that low–index facets of anatase TiO_2 follow the photoreactivity order of {001} < {101} < {010} for photocatalytic hydrogen evolution and·OH radical generation [103]. Similarly, a seeded growth technique also demonstrated that the {101} facets of anatase TiO_2 are more active than the {001} facets for photocatalytic water splitting [104]. Surprisingly, it was even found that the photocatalytic activity for H_2 production over the {111} facet exposed anatase TiO_2 is about 5, 9, and 13 times higher than that of the TiO_2 sample exposed with dominant {010}, {101}, and {001} facets, respectively [105]. However, most researchers ignored the synergetic effects of various co-exposed facets in one sample. More attention has to be paid to finding special facets rather than the balanced ratio of different exposed facets for the best photocatalytic efficiency of water splitting.

Recently, Yu's group found that an optimal ratio of the exposed {101} and {001} facets of TiO_2 played a significant role in the enhancement of photocatalytic performance for the reduction of CO_2 [106]. As shown in Fig. 5, the surplus electrons on the {101} facets will overflow onto the {001} facets and then have a fast recombination with the holes on the {001} facets if the percentage of {101} facets is too low to hold all the photoexcited electrons, this process will lead to a decrease in the photocatalytic activity. The results clearly showed that it is of great importance to find the balanced ratio of different exposed facets in achieving the best photocatalytic efficiency [107]. This finding may shed light on the design and fabrication of advanced nanosheet-based semiconductors for water splitting.

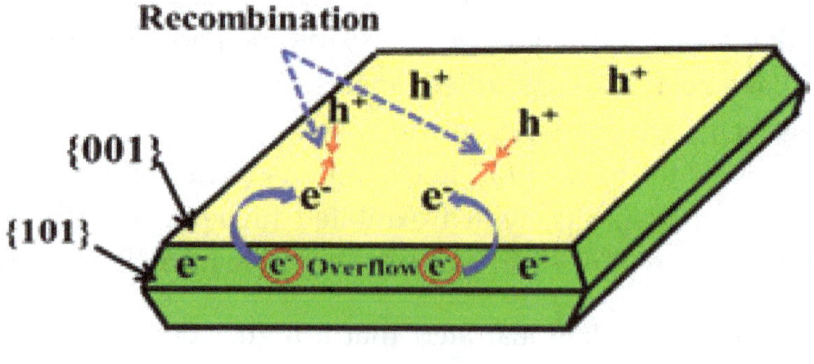

Figure 5. The electron overflow effect on the {101} facets of TiO_2 [107].

In 2011, crystal facet dependence of TiO_2 photocatalysis has been evidenced by using single-molecule imaging and kinetic analysis [108]. Single–particle spectroscopy (microscopy) has been used to explore the structural and kinetic features of "bulk" catalysis because of its high sensitivity and selectivity. This study demonstrated that the reaction sites for the effective reduction of the probe molecules were preferentially located on the {101} facets of the crystal rather than on the high surface energy {001} facets. This preference originated from the unique properties of the {101} facets in terms of their electron-trapping probability induced by the specific facet. This observation emphasizes the important role of the {101} surfaces as the reductive site in TiO_2 photocatalysis and is in agreement with the conclusion that the reactivity of the {001} facets towards oxidation is higher than that of the {101} facets.

This aforementioned investigation shows that the reactivity of a photocatalyst can be controlled by tuning the exposed facets. Another tool-morphology control, which means the formation of different surface facets with different surface atomic structures, also provides an effective method to tune the selectivity of the photocatalysts. For example, Liu et al. [95] reported a fluoride-mediated self-transformation method for fabricating hollow TiO_2 microspheres (HTS) with anatase polyhedra with about 20% exposed {001} facets. The fluorinated HTS exhibited preferential decomposition of methyl orange (MO) compared with methyl blue (MB). In contrast, surface-modified HTS that was either washed with NaOH or calcined at 600°C favored the decomposition of MB over MO. This example demonstrated the importance of the surface structure in modifying the catalytic selectivity of titania. Therefore, it is expected that, by controlling the exposed facets, we can design photocatalysts with both high reactivity and high selectivity. The photocatalysts that are terminated by specific facets allow the same adsorption states of reactant molecules and generate photoexcited electrons of similar energies on the specific facets. It is worth noting that these properties will be beneficial for the solar-induced selective photoconversion of carbon dioxides into specific valuable fuels because this process typically requires an undesired separation process and produces mixed hydrocarbons, including CH_4, CH_3OH, and HCOOH. It is believed that the breakthrough in making specific facets will intensify the development of selective organic transformations that are based on semiconductor photocatalysts.

High-index facets of nanomaterials usually have unique surface atomic structures, such as a high density of atomic steps, dangling bonds, kinks, and ledges, which can all act as active sites. Unfortunately, these unique surface atomic structures always have a high surface energy and high crystal growth rate, which is not naturally preferential growth and is easy to rapidly diminish during the crystal-growth process, so it is quite challenging to synthesize tailor–made crystals.

Yang and coworkers first reported the formation of anatase TiO_2 crystals that are exposed by high–index {105} facets [109]. They produced the product with well–faceted surface by a modified high–temperature gas–phase oxidation route with titanium tetrachloride ($TiCl_4$) as the Ti source. During the $TiCl_4$ oxidization process, the co–adsorption of oxygen, chlorine, or other related species will occur and may specifically lower the Gibbs free energy of the {105} facets thus the typical atomic configuration on the {105} facets can be stabilized and reserved. The unique stepped atomic configuration on the high–index {105} facets makes these materials promising candidates in the areas of renewable clean energy and environmental remediation.

Rutile is the thermodynamically stable phase of TiO_2 polymorphs, which can be obtained by typically three methods: (i) the hydrolysis of Ti precursors and subsequent crystallization; (ii) the post–transformation from anatase/brookite phase via thermal treatment (phase transformation temperature required depends on the particle size of TiO_2) [110] and (iii) mechanical processing [111]. Although rutile is considered to be less active in photocatalytic reactions compared to anatase, nanostructured rutile has also been used photocatalysis applications and in some cases show even higher activity than anatase. Band gap of rutile TiO_2 is 0.2 eV smaller than anatase one and further results in a wider absorption range, which may be the advantage of this phase.

Afterwards, various morphologies of rutile have been developed [112-119], with the nanorod being is a common morphology. The synthesis routes of such rutile nanorods with a high aspect ratio have been well documented in the literature [114,119,120-129]. Generally speaking, the presence of Cl ions as mineralizer in the synthesis system is favourable for rutile TiO_2, regardless of the source of Cl. In the case of the specific synthesis routes of controlling morphology of rutile, there are two representative examples demonstrate the formation of faceted rutile crystals. One is the rapid formation of self-assembled microspheres with rutile nanorods by microwave heating of $TiCl_3$ at 200°C for only 1 min [115]. The nanorods are exposed with {110} and {111} facets, but because of the extremely rapid growth rate, the surface is not smooth. Interestingly, the synthetic rutile nanorods have a smaller bandgap of 2.8 eV compared with the conventional 3.0 eV, which may facilitate the photocatalysis ability under visible light irradiation. The other one is reported by Kakiuchi et al. [116], who observed the dependence of degree of perfection of facets on hydrothermal temperature, where $TiCl_3$ was also used as a precursor together with NaCl additive. For example, at low temperature (80°C), only needle–like nanorods without well–recognized facets were formed. However, when elevating the temperature to 200°C, well–developed lateral {110} and top {111} facets can be observed. Apparently, this result indicates that a higher temperature is favorable for growing crystals with well-developed facets.

Compared with anatase and rutile, brookite phase TiO_2 has attracted little interest due to the generally considered lack of photocatalytic activity. However, increasing literatures have shown that brookite is also photocatalytically active and even has unique photocatalytic properties in some cases [130-133]. However, among the synthetic brookites, crystal facets are usually non-recognizable. Interestingly, Buonsanti et al. [134] developed a nonhydrolytic synthesis route to successfully prepare high–quality anisotropically shaped brookite nanorods with a length of 30–200 nm. These rods are determined to be dominantly enclosed with the longitudinal {210}/{100} and basal {001} facet, which is in agreement with the equilibrium shape of brookite crystals predicted from the Wulff construction.

2.2. α-Fe_2O_3

Hematite (α-Fe_2O_3) is the most thermodynamically stable form of iron oxide under ambient conditions and it is also the most common form of crystalline iron oxide. The iron and oxygen atoms are naturally arranged in the corundum structure, which is trigonal–hexagonal scalenohedral (3 2/m) with space group R–3c, lattice parameters a = 5.0356 Å, c = 13.7489 Å, and six formula units per unit cell [135,136]. It is easy to understand hematite's structure based on the packing of the anions, O^{2-}, which are arranged in a hexagonal closed–packed lattice along the

[001] direction. The cations (Fe^{3+}) occupy the two–thirds of the octahedral interstices (regularly, with two filled followed by one vacant) in the (001) basal planes, and the tetrahedral sites remain unoccupied. The arrangement of cations can also be considered as producing pairs of FeO_6 octahedra that share edges with three neighboring octahedra in the same plane and one face with an octahedron in an adjacent plane in the [001] direction (Fig. 6). The face–sharing is responsible for a trigonal distortion of the octahedra as the proximal iron atoms are repelled to optimize the crystal's Madelung energy. As a result, hematite exhibits a C_{3v} symmetry and there are two different Fe–O bond lengths (Figure 6). However, the electronic structures of the distorted FeO_6 octahedral are thought to be similar to undistorted clusters [133,136]. Hematite is antiferromagnetic at temperatures below 260 K and is a weak (parasitic) ferromagnet at room temperatures. While the magnetic properties of hemitate are not particularly dependent on its photo electrochemical performance, the iron spin configuration does influence the optoelectronic and carrier transport properties of hematite. The absorption of photons by hematite starts from the near–infrared spectral region where weak absorption bands (with absorption coefficients, a, of the order 10^3 cm^{-1}) are due to transition states electrons between two d orbital energy levels of the Fe^{3+} ion, which are split by an intrinsic crystal field [136,138]. Analysis by means of a Tauc plot shows the indirect nature of the band gap for the α-Fe_2O_3 involving d orbital to d orbital transition sand a direct transitions from O (2p) to Fe (3d), which occurs only for band gaps > 3.2 eV [139-141].

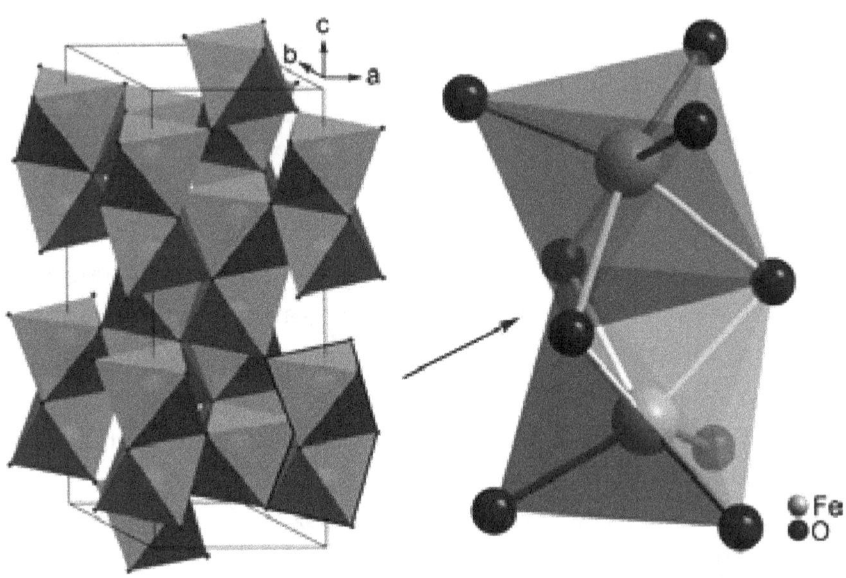

Figure 6. The unit cell (left) of hematite shows the octahedral face–sharing Fe_2O_9 dimers forming chains in the c direction. A detailed view (right) of one Fe_2O_9 dimer shows how the electrostatic repulsion of the Fe^{3+} cations produce long (light grey) and short (dark grey) Fe–O bonds [136,137].

Hematite (α-Fe_2O_3), an environmental friendly n-type semiconductor (Eg = 2.1 eV), has been widely used in many fields such as lithium–ion batteries [142], gas sensors [143–145], photocatalysis [146,147], water treatment [148] and water splitting for generating H_2. Hematite is one promising candidate for photocatalytic applications due to its narrow band gap of about 2.0–2.2 eV. Further, hematite absorbs light up to 600 nm, collects up to 40% of the solar spectrum energy, is stable in most aqueous solutions (pH > 3), and is one of the cheapest semiconductor materials available. Due to the band gap value α-Fe_2O_3 and the fact that it,s valence band edge is substantially lower than the water oxidation potential, it is a promising photoanode material for photoelectrochemical (PEC) water splitting. Photochemical water splitting involves a dispersed material in pure water and accordingly produces hydrogenand oxygen homogeneously throughout the solution [149]. The theoretical photocurrent density of α-Fe_2O_3 is ~12.6 mA/cm^2 under AM 1.5 G solar irradiation, and the solar energy conversion efficiency is ~15.5% in an ideal tandem PEC cell [150,151]. However, the photocatalytic performance of α-Fe_2O_3 is limited by certain factors such as high recombination rate of electrons and holes, low diffusion lengths of holes (2–4 nm), and poor conductivity, which led to both low efficiencies and a larger requisite over potential for photo-assisted water oxidation [152-156]. Many attempts have been made by researchers to overcome these anomalies of α-Fe_2O_3 such as lowering the recombination rate by forming nanostructures, enhancement in conductivity by doping with suitable metals and improving the charge transfer ability [157,158]. Apart from water-splitting applications, the photocatalytic activity of hematite can be used for the elimination of organic compounds in water treatment applications.

As the surface area plays an important role in determining the photocatalytic activity of materials, researchers have attempted to reduce the size of photocatalytic materials and enhance the photocatalytic properties of these materials by producing hematite in a nanoscale powder form. Many methods have been followed to synthesize α-Fe_2O_3 in a nanocrystalline form and in different shapes including hydrolysis [159], co-precipitation [160,161], hydrothermal methods [162-164], solvothermal methods [165,166], ionic liquid-assisted synthesis [167], thermal decomposition [168], combustion methods [169], and a combination reflex condensation and hydrothermal method [170].

Hosseinian et al. [171] synthesized nanostructured iron oxide of different morphologies and different phase compositions (α-Fe_2O_3 and Fe_3O_4) by a solid–state reaction (SSR) route. The photocatalytic activity was checked with respect to degradation of rhodamine B(RhB), and it was observed that the samples containing a mixture of α-Fe_2O_3 and Fe_3O_4 showed better photocatalytic activity than that of the pure α-Fe_2O_3. The higher photocatalytic activity observed for a mixed-phase sample was attributed to the higher transfer of electrons and holes generated during the photoreaction of α-Fe_2O_3 to the valence band of Fe_3O_4, which limits the recombination rates [172]. Yang et al. [173] synthesized α-Fe_2O_3 nanoparticles of uniform size (170nm to 2μm) by a hydrothermal route to study both magnetic as well as photocatalytic properties. The α-Fe_2O_3 powders with the smaller crystallite sizes show the highest photocatalytic degradation efficiency than that of the powders with larger crystallite sizes. Further, all the samples showed higher efficiency for degradation of the dye than that of the commercially available Degussa P25. Apte et al. [169] synthesized nano structured α-Fe_2O_3 powders in size ranging 25–55nm and their photocatalytic activity was analyzed with respect to the decomposition of hydrogen sulfide (H_2S) gas. α-Fe_2O_3 (necked structures) showed good photocatalytic properties and production of H_2. Zhou et al. [174] synthesized nanorods of α-Fe_2O_3 by thermal dehydration and compared the photocatalytic activity with microrods. The authors reported a higher degradation rate for rhodamine B(RhB) for nanodimensional α-Fe_2O_3 than that of the corresponding micron-sized rods. Higher Fe–O bond stretching frequencies were proposed as one of the key factors behind the enhanced photocatalytic activity. Particle size, composition, porosity, and the local structures are also the key factors that affect the photocatalytic properties of materials. Townsend et al. [159] compared the photocatalytic activity of three forms of Fe_2O_3 including bulk (crystallite size 120 nm), ultrasonicated bulk (crystallite size 40 nm), and nanopowders of α-Fe_2O_3 (crystallite size 5.4 nm). They found that the rate of oxygen evolution is higher when the crystallite size becomes smaller, and the highest rate was reported for α-Fe_2O_3 nanopowders (1072 μmol/h g). In the case of α-Fe_2O_3 nanopowders, the hole diffusion length is comparable to the crystallite size, which results in more availability of holes to react with water. Dang et al. [160] reported the effects of calcination temperature, reaction temperature, amount of catalyst, and duration of reaction on the catalytic properties. They reported an increase in photocatalytic activity with increasing calcination temperature, reaction temperature, and catalytic amount up to a certain extent, after which the activity decreases. Similar effects were also reported by Pawar et al. [175] for α-Fe_2O_3 nanoparticles synthesized by a sol-gel technique followed by the heat treatment at different calcination temperatures. The efficiency of the catalyst was analyzed with respect to various experimental variables such as calcination temperature, pH, light intensity, and concentration of dye and catalyst. Samples calcined at 600°C show the highest photocatalytic activity because of the formation of the more dominant α-Fe_2O_3 phase. The photocatalytic properties were analyzed for the 3–10 pH range, and the reactions at higher pH conditions showed better photocatalytic properties. In basic pH conditions, formation of OH· radical is more favored and electrostatic abstractive effects between cationic malachite green dye and negatively charged surface of α-Fe_2O_3 increases, which results in a higher probability of dye degradation. Light intensity shows a linear effect on the photocatalytic properties of α-Fe_2O_3 due to the increased availability of photons for the reaction. Similar effects were also reported by Liu et al. [176] for α-Fe_2O_3 nanorods. These authors examined the effect of the amount of catalyst and initial dye concentration on the photocatalytic properties. The optimum catalyst amount was reported to be 50 mg/L to achieve the highest photocatalytic activity. However, the photocatalytic activity degrades with increasing dye concentration. This effect was justified in terms of a decrease in transparency with an increase in dye and catalyst concentration after a particular value.

In a photoreaction, the porosity of the catalyst plays a major role in enhancing the photocatalytic properties. Sundarmurthy et al. [177] synthesized 1D α-Fe_2O_3 nanobraids and nanoporous structures by electrospinning to analyze the photocatalytic properties. The nanostructures show superior photocatalytic activity for the degradation of Congo red dye (CR) in a small fraction of time due to the porous surface and nanosized crystallites of α-Fe_2O_3, which provide more active catalytic centers and allow effective interaction between organic dye and α-Fe_2O_3, thereby enhancing photocatalytic degradation performance. α-Fe_2O_3 porous structures were prepared by Zhang et al. [162] and the photocatalytic activity was analyzed by the degradation of methylene blue(MB). They analyzed the effect of porosity and the amount of catalyst on photocatalytic activity. It has been reported that an optimized amount of catalyst (20 mg) is required for getting the highest rate of degradation of MB, less or more than this amount leads to lower photocatalytic activity. Large amounts of catalyst result in lesser illumination, and when the amount of catalyst is insufficient, the active sites are not sufficient to degrade the organic dye. Geng et al. [178] followed a number of Ni^{2+}/surfactant system routes for synthesizing α-Fe_2O_3 with a porous structure and rough surface which shows better photocatalytic properties than that of the α-Fe_2O_3 nanoparticles in the degradation of MB as a result of higher surface area. Gang et al. [147] prepared α-Fe_2O_3 micro/nano spheres synthesized by hydrothermal synthesis followed by the thermal treatment. The micro/nano spheres show a better dye degradation efficiency than that of the nanopowders. The calculated reaction rate for spherical structures is more than twice than that of the reaction rate of nanopowders and 12 times the reaction rate of the micron-sized powders. The better photocatalytic activityis the result of the higher specific surface area and porous structures.

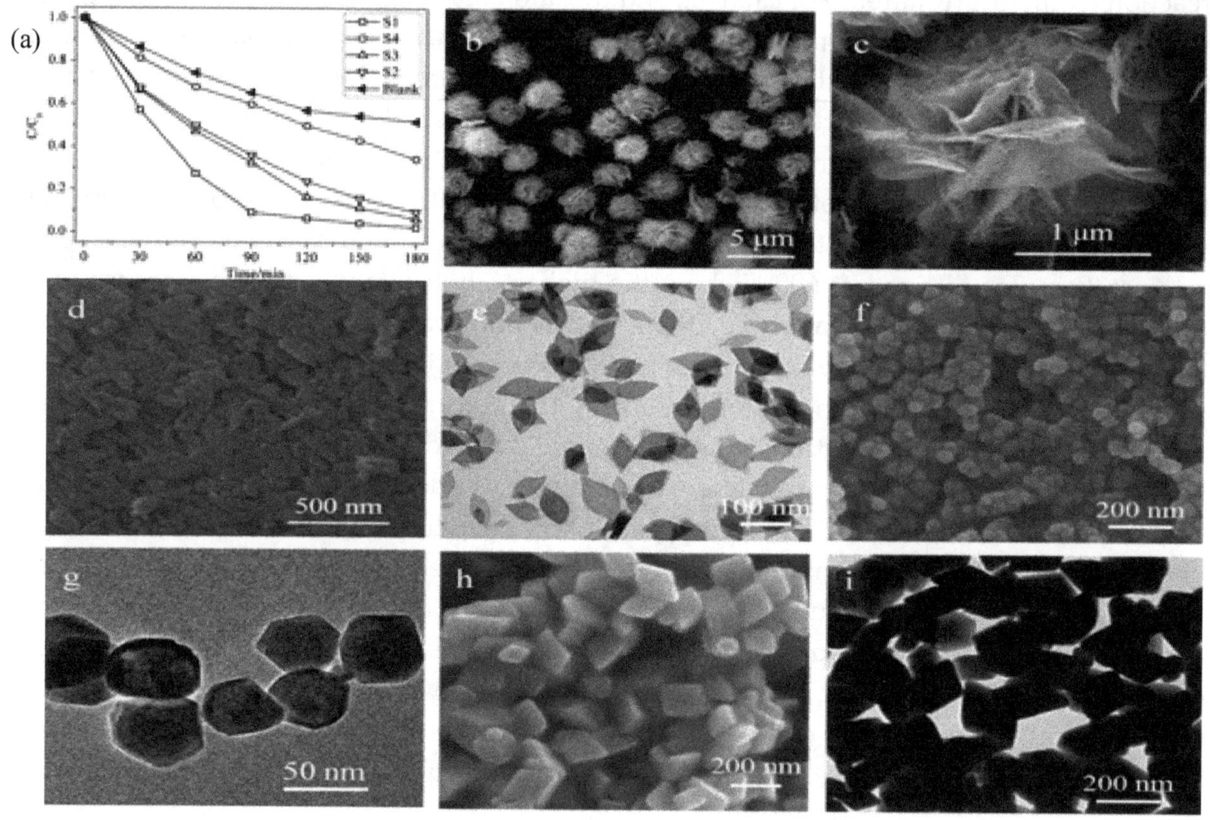

Figure 7. Photocatalytic degradation rate of RhB over the α-Fe_2O_3 nanostructures under visible light illumination in the presence of H_2O_2 additive (a),and SEM/TEM images of the α-Fe_2O_3 nanostructures: (b,c)S1;(d,e)S2;(f,g)S3;and(h,i)S4 [179].

Xu et al. [179], Zhou et al. [180] and Bharathi et al. [170] reported the effect of the surface morphology of α-Fe_2O_3 on its photocatalytic activity. α-Fe_2O_3 nanostructures with different morphologies such as microflowers, nanospindles, nanoparticles and nanorhombohedra were synthesized (Fig. 7) [179]. The photocatalytic activity was analyzed by monitoring the degradation of RhB in the presence of the catalyst. The best photocatalytic activity was observed for the samples with highest surface area and porosity. Similar surface area effects were also reported by Cheng et al. [181] for flower-like α-Fe_2O_3 nanostructures synthesized by a biphasic interfacial reaction route. The photocatalytic properties of α-Fe_2O_3 were evaluated by measuring the degradation of RhB. The results were compared with the commercial α-Fe_2O_3

powders and nanoflowers were found to have a better photocatalytic property than the commercial powders. The enhancement was related to the increase in crystallinity and increase in the surface area, which is also supported by results of other authors for TiO_2 [182] and Fe_2O_3 [183]. Similar surface area effects were also reported by Cao et al. [184], Xu et al. [166], and Li et al. [163] for α-Fe_2O_3 hollow microspheres prepared by solvothermal and hydrothermal methods. The photocatalytic activity was analyzed by the degradation of salicylic acid. The hollow spheres associated with nanosheets show better photocatalytic activity than that of then anoparticles of α-Fe_2O_3. Similar results were also reported by Majiet al. [168], where α-Fe_2O_3 powders prepared at 500°C show better photocatalytic activity for the degradation of rose Bengal dye than that of the powders prepared at 600°C and commercially available TiO_2(Degussa–25) as a result of higher surface area. α-Fe_2O_3 hollow spindles and spheres were prepared by Li et al. [164] and Xu et al. [167], respectively. These authors reported an enhancement in photocatalytic degradation efficiency as a result of the enhancement in specific surface area, which results in more unsaturated surface coordination sites exposed to the solution. The hollow microsphere facilitates more electron-hole transport and lowers the recombination rate. Hollow microspheres allow multiple reflections of visible light within the interior that encourage a more efficient use of the light source and enhance light–harvesting, leading to an increased quantity of ·OH available to participate in the photo-catalytic reaction. Along with this, the hollow spheres also provide ideal channels for the dye molecules and increase the probability of interaction.

Apart from crystallite size, the orientation of crystallites also plays a major role in enhancing the photocatalytic properties. This effect has been reported by Wu et al. [185], in which the authors prepared α-Fe_2O_3 nanocubes by a solvothermal method and reported a higher photocatalytic property for the {104} planes than that of the samples with {012} planes. The photocatalytic properties involve Fenton's reaction. The amount of Fe^{3+} on the surfaces of the catalyst play a very important role in the Fenton reaction in which the reduction of Fe^{3+} to Fe^{2+} generates hydroxyl radicals (·OH) [186]. It has been reported by Lv et al. [187] that {104} planes of α-Fe_2O_3 contain 10.3 atoms/nm^2 of exposed Fe^{3+} ions, whereas the {012} planes contain 7.33 atoms/nm^2 of exposed Fe^{3+} ions. This explains the higher reactivity of {104} planes than that of the {012} planes. Along with the surface morphology, oxygen pressure and amount of the catalyst also play a major role in enhancing the photocatalytic properties. Isaev et al. [188] reported an enhancement in the photocatalytic activity with an increase in the quantity of Fe_2O_3 up to a certain point, after that, the photocatalytic activity is decreased. Similarly, the authors reported an enhancement in dye degradation with increased oxygen content. The reason behind the enhancement in photocatalytic behavior is due to the formation of more oxygen-containing active species such as HO·, O_2·, and HO_2· oxidizing species. Zhou et al. [189] investigate visible–light–induced photodegradation of model dye rhodamine B (RhB) in the presence of hydrogen peroxide (H_2O_2) over hematite architectures, namely 1D nanorods, 2D nanoplates, and 3D nanocubes (Fig. 8), and the reactivity trend can be rationalized as exposed facets in the order {110} > {012} >> {001}. This photocatalytic activity order can be well explained by different facets of α-Fe_2O_3 surface atomic and electronic structures.

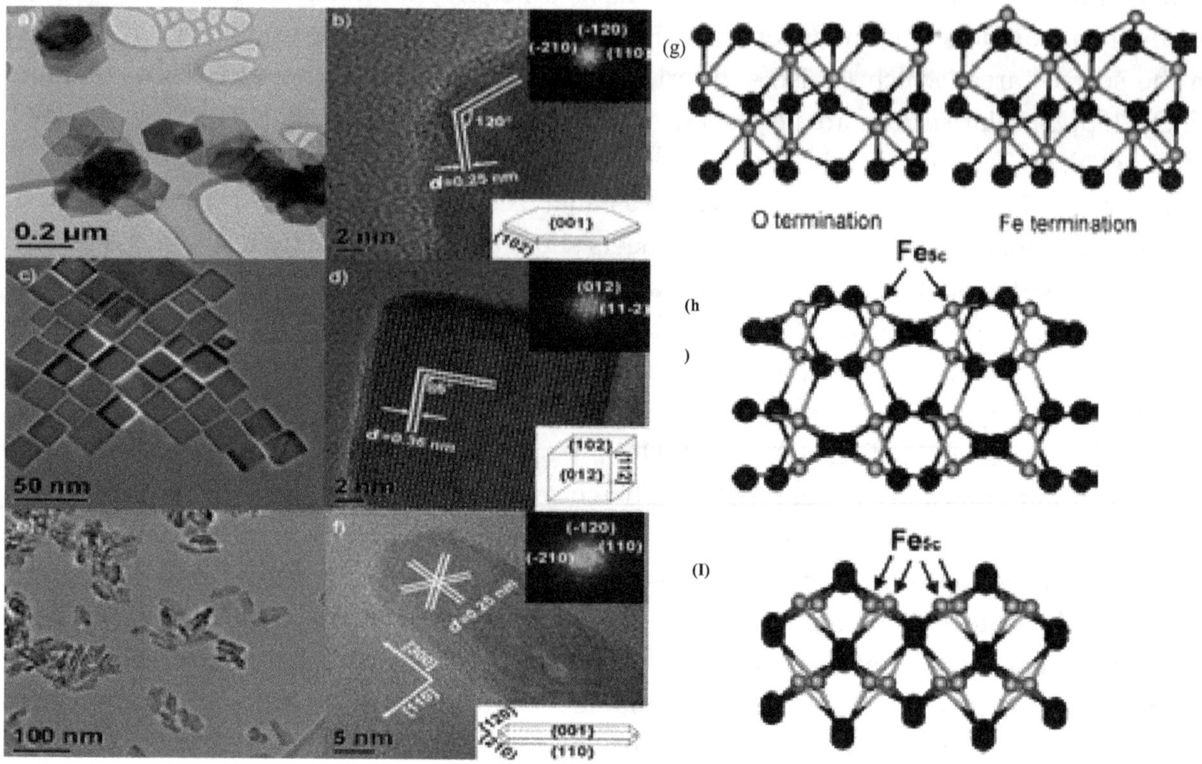

Figure 8. Representative morphologies and structures of α-Fe_2O_3 architectures.(a) TEM image and (b) HRTEM image of 2D α-Fe_2O_3 nanoplates. Insets: FFT pattern and drawing of a plate. (c) TEM image and (d) HRTEM image of 3D α-Fe_2O_3 nanocubes. Insets: FFT pattern and drawing of a cube. (e) TEM image and f) HRTEM image of 1D α-Fe_2O_3 nanorods. Insets: FFT pattern and drawing of a rod. Side views of surface terminations of α-Fe_2O_3. (g) {001}, (h) {012}, and (i) {110}. Large black spheres are oxygen and small gray spheres are iron. The coordinatively unsaturated iron atoms on the {012} and {110} surfaces are shown by arrows [189].

3. Ternary oxide system

3.1. $BiVO_4$

Bismuth vanadate ($BiVO_4$), which is an n–type semiconductor, has been identified as one of the most promising photocatalytic materials. As it is well known, $BiVO_4$ exists in three polymorphs of monoclinic scheelite, tetragonal scheelite, and tetragonal zircon structures, with bandgaps of 2.4, 2.34, and 2.9 eV, respectively. $BiVO_4$ exists naturally as the mineral pucherite with an orthorhombic crystal structure [190]. However, $BiVO_4$ prepared in the laboratory does not adopt the pucherite structure but crystallizes either in a scheelite or a zircon-type structure (Fig. 9) [191,192]. The scheelite structure can have a tetragonal crystal system (space group: $I4_1/a$ with $a = b = 5.1470$ Å, $c = 11.7216$ Å) or a monoclinic crystal system (space group: $I2/b$ with $a = 5.1935$ Å, $b = 5.0898$ Å, $c = 11.6972$ Å, and $b = 90.3871$) [192,193] while the zircon-type structure has a tetragonal crystal system (space group: I_41/a with $a = b = 7.303$ Å and $c = 6.584$ Å) [192,194].

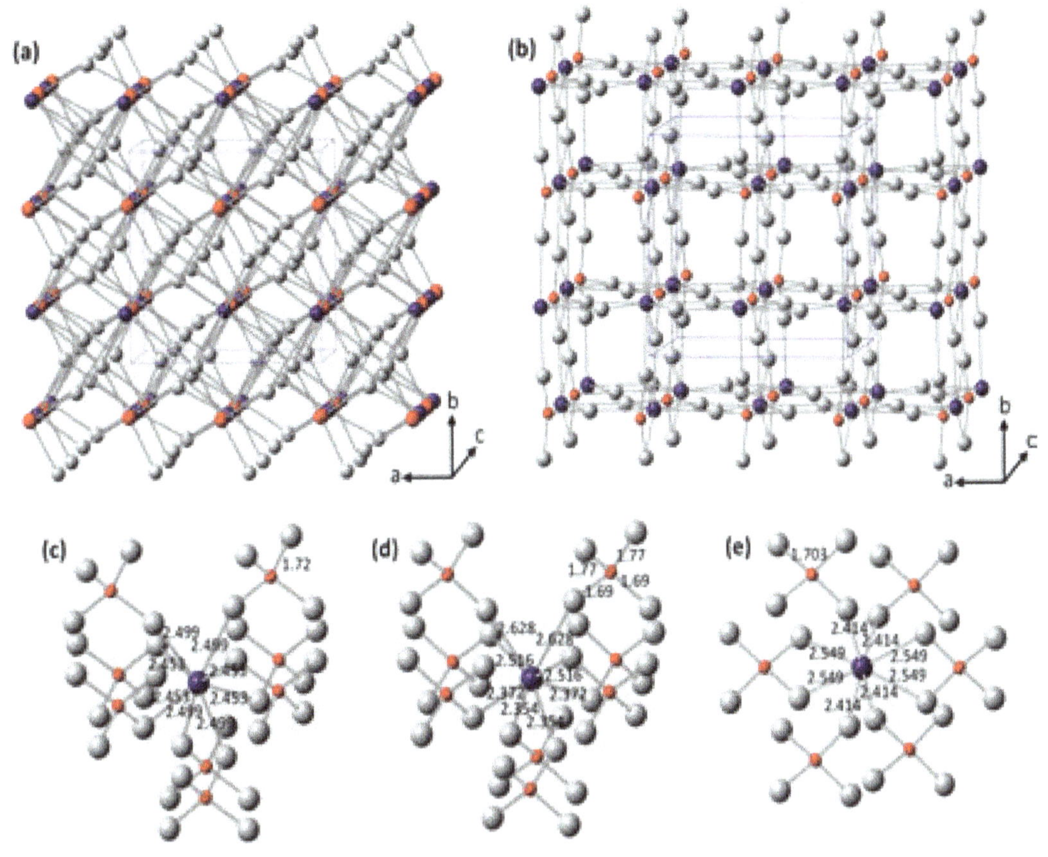

Figure 9. Crystal structures of (a) tetragonal scheelite and (b) zircon-type $BiVO_4$ (red: V, purple: Bi, and gray: O). The crystal structure of monoclinic scheelite is very similar to what is shown in (a) with the exception being the subtle changes in atomic positions of Bi, V, and O. Local coordination of V and Bi ions in (c) tetragonal scheelite, (d) monoclinic scheelite, (e) and zircon-type $BiVO_4$ structure with bond lengths shown in Å [192,194].

In the scheelite structure, four O atoms coordinate each V ion within a tetrahedral site and eight O atoms from eight different VO_4 tetrahedral units coordinate each Bi ion [192,193]. Fig. 9(a) shows the four-coordinated V center and the eight-coordinated Bi center alternating along the [001] direction. Two Bi centers and one V center coordinate each O atom in this structure, and a three-dimensional structure was formed by holding the Bi and V centers. The only difference between the tetragonal and monoclinic scheelite structure is that the local environments of V and Bi ions are more significantly distorted in the monoclinic structure, which removes the fourfold symmetry necessary for a tetragonal system. For example, in the tetragonal scheelite, all four V–O bond lengths were equal (1.72 Å), while in a monoclinic scheelite structure, there are two different V–O bond lengths(1.77 Å and 1.69 Å). In the same manner, in the tetragonal scheelite structure, only two very similar Bi–O distances exist (2.453 Å and 2.499 Å), while in the monoclinic scheelite structure, the Bi–O distances change significantly (2.354 Å, 2.372 Å, 2.516 Å and 2.628 Å) [192,193]. The significant distortion of the Bi–polyhedra indicates that the Bi 6s alone is more sterically expressed in the monoclinic scheelite structure.

It should be noted that the monoclinic scheelite structure of $BiVO_4$ was originally reported with the space group *I*2/b, which is a nonstandard space group [192,193]. Some recent studies of $BiVO_4$ have used a standard space group *C2/c*, which is converted from *I2/*b. Changes in the crystallographic axes via the conversion of a monoclinic *I*–centered (body-centered) cell to a monoclinic C-centered cell are shown in Fig. 10 [195]. With this cell conversion, the new cell parameters for *C2/c* are a' = 7.2472 Å, b' = 11.6972 Å, c' = 5.0898 Å, and β' = 134.225°. The choice of the *I*–centered monoclinic cell has the advantage of easily showing its structural relationship to the tetragonal scheelite structure that was reported in a body–centered space group, $I4_1/a$, using the identical unit cell choice and crystallographic axes. Since both *I2/b* and *C2/c* space groups, which have different unit cell choices and crystallographic axes, are commonly used to describe the monoclinic scheelite structure of $BiVO_4$, it is necessary to clarify the space group used when referring to specific atomic planes or crystal directions as well as the *hkl* indices of X-ray diffraction peaks in order to prevent any possible confusion [192].

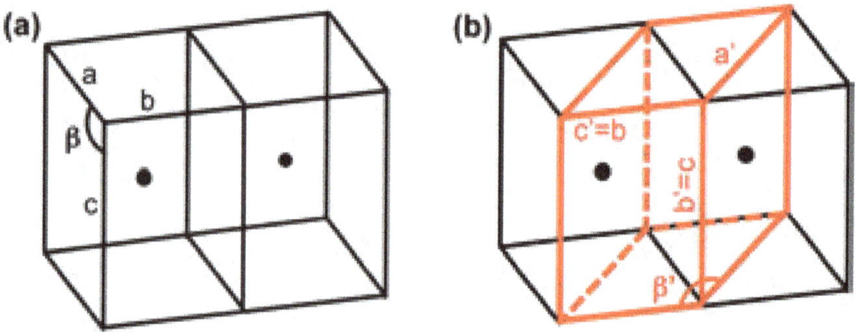

Figure 10. Cell conversion of (a) *I*–centered monoclinic to (b) C–centered monoclinic. *a, b, c,* and *β* represent the unit cell parameters for the *I*–centered cell and a, b0, c0, and b0 for the C–centered cell [195].

In the zircon–type structure, V is still stabilized by four O atoms and Bi is coordinated by eight O atoms. However, since two VO_4 units provide two O atoms to Bi, each Bi is surrounded by only six VO_4 units, as shown in Fig. 9 (e). To form a 3D structure, two Bi centers and one V center are connected by all oxygen atoms, which holds the V and Bi centers together [192].

It was reported that the low temperature synthesis (e.g., precipitation at room temperature) can form a zircon–type structure [192,196,197]. However, in this process, kinetics plays a critical role in the determination of final products, so the structure type obtained at low temperatures may change with different synthesis methods used and detailed conditions. A phase transition from tetragonal zircon to monoclinic scheelite was reported to occur irreversibly at 670–770 K [192,196]. Among scheelite structures, the tetragonal phase is a high temperature phase and the phase transition between monoclinic scheelite $BiVO_4$ and tetragonal scheelite $BiVO_4$ was observed to occur reversibly at 528 K [192,196].

In the zircon–type $BiVO_4$, the charge-transfer transition from O 2p orbitals to empty V 3d is mainly responsible for the bandgap transition. In the scheelite structure, the bandgap is reduced because the 6s state of Bi^{3+} appears above the O 2p and the transition from Bi $6s^2$ (or hybrid Bi $6s^2$–O 2p orbitals) to the V 3d becomes possible. Among scheelite $BiVO_4$ structures, Tokunaga et al. reported that monoclinic scheelite structure shows much higher photocatalysis activity for the photocatalytic water oxidation compared with tetragonal scheelite structure

[192,198]. The bandgap energies of the tetragonal and monoclinic scheelite $BiVO_4$ shows little difference and the more severe distortion of the metal polyhedra present in the monoclinic scheelite $BiVO_4$is the reason why the photocatalytic performance is different [192,198]. As discussed earlier, the local environment of Bi in the monoclinic scheelite structure is much more distorted than that in the tetragonal scheelite structure (Fig. 9 (c) and (d)).

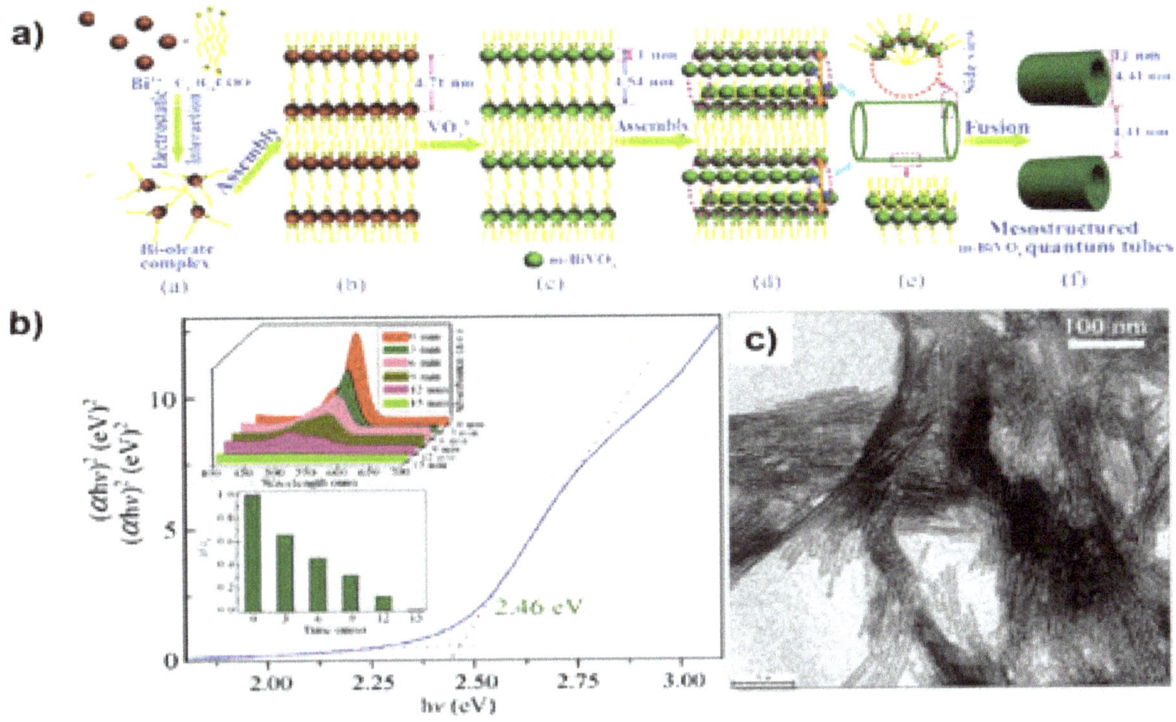

Figure 11. (a) Mechanism for the formation of $BiVO_4$ quantum tubes. (b) Optical absorption edge of $BiVO_4$ quantum tubes (top and bottom insets:photodegradation of RhB vs. irradiation time under visible light). (c) TEM image of $BiVO_4$ quantum tubes after the photodegradation [205]

Morphology control is an efficient method to facilitate carrier transportation and light harvesting, accelerate charge movement within the material structure and assist the collection and separation of electron-hole pairs at the interface of the materials [199-202]. Controllable synthesis of $BiVO_4$ with controlled morphologies has been notable developed (such as nanorods, nanowires, nanotubes (NTs), nanobelts, nanoellipsoids, hollow spheres, and even some hierarchical architectures) and corresponding morphology-dependent photocatalytic properties have also been extensively studied [203-211]. For example, Tada et al. first fabricated $BiVO_4$ nanorods (NRs) using polyethylene glycol (PEG) as a shape-directing agent [203]. Yu et al. developed a template-free solvothermal method to synthesize $BiVO_4$ nanotubes (NTs) [203]. Xie et al. reported a novel assembly-fusion strategy for the synthesis of $BiVO_4$ quantum tubes with an ultra-narrow diameter of 5 nm, ultrathin wall thickness down to 1 nm, and exposed {010} facets (Fig. 11 (a), (c)) [206]. As the increase of the reaction time, optical absorption edge and band energy of the $BiVO_4$ quantum tubes are significantly blue-shifted compared with bulk $BiVO_4$, which is due to the well-known quantum size confinement effect (Fig. 11 (b)).

Nanosized building blocks, such as nanowires, nanobelts, nanosheets, and nanotubes possess interesting properties, and the self-assembling of them into hierarchical architectures is much more interesting and has attracted great attention [199]. Liu et al. and Chen et al. synthesized $BiVO_4$ porous hollow microspheres composed of single–crystalline nanosheets using a solvothermal–induced self-assembling method (Fig. 12 (a), (b)) [207,209]. These hollow microspheres exhibited excellent photocatalytic activity due to the increased specific surface area and light harvesting ability. Xie et al. also reported the multi-responsive function of ellipsoidal $BiVO_4$ assembled from many small nanoparticles with major exposed {101} facets [210]. Similarly, Zhao et al. synthesized uniform hyperbranched $BiVO_4$ via a surfactant–free hydrothermal route (Fig. 12 (c)) [211]. The crystal consists of four trunks with branches distributed on opposite sides, this unique structure is beneficial from the different growth rates along a, b, and c axes: preferentially along the [100] direction at the beginning and subsequently along the [010] and [001] directions. The loosely packed building units of the hyperbranched structure exhibits excellent photocatalytic activity, because (i) the small crystal size allows the inside generated electron-hole pairs efficiently transporting from inside out to the surface and (ii) the large surface area provides abundant active sites for the photocatalytic reaction and promotes light harvesting as well as reactant adsorption.

One of the main reasons for the charge recombination in $BiVO_4$ is the long diffusion length of the photo–induced electrons [212-214]. Tailoring porous $BiVO_4$, especially ordered porous structures, can shorten the diffusion length and thus facilitate charge migration, providing a readily accessible channel and increasing the adsorption of reactants and the supply of more surface active sites [214,210]. Yu et al. reported that ordered mesoporous $BiVO_4$ shows a higher photoactivity than conventional $BiVO_4$, and this mesoporous $BiVO_4$ was fabricated by nanocasting using mesoporous silica KIT–6 as the replica parent template (Fig. 13 (a)) [210]. Ordered macroporous $BiVO_4$ with controllable dual porosity was synthesized by Xie et al. for efficient solar water splitting and the relationship between the geometrical characteristics and the charge migration was also demonstrated (Fig. 13 (b)) [215]. There are mainly two factors that determined by the geometrical characteristics of periodically ordered macroporous structures (Fig. 13 (b), (i)): the diameter of the macropores surrounded by the final skeletal walls (denoted as D1) and the diameter of the pores between neighboring macropores (denoted as D2). Previously, Lee et al. observed an efficient photo-induced charge drift mobility within the proper D1 size [216]. Based on this, Xie et al. further synthesized ordered macroporous $BiVO_4$ architectures with controllable dual porosity (aforementioned as D1 and D2) via a modified colloidal crystal templating method (Fig. 13 (b), (i) and (ii)), and verified that charge migration in periodically ordered macroporous architectures has a strong dependence on D1 and D2 (Fig. 13 (b), (iii)) [215]. On the one hand, no matter in the bulk and on the surface, it is believed that a smaller D2 is favorable for charge migration. On the other hand, a smaller D1 blocks bulk charge migration but facilitates surface charge migration.

The specific crystal facet determines the surface active sites and even the electronic structure, as a result crystal facets play a critical role in photocatalysis. [198,217], and consequently, it is of great importance to develop the crystals exposed with highly reactive facets [217-223]. Xi et al. synthesized well-defined $BiVO_4$ nanosheets exposed with {001} facets using a straightfor-

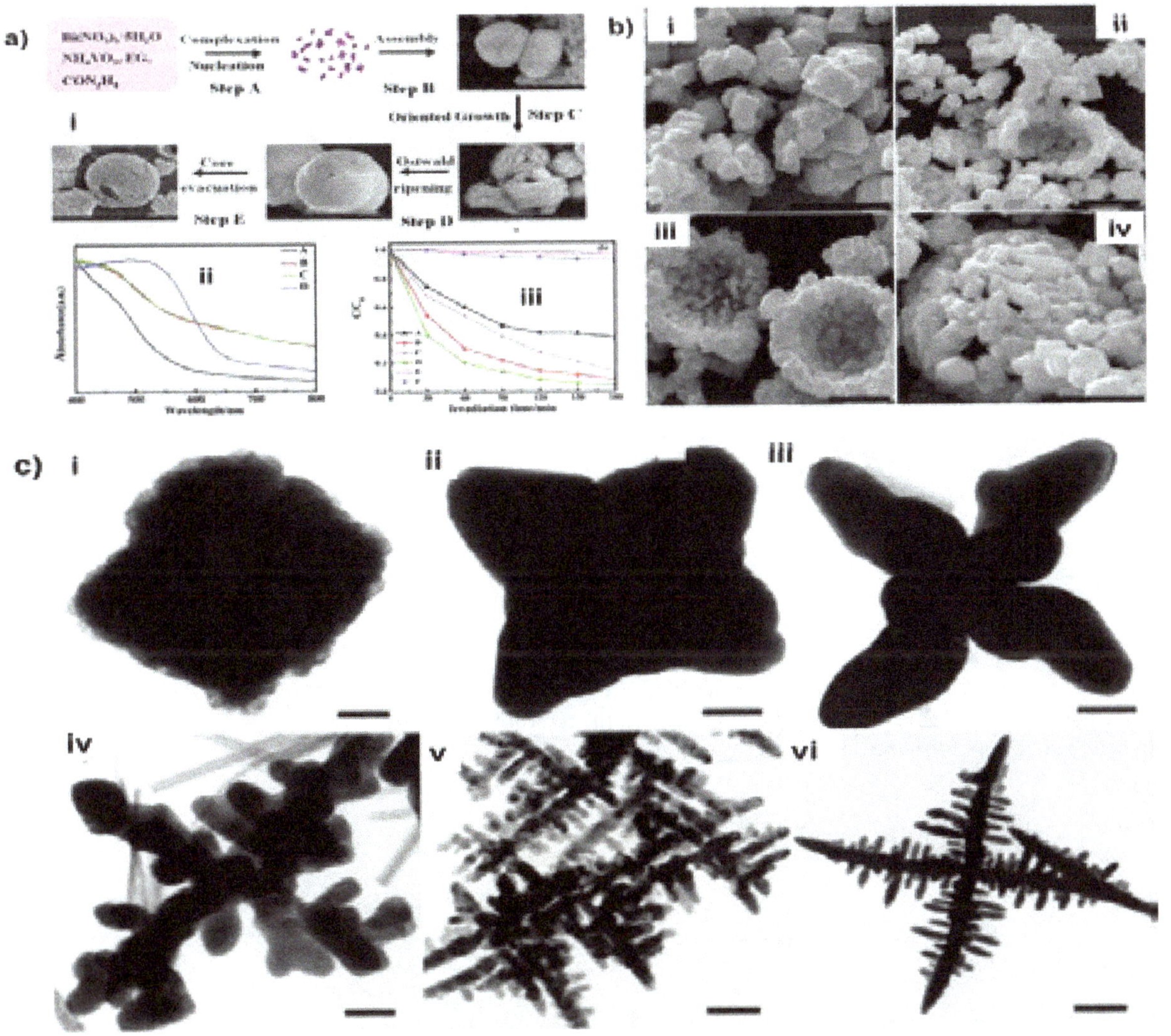

Figure 12. (a) Formation mechanism, UV–Vis absorption, and RhB photodegradation of hollow $BiVO_4$ microspheres [207]. (b) Morphology evolution of $BiVO_4$ hollow spheres via a hydrothermal method using urea as the guiding surfactant (I: 2 h; II: 4 h; III: 8 h; IV: 12 h; V: 24 h; scale bar is 2 μm) [208]. (c) Morphology evolution of hyperbranched $BiVO_4$ at intervals of 10 min (I), 20 min (II), 30 min (III), 45 min (IV), 1 h (V), and 3 h (VI), respectively (the scale bars are 100, 200, 200, 200, 400 and 500 nm, respectively) [210].

ward hydrothermal route without any template or organic surfactant (Fig. 15 (a)) [218]. Typically, $BiVO_4$ crystals show a regular decahedron shape with controllable exposed facets of {010}, {011}, {110} and {111}, as shown in Fig. 14. Li et al. Synthesized $BiVO_4$ with a highly exposed (010) facet using $TiCl_3$ as a directing agent, and correlated this to the high activity in O_2 evolution on $BiVO_4$ (Fig. 15 (b)) [219]. Inspired by this work, facet-dependent photocatalytic activity for water oxidization on $BiVO_4$ was investigated by density functional theory (DFT) calculations, particularly between the (010) and (011) facets (Fig. 14 (c)) [220]. The (010) facet has a higher activity compared with the (011) facet due to its higher charge carriers mobility, easier adsorption of water, and lower overall potential energy of O_2 evolution.

Recently, many studies have reported that photo–induced electrons and holes may be drifted to different crystal facets [217,221-223], which means photo-reduction and oxidation may

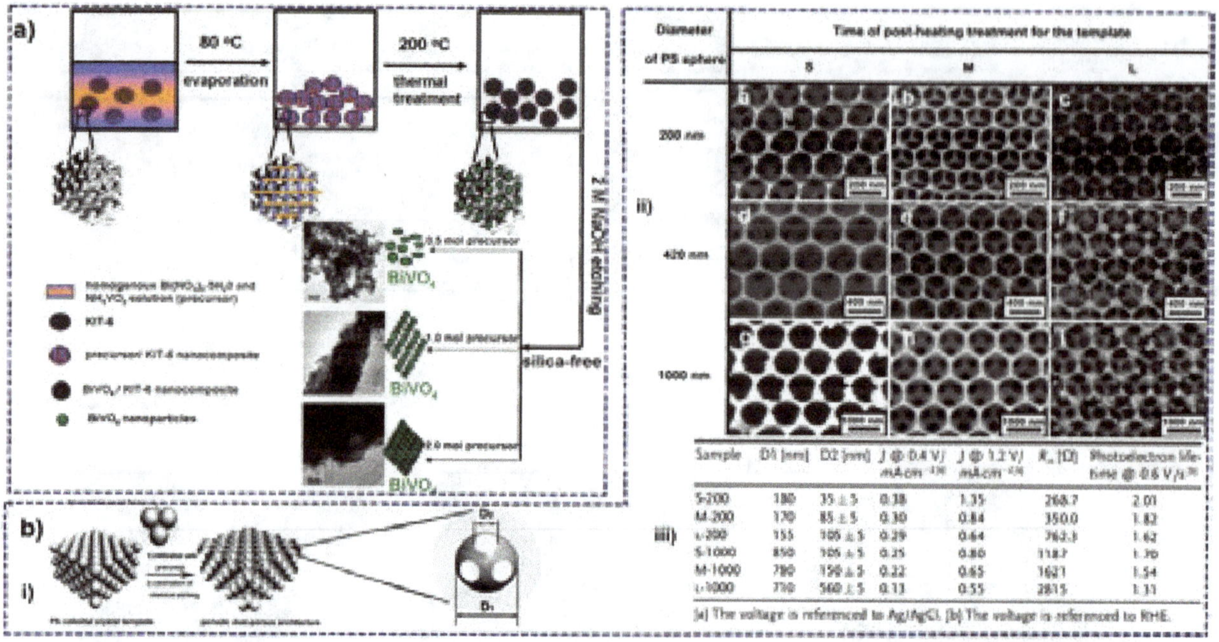

Figure 13. (a) Proposed process for the fabrication of ordered mesoporous $BiVO_4$ [211]. (b) (i)Schematic representation of dual porosity in periodically ordered porous $BiVO_4$; (ii) typical SEM images of corresponding $BiVO_4$; (iii) relationship between PEC performance and dual porosity [216].

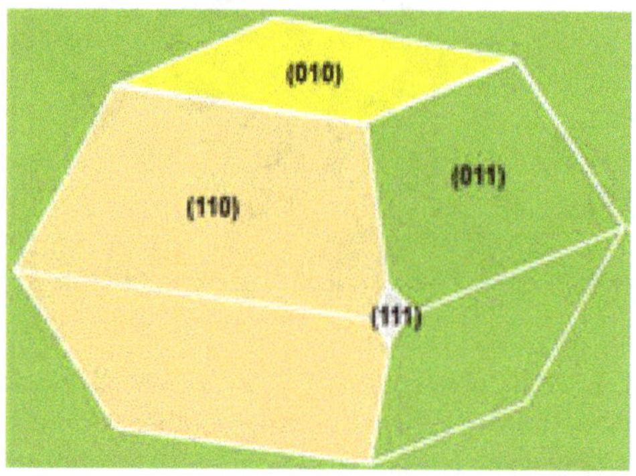

Figure 14. Typical crystal of $BiVO_4$ exposed with the {010}, {011}, {110} and {111} facets.

happen preferentially on different facets. Therefore, the cooperation of different facets is very important to obtain high quantum efficiency. Using photochemical labeling, Li et al. discovered that photo-excited electrons-driven reduction reaction (Pt–photodeposition) and photo-excited holes-driven oxidation reaction (MnO_x-photodeposition) take place on the {010} and {110} facets, respectively (Fig 16a) [222], which implies that the photo-induced electrons and holes move to the {010} and {110} facets, respectively. Notably, it provides a very useful inspiration to selectively deposited co-catalyst on specific facets via photodeposition (Fig. 16 (b)) [223]. Using this concept, the photocatalyst with Pt on the {010} facets and MnO_x on the {110} facets exhibits a much higher activity in both photocatalytic and PEC water oxidation, compared with the counterparts with randomly distributed Pt and PbO_2 co–catalysts (Fig. 16 (c), (d)). The coupling of co–catalysts on selected semiconductor facets may open up a new strategy for developing highly efficient photocatalysts.

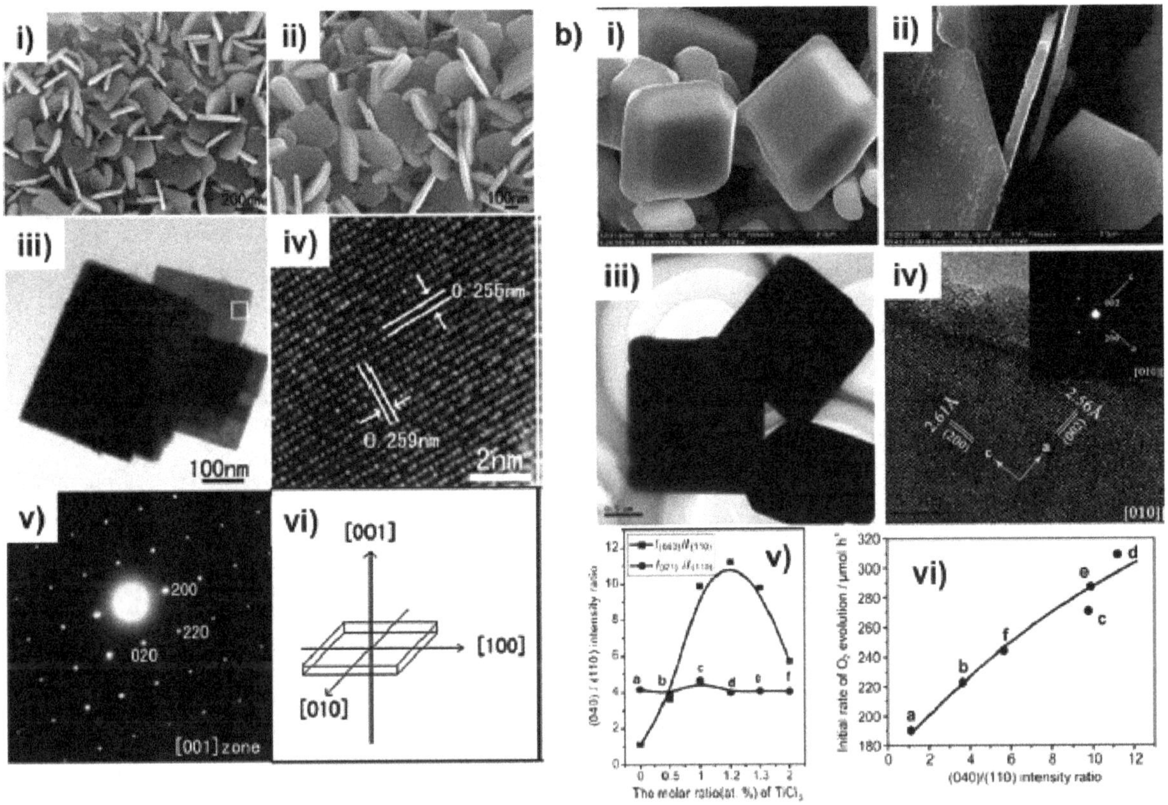

Figure 15. (a) SEM and HRTEM images of $BiVO_4$ nanoplates exposed with the {001} facets [213]. (b) Facet(010/110)-dependent photoactivity of oxygen evolution on $BiVO_4$ [214].

3.2. Bi_2WO_6

The Aurivillius family have a general formula of $Bi_2A_{n-1}B_nO_{3n+3}$ (A = Ca, Sr, Ba, Pb, Bi, Na, K and B = Ti, Nb, Ta, Mo, W, Fe), and Bi_2WO_6 is the simplest member of this family (where n = 1) and usually have the layer structures and unique properties [224]. Fig. 17 shows a schematic structure of the Bi_2WO_6 crystalline with orthorhombic structures constructed by alternating $(Bi_2O_2)_n^{2n+}$ layers and perovskite–like$(WO_4)_n^{2n-}$ layers [225]. More recently, many Aurivillius-based compounds have been reported which exhibit interesting properties suitable for photocatalytic applications. Of these, Bi_2WO_6 is the simplest and probably the most studied example within this family. In this bismuth tungstate, the perovskite–like structure is defined by WO_6 units which form a layer perpendicular to the (100) direction and sandwiched between the $(Bi_2O_2)^{2+}$ units. Layers sandwiched structure favors the efficient separation of photogenerated electron-hole pairs and then improves the photocatalytic activity, which can be ascribed to the formed internal electric fields between the slabs [226,227]. Due to its preferable band composition and unique layered structure, Bi_2WO_6 possesses several advantages as photocatalysts over the competing materials, especially in the view of practical applications, including its desirable visible– light absorption, relatively high photocatalytic activity and good stability.

Bi_2WO_6 consists of accumulated layers of corner-sharing WO_6 octahedral sheets and bismuth oxide sheets [228,229]. The conduction band of Bi_2WO_6 is composed of the W5d orbital; its valence band is formed by the hybridization of the O2p and Bi6s orbitals, which not only makes

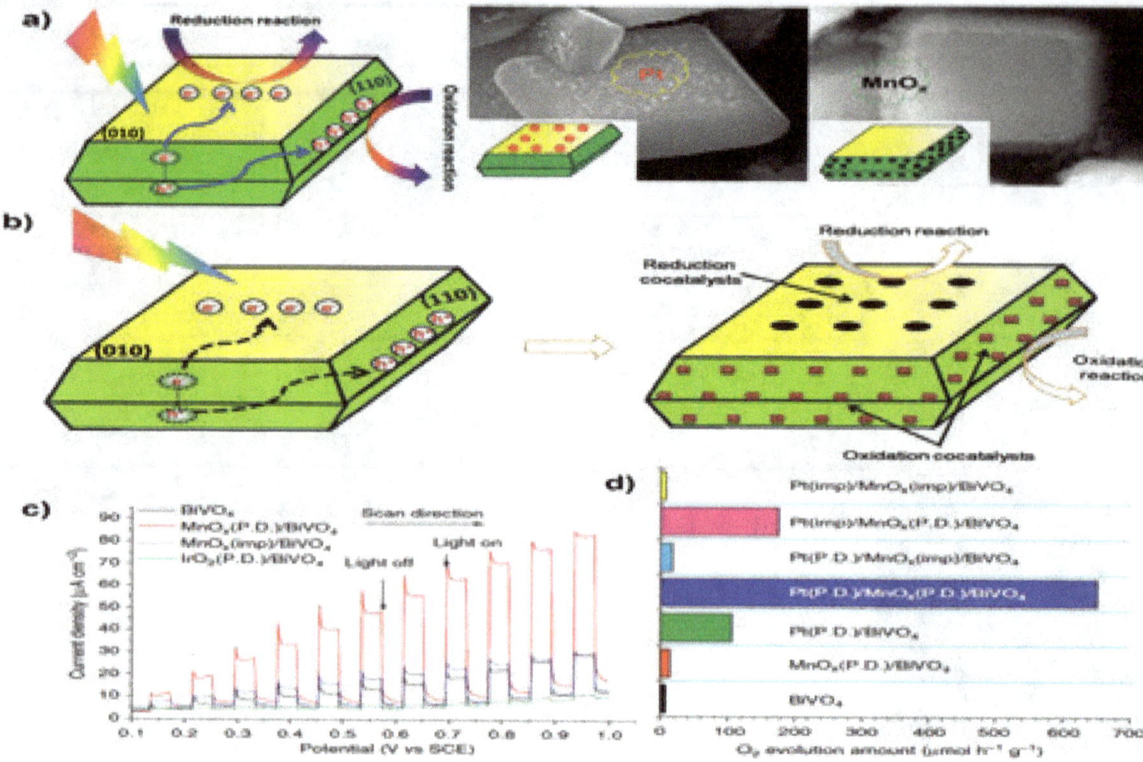

Figure 16. (a) Charge separation between the {010} and {110} facets confirmed by Pt and PbO_2 photodeposition on $BiVO_4$ [218]. (b) Selective deposition of dual redox co-catalysts on specific facets of $BiVO_4$ [219]. (c, d) Photoelectrocatalytic and photocatalytic water oxidation activity of $BiVO_4$ with selectively deposited co-catalysts on specific facets and randomly distributed co-catalysts [220].

the VB largely dispersed and thus results in a narrowed band gap of Bi_2WO_6 (2.8 eV) capable of absorbing visible light (λ > 400 nm), but also favors the mobility of photogenerated holes for specific oxidation reactions [230]. Such a band structure indicates that charge transfer in Bi_2WO_6 upon photoexcitation occurs from the O2p + Bi6s hybrid orbitals to the empty W5d orbitals, as illustrated in Fig. 18 [231].

As early as in 1999, the solid–state method was first used by Kudo et al. to synthesize Bi_2WO_6 photocatalyst [232], but the particle sizes of the product are in micrometers and the specific surface area is very small, which greatly limit its application in the photocatalysis. In the aim of obtaining micro or nanosized Bi_2WO_6 structures with enhanced photocatalytic activity, several groups have developed many advanced synthetic methods including sol–gel method [233], combustion synthesis method [234], ultrasonic method [235], co-precipitation method [236], sol-gel method/calcining method [237], and hydro/solvothermal method [238-245].

Bi_2WO_6 micro/nano–structures with diverse shapes exhibit different photocatalytic activities, and currently some of them have been used not only for the photodegradation of other organic pollutants but also for the photocatalytic disinfection. In 2005, Zhu's group have developed a Bi_2WO_6 nanoplates [246,247] applied in the photodegradation of rhodamine B (RhB) under visible–light irradiation. Notably, the photocatalytic reaction constant (k) of the best quality Bi_2WO_6 nanoplates is three times higher than that of the sample prepared by solid–state reaction [246]. In addition, they found a significantly pH-dependence of the photo-assisted

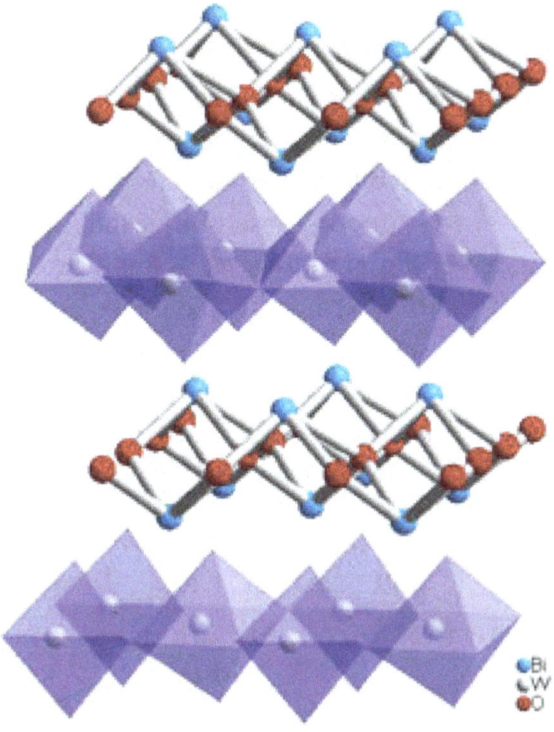

Figure 17. Structure of Bi_2WO_6 showing the WO_4^{2-} and $Bi_2O_2^{2+}$ layers [225].

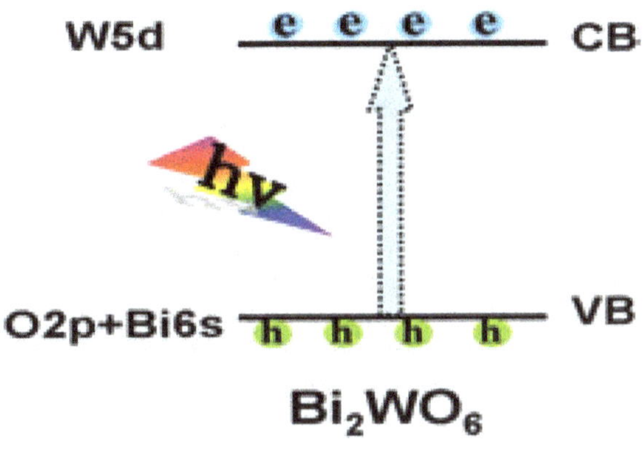

Figure 18. Band structure of the Bi_2WO_6 photocatalyst. [231]

degradation of RhB in aqueous Bi_2WO_6 as the pH varies from 5.03 to 9.89, where the highest degradation rate was achieved at pH 6.53. It is proposed that the pH of the solutions can affect the mode and extent of adsorption of RhB on the Bi_2WO_6 surface and further the transformation rate of RhB indirectly. They further used the total organic carbon measurement to determine the high mineralized degree of RhB [247]. Further studies on the mechanism shows that a photocatalytic process and a photosensitized process is involved in the the Bi_2WO_6–assisted photodegradation of RhB [247]. However, the contribution of RhB photodegradation driven by the light–excited RhB was much slower than by the light–excited Bi_2WO_6. The experimental results show that only 19% of RhB was degraded by photosensitized action, while 81% of RhB was degraded by a photocatalytic process [247].

In order to further improve the photocatalytic activity of Bi_2WO_6, some groups have developed Bi_2WO_6 nanoplates superstructures [248,249-253]. Zhang group also prepared Bi_2WO_6 micro/nanostructures, including nanoplates, tyre/helixlike, disintegrated–flower–like and flower–like superstructures [238,239]. The photodegradation results of RhB show that these Bi_2WO_6 micro/nanostructures exhibit different photocatalytic activities under visible-light ($\lambda > 400$ nm) irradiation, as shown in Fig. 19 (a). Among these photocatalysts, the uncalcined flower-like Bi_2WO_6 superstructure prepared with pH = 1 processes an improved photocatalytic performance, which can degrade 84% of RhB in 60 min [238]. Besides, the photocatalytic performance can be further improved by the calcination process, and the result calcined flower–like Bi_2WO_6 superstructure has a higher photocatalytic activity, which can degrade 97% of RhB in 60 min (Fig. 19 (b)). This performance is also superior to other traditional photocatalysts such as TiO_2 (P25) and bulk SSR-Bi_2WO_6 powder prepared by solid–state reaction [239].

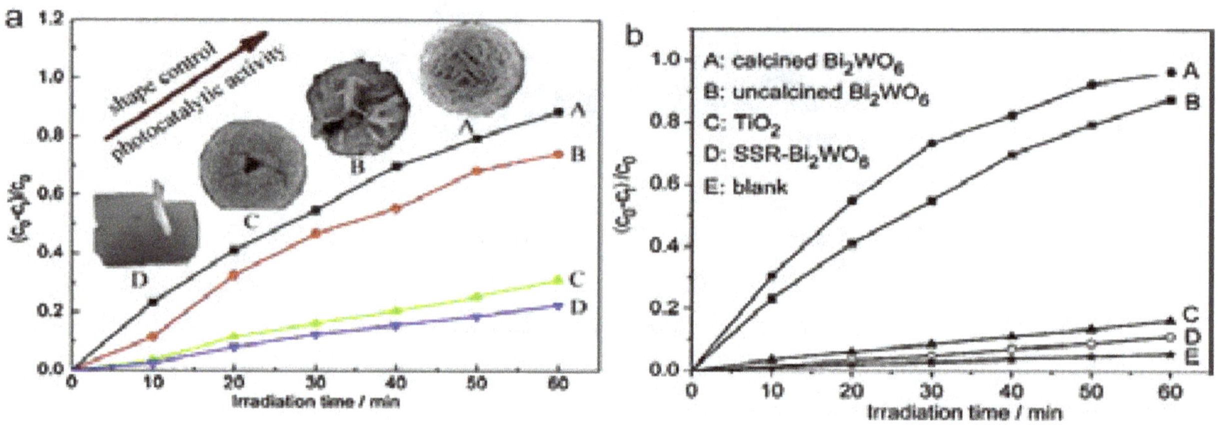

Figure 19. (a) The photodegradation efficiencies of RhB as a function of irradiation time by different Bi_2WO_6 nano/micro–structures: (A) the uncalcined flower–like Bi_2WO_6 superstructure prepared with pH = 1, (B) the uncalcined disintegrated flower-like Bi_2WO_6 superstructure prepared with pH = 2.5, (C) the calcined tyre/helix–like Bi_2WO_6 superstructures prepared with pH = 1 and P123, (D) the uncalcined Bi_2WO_6 nanoplates prepared with pH = 7.5; [239] (b) the photodegradation efficiencies of RhB as a function of irradiation time by photocatalyst samples: (A) the calcined flower-like Bi_2WO_6 superstructure, (B) the uncalcined flower-like Bi_2WO_6 superstructure, (C) TiO_2 (P25), (D)bulk SSR-Bi_2WO_6 powder, and (E) blank [238].

The novel flower-like superstructures of the uncalcined or calcined Bi_2WO_6 is mainly responsible for the highly improved photocatalytic activity. At the same time, as shown in SEM images (Fig. 20), there are plenty of meso– or macro–diameter sized pores in the flower-like superstructures, which can be considered as electron transport paths that also contributes to the photocatalysis process. [254]. It is generally believed that it is an integral part of the architectural design if the reactant molecules can easily move in or out of the nanostructured materials, the efficiency of the photocatalysis can be improved, and here, meso– or macro–diameter sized pores provides the important transport paths [254,255]. They also believe that the introduction of textural transport paths in the uncalcined or calcined Bi_2WO_6 superstructures facilitate the reactant molecules to easily incorporate with the reactive sites on the framework walls of photocatalysts, which leads to excellent photocatalytic performance for the degradation of RhB [239]. On the other hand, fewer defects, which acting as electron-hole recombination centers, can be significantly reduced by improved crystallinity of Bi_2WO_6 through the calcination process of Bi_2WO_6 [238]. This has been proved by Amano et al. [256] who experimentally investigated the influence of crystallization on the lifetime of photoexcited electrons from Bi_2WO_6. The recombination rate of electrons decay with holes can be characterized by the intensity of transient IR absorption after a 355 nm laser pulse [256]. If an appreciable absorbance at 100 μs in Bi_2WO_6 crystalline can be observed, it means a slow recombination rate and a long lifetime of photogenerated carriers, which is beneficial for driving appreciable photocatalytic reactions. However, no transient absorption for amorphous Bi_2WO_6 samples was observed implying a fast recombination of electron-hole pairs, leading to negligible photocatalytic activity. Therefore, the higher photocatalytic activity of the calcined Bi_2WO_6 is explicable in several cases [238,251,256].

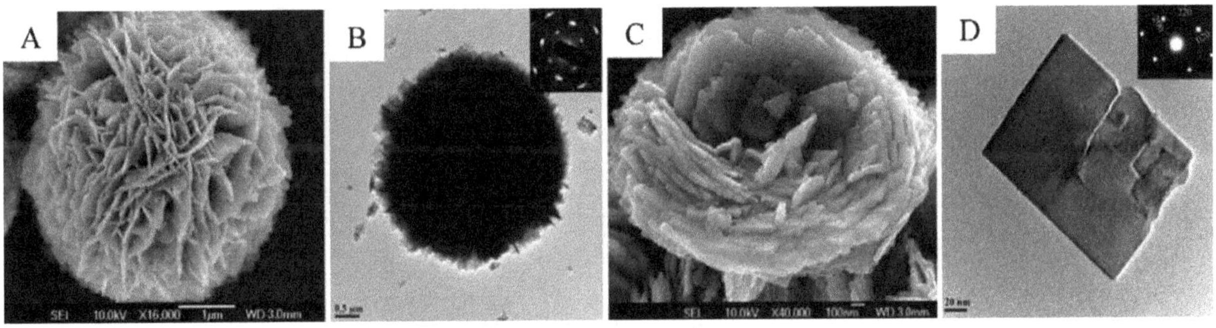

Figure 20. SEM (A) and TEM (B) images of an individual flower-like Bi_2WO_6 superstructure (inset: SEAD pattern recorded at the corner of this individual sphere); (C) SEM image of a broken Bi_2WO_6 sample; (D) TEM image of a peeled fragment (inset: SEAD pattern recorded at this individual nanoplate) (conditions: pH = 1, hydrothermally treated at 160 °C for 20 h, no surfactant, uncalcined) [238].

It is worth noting that Amano et al. [256] have demonstrated a high photocatalytic activity of crystalline Bi_2WO_6 under visible-light ($\lambda > 400$ nm) irradiation for oxidative decomposition of gaseous acetaldehyde (AcH) to produce CO_2, however, amorphous Bi_2WO_6 sample exhibits negligible photocatalytic activity under same condition. Because the mineralization of colorless AcH does not involve a dye–sensitized process, this result provides an solid conclusion that crystalline Bi_2WO_6 has excellent visible–light–driven photocatalytic activity. Furthermore, the photocatalytic activity of crystalline Bi_2WO_6 has been finely evidenced by its diffuse reflectance photoabsorption spectrum and action spectrum, that is, 8% apparent quantum efficiency at wavelength of 400 nm [256]. In the future, much more clear evidence needs to be provided to explore the mechanism of visible–light–driven photocatalytic activity of Bi_2WO_6. [256].

Recently, Amano et al. [252] reported the preparation of Bi_2WO_6 superstructures with similar hierarchical architecture, secondary particle size, crystalline shape, exposed crystalline lattice planes, and crystalline content. The only difference is that as the increasing hydrothermal reaction temperature, their specific surface areas of the products were different due to the increase thickness of crystalline rectangular platelets. The specific surface area of the product is very important, because when levels of crystalline content of Bi_2WO_6 flake-ball particles is similar, the higher the specific surface area is, the better the photocatalysis ability it shows.

This proportional relation could be explained by the fact that the initial rate of AcH decomposition was expressed by first-order kinetics with respect to the amount of surface-adsorbed AcH, which is proportional to the specific surface area of Bi_2WO_6 samples.

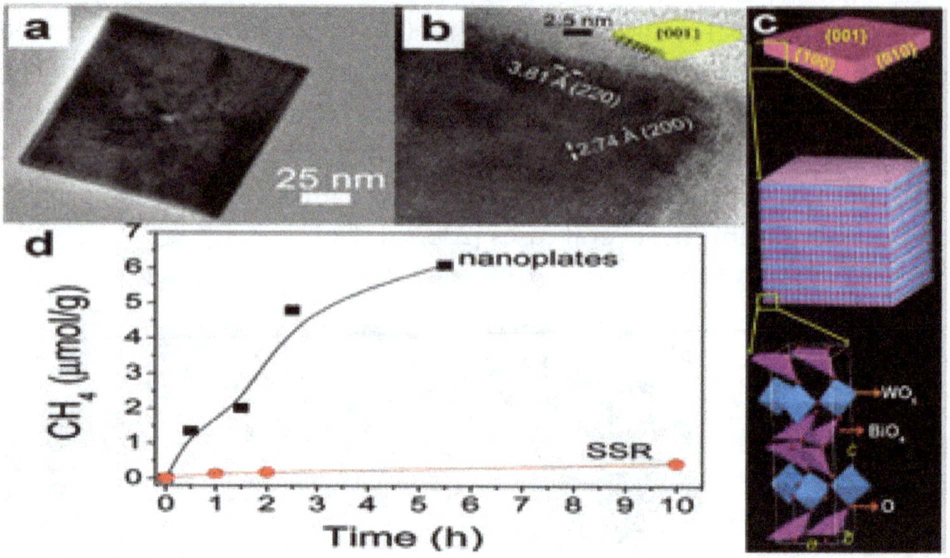

Figure 21. (a) TEM, (b) HRTEM, and (c) structural model of Bi_2WO_6 square nanoplates. (d) CH_4 generation over nanoplates and the SSR sample under visible–light irradiation ($\lambda > 420$ nm) [257].

In 2011, Zou and coworkers reported a remarkable increase in the CO_2 reduction with water to yield CH_4 over Bi_2WO_6 square nanoplates (Fig. 21 (a)–(c)) under visible-light irradiation as compared with yield over Bi_2WO_6 made by solid-state reaction (SSR) [257]. In detail, the CH_4 production rate increases from 0.045 mmol g^{-1} h^{-1} for the SSR sample to 1.1 mmol g^{-1} h^{-1} for the nanoplate catalyst (Fig. 21 (d)) [257]. Considering that the band gap of B_2WO_6 nanoplates and SSR sample is very close, geometrical factors of the photocatalyst is mainly responsible for the photoactivity enhancement. Firstly, reducing lateral dimension of the nanoplate to the nanometer scale offers a higher specific surface area. Secondly, the ultrathin geometry of the nanoplate facilitates the transfer of the charge carriers from the bulk onto the surface, where they participate in the photoreduction reaction. Thirdly, the preferentially exposed (001) crystal plane of the nanoplates is more effective than other crystal planes [257].

3.3. BiOX (X = Cl, Br, and I)

Another new class of interesting layered materials, Bismuth oxyhalides (BiOX; X = Cl, Br, and I), shows promising photocatalytic energy conversion and environment remediation ability, because of their unique layered–structure–mediated fascinating physicochemical properties and suitable band structures, along with their high chemical and optical stability, nontoxicity, low cost, and corrosion resistance [258-260]. The layered BiOX (X = Cl, Br and I) semiconductor materials, as members of the Sillen–Aurivillius family, have a tetragonal PbFCl–type structure (space group *P4/nmm*), which consists of [X–Bi–O–Bi–X] slices stacked together by the nonbonding (van der Waals) interactions through halogen atoms along the *c*-axis [261]. In each

[X–Bi–O–Bi–X] layer, a bismuth center is surrounded by four oxygen and four halogen atoms, creating an asymmetric decahedral geometry. The covalent bonds is the interaction bond within the [Bi_2O_2] layers, whereas the [X] layers are stacked together by van der Waals, forces (nonbonding interactions) between the X atoms along the *c*-axis. The strong intralayer covalent bonding and the weak interlayer van der Waals interaction can induce the formation of layered structures. For BiOX crystals, the valance band maximum mainly comprises of O2p and X np states (n = 3, 4, and 5 for X = Cl, Br and I, respectively) and the Bi 6p states dominate the conduction band minimum [262-266]. As the atomic numbers of X increases, the contribution of X ns states increases remarkably, and the dispersive characteristic of band energy level becomes more and more striking, thereby narrowing the band gap. Taking BiOCl, for example, as illustrated in Fig. 22, BiOX (X = Cl, Br, I) are characterized by the layered structure that are composed of [Bi_2O_2] slabs interleaved with double halogen atom slabs along the [001] direction. A highly anisotropic structural, electrical, optical, and mechanical properties of this material origin from the nature of its strong intralayer covalent bonding and the weak interlayer van der Waals, interaction, this unique structure allows BiOX to apply in many promising potential applications including photocatalytic wastewater and indoor-gas purification, water splitting, organic synthesis, and selective oxidation of alcohol [267-277].

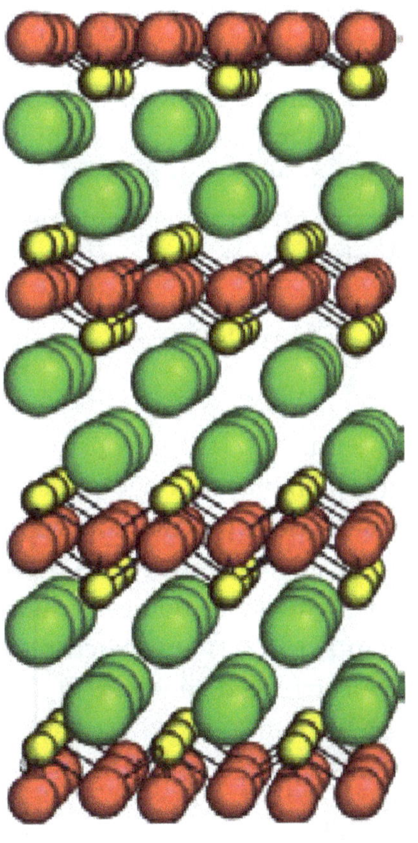

Figure 22. The schematic diagram of crystal structure of BiOCl (green, Cl atoms;yellow, Bi atoms; red, O atoms) [278].

One–dimensional nanostructures (1D), which refers to the materials with nanoscaled thickness and width, while the length can be several micrometers or longer, is considered to be promising in photocatalysis application. The prolonged length scale may allow the 1D nanomaterials to contact the macroscopic world for various measurements [279,280]. Besides, the high aspect ratio of 1D nanostructured semiconductors also facilitates the fast photoexcited electron-hole separation, which is favorable for highly efficient photocatalytic reactions.

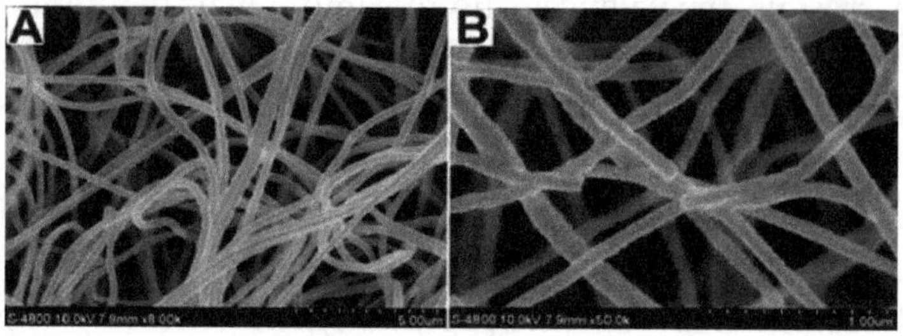

Figure 23. SEM images of the prepared (A) PAN/$BiCl_3$ nanofibers and (B) BiOCl nanofibers [281].

BiOX (X = Cl, Br and I) material is naturally preferentially grow into nanoplates/sheets with 2D features due to its highly anisotropic layered structures. As a result, hard templates for the synthesis of the 1D bismuth oxyhalide nanostructures is commonly used, becaused the template can be easily removed by subsequent thermal or chemical treatments [281-283]. For example, Liu et al. [281] developed the electrospinning method to synthesize BiOCl nanofibers, as shown in Fig. 23. After thermal removal of the polyacrylonitrile (PAN) template at 500°C for 10 h, they can obtain the BiOCl nanofibers with diameters ranging from 80 to 140 nm. Interestingly, the as-prepared BiOCl nanofibers showed high activity towards rhodamine B (RhB) degradation under the UV irradiation, and the photodegradation rate was found to be about three times faster than that of Bi_2O_3 nanofibers obtained in the same way. In addition to PAN, some other templates involving activated carbon fibers (ACFs) [282] and anodic aluminium oxide (AAO) [283] have also been used to prepare BiOCl nanofibers/nanowire arrays, which displayed efficient photocatalytic performance in the degradation of organic dyes.

In the past few years, 2D nanomaterials, such as graphene, transition metal dichalcogenides and layered double hydroxides (LDHs), have gained great attention for their extraordinary physical/chemical features and promising applications in a great deal of applications [284-287]. Intrinsically, van der Waals bonds or electrostatic forces between the layer structure in such 2D nanostructures is the origin of its lamellar structure. Similarly, the layered structure makes BiOX (X = Cl, Br and I) tend to the intrinsic 2D nanostructures, such as nanoplates, nanosheets and nanoflakes. The formed intra-electric field between [Bi_2O_2] layers and halogen atom layers could accelerate the transfer of the photo-induced carriers and enhance the photocatalytic activity of BiOX (X = Cl, Br and I) [288].

To date, numerous synthetic methodologies have been exploited for the preparation of 2D BiOX nanomaterials, such as hydrolysis, [288-292] hydrothermal/solvothermal synthesis, [293-295] and thermal annealing [296]. For instance, recently, Zhang's group [293] has synthesized 2D BiOCl nanosheets with predominantly exposed {001} and {010} facets by selective addition of the mineralizing agent NaOH. Interestingly, BiOCl nanosheets with exposed {001} facets displayed higher UV–induced photocatalytic degradation of MO dye, while the counterpart with exposed {010} facets exhibited higher degradation activity under visible light. On the one hand, the generated internal electric field along the [001] direction is more favorable for direct semiconductor photoexcitation under UV irradiation as shown in Fig. 24, which was also confirmed by the higher photocurrent of {001} facets than that of {010} facets from the transient photocurrent responses. On the other hand, compared with {001} facets, the larger surface area and open channel feature of {010} facets facilitate the adsorption of dye molecules, which further results in its better indirect dye photosensitization performance under visible light irradiation.

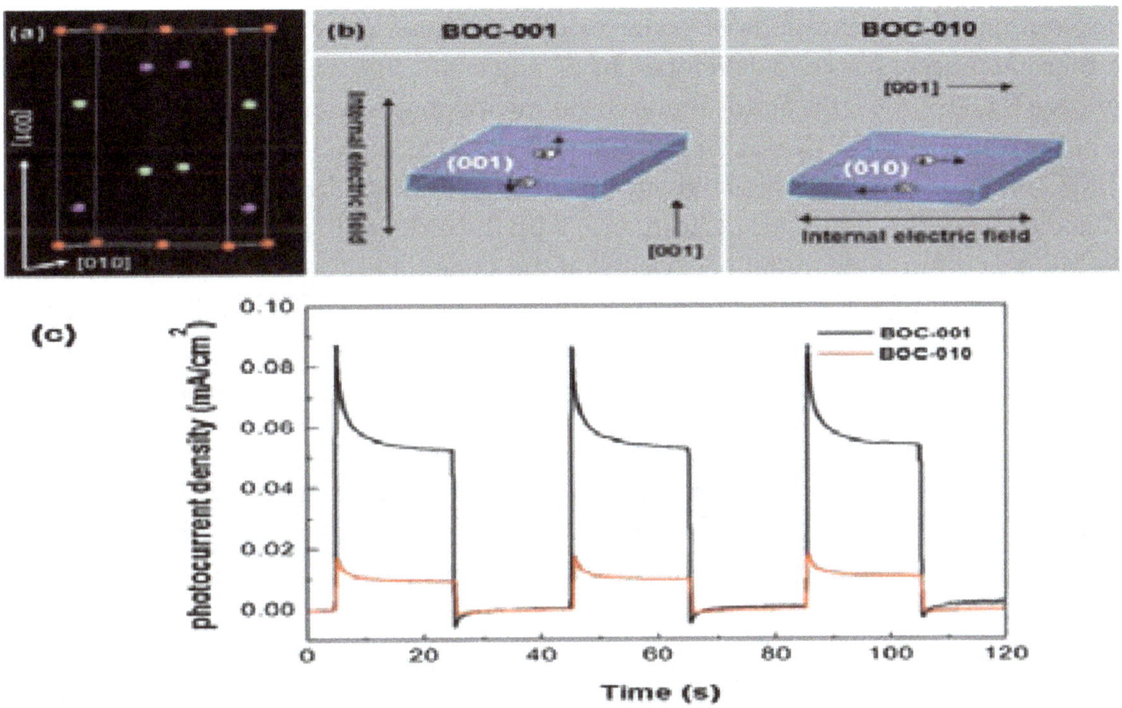

Figure 24. (a) Crystal structure of BiOCl. (b) Model showing the direction of the internal electric field in each of the BiOCl nanosheets. (c) Photocurrent responses of the BiOCl nanosheets in 0.5 M Na_2SO_4 aqueous solutions under UV-vis irradiation [293].

Researches have also using density functional theory (DFT) computations to reveal the nature of such a facet-dependent photocatalytic property in BiOX (X = Cl, Br and I) [297]. The halogen X–terminated {001} facets shows great thermodynamic stability and could efficiently separate photo-generated electron–hole pairs, whereas the formation of deep defect levels in the band gap of BiX–terminated {110} and other facets with surface O vacancies are bad for the carrier separation. This finding reveals the insight into the fundamental facet determined photocatalysis of BiOX (X = Cl, Br and I), which explains the superior photocatalytic performance of BiOX (X = Cl, Br and I) nanosheets with higher percentage of {001} facets than those with lower ones [289,290,296].

Self-assembly is a strong tool in nanotechnology fabrication of making low dimensional (e.g., 1D nanorods, 2D nanosheets, etc.) materials into their higher–dimension (3D) multifunctional superstructures, which plays a major role in material synthesis and device engineering and has been paid much attention recently [298-302]. Comparison with 1D and 2D nanostructures, 3D hierarchical nano/microstructures, which integrate the features of the nanoscale building units and their assembled architectures, are more attractive for solar energy storage and conversion [300-302]. Furthermore, 3D architectures could endow the BiOX (X = Cl, Br, I) semiconductors with improved light harvesting, shortened diffusion pathways, faster interfacial charge separation and more reactive sites, thus enhancing their photocatalytic efficiencies.

Hydro/solvothermal routes are definitely the most robust method among the methodological synthesis of the 3D BiOX (X = Cl, Br and I) hierarchical assemblies, [303-322] which are usually carried out at critical conditions of water or other organic solvents. In 2008, a generalized solvothermal process has been developed by Zhang et al. [303], who use ethylene glycol (EG) to prepare BiOX (X = Cl, Br, and I) hierarchical microspheres from 2D nanoplates. The band gaps of the resulting BiOX (X = Cl, Br and I) samples are calculated to be 3.22, 2.64, and 1.77 eV for BiOCl, BiOBr, and BiOI, respectively. Under visible-light irradiation, the BiOI sample exhibited the best photocatalytic performance with the order of BiOI > BiOBr > BiOCl evaluated by MO dye solution degradation. Almost at the same time, Tang et al. [304] also prepared 3D microspherical BiOBr architectures assembled by nanosheets through EG–assisted solvothermal synthesis. The band gap of the BiOBr architectures is 2.54 eV, so it shows higher photocatalytic activity for MO decomposition under visible–light irradiation than the BiOBr bulk plates.

How to realize the hierarchical architectures in the microstructure modulation of BiOX (X = Cl, Br and I) nano/microstructures with hollow voids draws much attention due to their better penetrability and higher light utilization. Recently, Huang group [313] has developed a method to synthesize uniform BiOBr hollow microspheres in the presence of 2-methoxyethanol solvent a mini-emulsion-mediated solvothermal route. The size of the BiOBr hollow microspheres is in the in the range of 1-2 mm and shell thickness of about 100 nm, which are composed of numerous interlaced 2D nanosheets. As demonstrated in Fig 25, by observing a Tyndall effect of the precursor suspension, the author confirmed that the 1–hexadecyl–3–methylimidazolium bromide ionic liquid ([C16Mim]BrIL) can not only serve as a Br source but also create a colloidal mini-emulsions. The diameter of the BiOBr hollow microsphere is determined by the size of the of the emulsion because the reaction takes place at the phase interface edge of the mini–emulsion rather than in the itself. Under visible-light irradiation, such BiOBr hollow microspheres displayed superior photocatalytic activity in degradation of RhB dye and reduction of Cr^{VI} ions to the samples with micro-flower shape. Xia and co-workers [319] prepared BiOI hollow microspheres by the EG–assisted solvothermal method using 1-butyl-3-methylimidazoliumiodine ([Bmim]I) IL as the reactive templates and I source. Under visible-light irradiation, such 3D BiOI hollow microspheres exhibited higher photocatalytic activity toward MO degradation than that of 2D BiOI nanoplates. Besides the halide ion-containing ILs [304,304,312,314,317,322], surfactants such as poly(vinylpyrrolidone) (PVP) [308] and hexadecyltrimethylammonium bromide (CTAB) [311,312,314,316,318,319] have been used to tailor the self-assembly process of the BiOX (X = Cl, Br, I) hierarchical architectures, and in particular CTAB could act as reactive template to provide Br - ions for BiOBr.

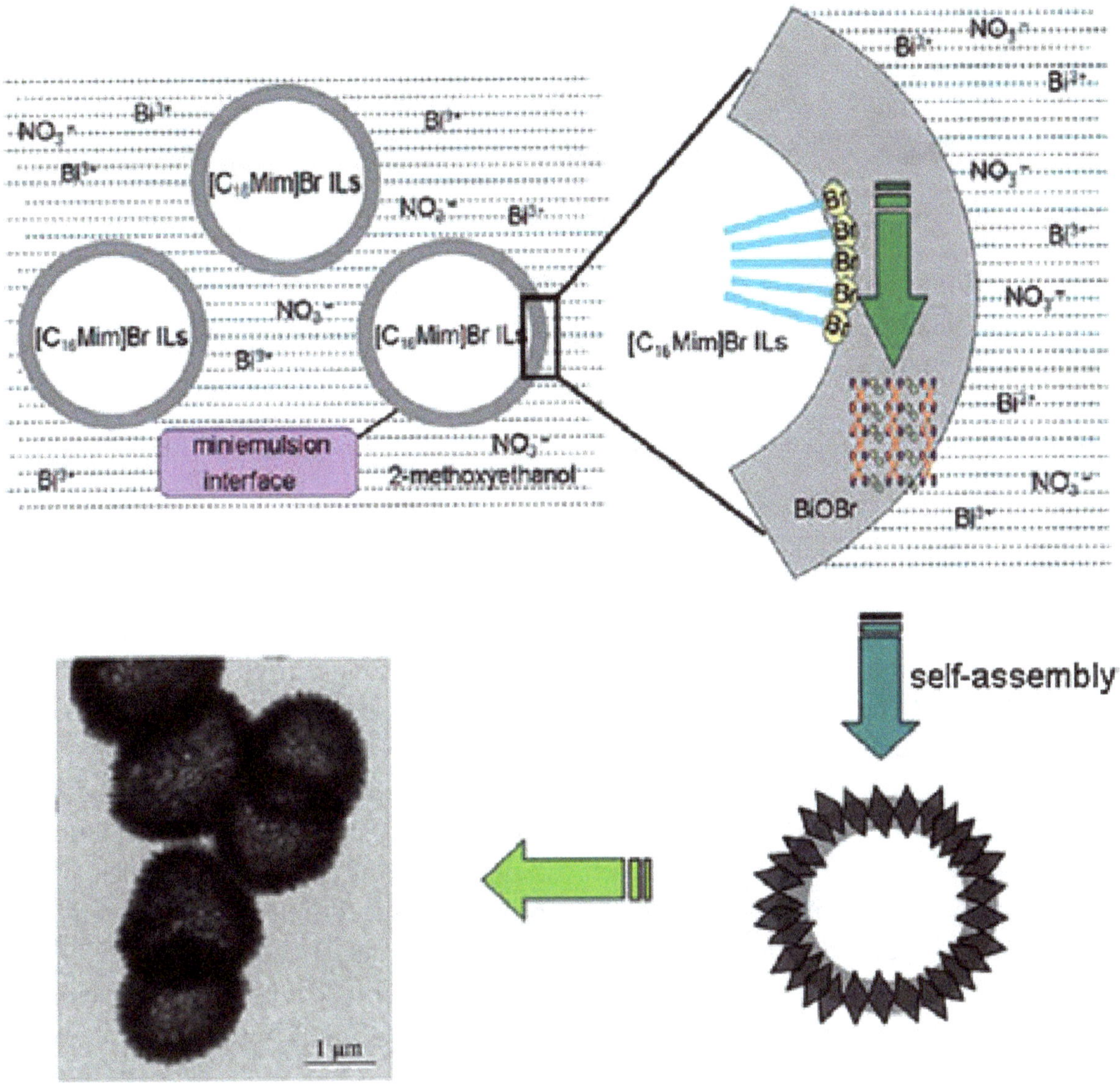

Figure 25. The schematic formation process of the BiOBr hollow microspheres by the mini-emulsionmediated solvothermal route [313].

Apart from the hydro/solvothermal syntheses, other synthetic procedures are also used to synthesize the ordered superstructures of BiOX (X = Cl, Br and I) semiconductors, such as hydrolysis [323,324], direct precipitation [325,326], sonochemical route [327,328], refluxing method [329], chemical bath [330] and solution oxidation process [331]. For example, Xiong and co-workers [331] reported a rapid in situ oxidation process to fabricate 3D flower-like BiOCl hierarchical nanostructures by reacting metallic Bi nanospheres and $FeCl_3$ aqueous solution at room temperature. As illustrated in Fig. 26, in the presence of Cl- ions, the redox potential of Bi species could be reduced from +0.308 V (Bi^{3+}/Bi vs. SHE) to +0.16 V (BiOCl/Bi). Therefore, the high redox potential of Fe^{3+} ($E(Fe^{3+}/Fe^{2+})$ = +0.771 V) could oxidize the surface of Bi nanospheres into the final 3D BiOCl hierarchical nanostructures. Compared with the commercial BiOCl sample, such flower-like BiOCl nanostructures obtained displayed much better RhB photodegradation activity and higher photoelectric conversion performances.

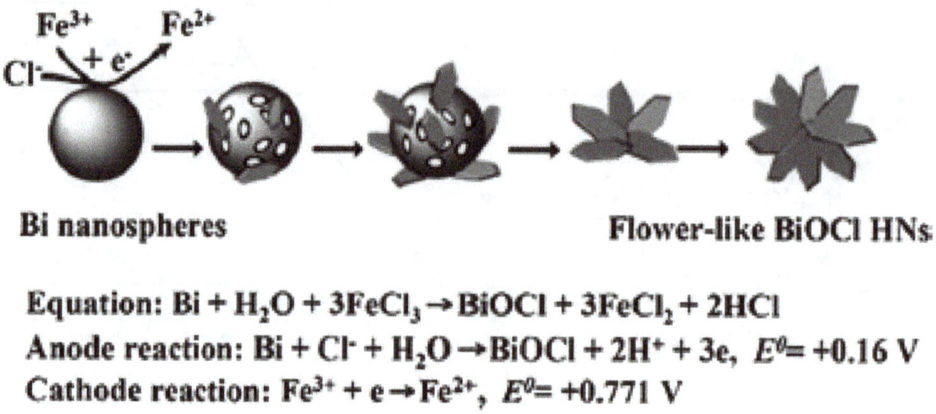

Figure 26. Schematic illustration of the fabrication of flower–like BiOCl hierarchical nanostructures by an in situ oxidation process [331].

Interestingly, Zhang [332] found that the photoactivity of BiOCl nanosheets shows a highly exposed facet-dependent effcet. The BiOCl nanosheets with exposed {001} facets showed higher direct semiconductor photoexcitation activity towards pollutant degradation due to both the surface atomic structure and suitable internal electric fields under UV light irradiation. Under visible light, highly exposed {010} facet BiOCl nanosheets shows superior indirect dye photosensitization activity for methyl orange degradation, which is due to the larger surface area and open channel characteristic of BiOCl nanosheets. It is belived that the enlarged surface area and open channel could enhance the adsorption capacity of methyl orange molecules as well as provide more contact sites between the photocatalyst and dye molecules, thereby facilitating the indirect dye photosensitization process because more efficient electron injection from the photoexcited dye into the conduction band of the catalyst happened (Fig. 27). These findings not only clarified the origin of facet-dependent photoreactivity of BiOCl nanosheets but also provided effective guidance for the design and fabrication of highly efficient bismuth oxyhalide photocatalyst.

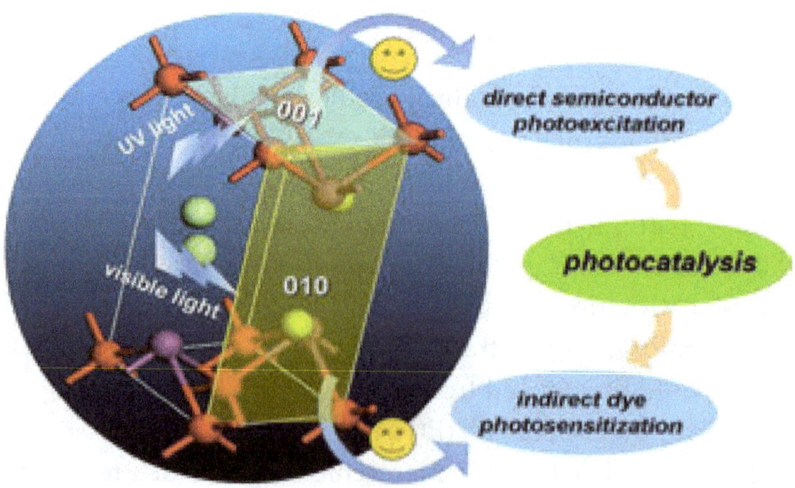

Figure 27. Schematic illustration of facet-dependent photoreactivity of BiOCl single-crystalline nanosheets [332].

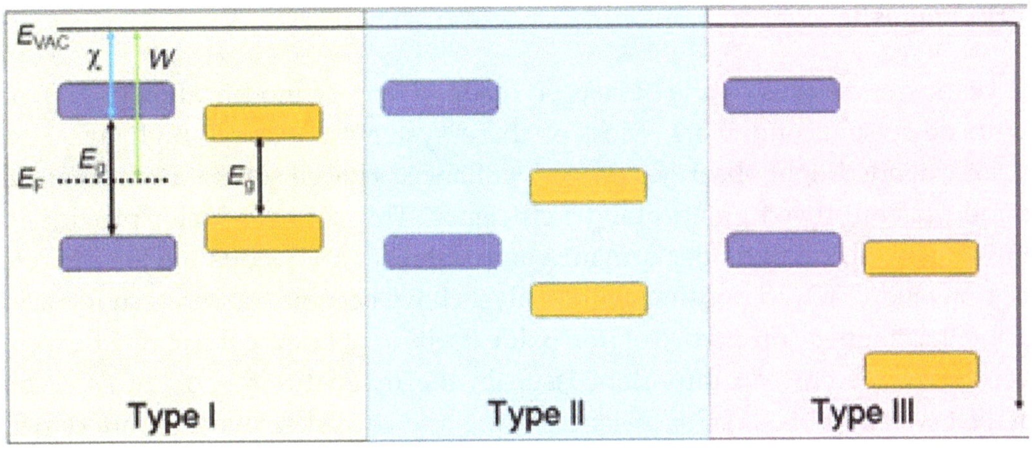

Figure 28. Schematic diagram of three types of semiconductor heterostructures [338].

It is speculated by Ye et al. that oxygen vacancy induced by UV light could yield an intermediate state between the valence and conduction bands to narrow the band gap, which may make oxygen–deficient BiOCl a promising alternative for the visible–light–driven photocatalytic reaction [333,334]. Zhang group recently found that oxygen vacancies of BiOCl can be created by the reductive ethylene glycol because it could easily react with the oxygen –terminated (001) surface at 160°C, which is evidenced by the electron spin resonance (ESR) spectra [335]. The resulting oxygen vacancies not only extended the light-response edge up to 650 nm but also enabled the effective capture of photoinduced electrons and molecular oxygen to generate superoxide anion radicals, both of which are of great important for realizing high photocatalysis efficiency of the photocatalyst. Recently, Xie et al. demonstrated that with the reduced thickness of the {001} facet–dominant BiOCl nanosheets to the atomic scale, the defects mainly change from isolated defects to triple vacancy, which could significantly promote the sunlight–driven photocatalytic activity of BiOCl nanosheets. The enhanced adsorption capability, the separation of electron-hole pairs and the generated reductive photoexcited electrons is mainly responsible for this improvement [261].

4. Composites systems

In contrast to one individual semiconductor photocatalyst, semiconductor composites are more intriguing for their interfacial heterostructures, which are formed at their junctures and have an important effect on their photocatalytic performances. There are usually three types of band positions in semiconductor heterojunctions: straddling gap (type I), staggered gap (type II) and broken gap (type III), as presented in Fig. 28. Among them, semiconductor composites with the staggered gap (type II) have drown much attention in the field of heterogeneous photocatalysis [336,337]. In this system, the photoinduced electrons and holes can be easily separated at the interface of the two semiconductors via effective interfacial charge transfer, thereby enhances the photocatalytic performance of the semiconductor composites.

4.1. Binary composite

In recent years, tremendous efforts have been made in surface modification of TiO_2 nanomaterials with other semiconductors. Most of these systems possess a high dye adsorption capacity, an extended light absorption range, enhanced charge separation, promoted mass-transfer and thus improved photocatalytic efficiency. This semiconductor provides the best compromise between catalytic performance and stability in aqueous media. Therefore, the magnetic iron oxide/TiO_2 composite photocatalysts have become the research focus in recent years. Using the magnetic properties of iron oxide itself for obtaining the magnetic recoverable photocatalyst has become an important issue in the magnetic iron oxide/TiO_2 composite photocatalyst system [339-342]. For instance, Wang and coworkers have reported the fabrication of core–shell Fe_3O_4@SiO_2@TiO_2 microspheres through a wet–chemical approach. The microspheres possess both ferromagnetic and photocatalytic properties. The TiO_2 nanoparticles on the surfaces of the microspheres degraded organic dyes under the illumination of UV light. Furthermore, the microspheres were easily separated from the solution after the photocatalytic process due to the ferromagnetic Fe_3O_4 core. The photocatalysts were recycled for further use and the degradation rate of methyl orange still reached 91% after six cycles of reuse [343]. As shown in Fig. 29, Chalasani and Vasudevan have demonstrated water-dispersible photocatalytic Fe_3O_4@TiO_2 core–shell magnetic nanoparticles by anchoring carboxy–methyl beta–cyclodextrin (CMCD) cavities to the TiO_2 shell, and photocatalytically destroyed endocrine-disrupting chemicals, bisphenol A (BPA) and dibutyl phthalate, present in water. The particles, which were typically 12 nm in diameter, were magnetic and removed from the dispersion by magnetic separation and then reused. The concentration of BPA solution was determined by liquid chromatography, and then irradiated under UV light for 60 min. After photodegradation of BPA, the CMCD–Fe_3O_4@TiO_2 nanoparticles that were separated from the mixtures by a magnet, and can be reused for the photodegradation of newly prepared BPA solutions. The recycle photocatalytic performance of CMCD–Fe_3O_4@TiO_2 for the photodegradation of BPA was excellent and stable, retaining 90% efficiency after 10 cycles [345]. For obtaining the magnetically recovered photocatalysts, Fe_3O_4 and γ-Fe_2O_3 were often employed due to their higher saturation magnetization and good magnetic separation ability.

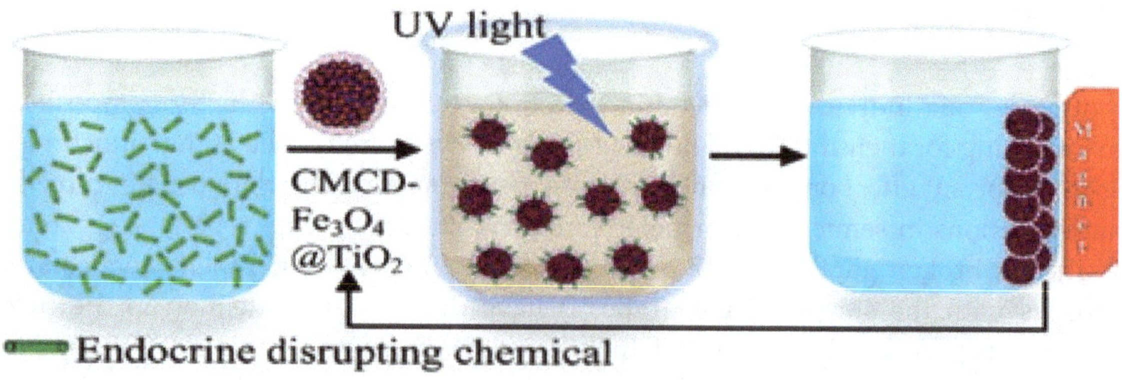

Figure 29. Scheme for the reuse of cyclodextrin-functionalized Fe_3O_4@TiO_2 for photocatalytic degradation of endocrine-disrupting chemicals in water supplies [344].

On the other hand, α-Fe_2O_3 has often been introduced into the magnetic iron oxide/TiO_2 composite photocatalyst in order to use its narrow band gap properties and to obtain magnetic iron oxide/TiO_2 composite heterostructures [344-348]. For example, Peng and coworkers have synthesized Fe_2O_3/TiO_2 heterostructural photocatalysts by impregnation of Fe^{3+} on the surface of TiO_2 and annealing at 300°C, the composites possess different mass ratios of Fe_2O_3 vs. TiO_2. The photocatalytic activities of Fe_2O_3/TiO_2 heterocomposites, pure Fe_2O_3 and TiO_2 were studied by the photocatalytic degradation of Orange II dye in aqueous solution under visible-light ($\lambda > 420$ nm) irradiation. The Fe_2O_3/TiO_2 heterogeneous photocatalysts exhibited an enhanced photocatalytic ability for Orange II, higher than either pure Fe_2O_3 or TiO_2. The best photocatalytic performance for Orange II could be obtained when the mass ratio in Fe_2O_3/TiO_2 is 7 : 3. The results illustrate that the generation of heterojunctions between Fe_2O_3 and TiO_2 is key for improving movement and restraining the recombination of photoinduced charge carriers, and finally improving the photocatalytic performance of Fe_2O_3/TiO_2 composites [348]. Recently, Palanisamy and coworkers have prepared Fe_2O_3/TiO_2 (10, 30, 50, 70 and 90 wt% Fe_2O_3) photocatalysts by a sol-gel process. Mesoporous Fe_2O_3/TiO_2 composites exhibited excellent photocatalytic degradation ability for 4–chlorophenol in aqueous solution under sunlight irradiation. The author claimed that the photogenerated electrons in the VB of TiO_2 are transferred to Fe(III) ions resulting in the reduction of Fe(III) ions to Fe(II) ions. Thus, the photoinduced holes in the VB of Fe_2O_3/TiO_2 cause an oxidation reaction and decompose the 4-chlorophenol to CO_2 and H_2O. Meanwhile the transferred electrons in Fe(III) ions could trigger the reduction reaction [349].

Wang's group successfully synthesized Bi_2WO_6-TiO_2 hierarchical heterostructure through a simple and practical electrospinning-assisted route (Fig. 30 (A) and (B)) [350]. As shown in Fig. 30 (A), Bi_2WO_6 nanoplates grew aslant on the primary TiO_2 nanofibers. These three dimensional (3D) Bi_2WO_6–TiO_2 hierarchical heterostructures exhibited enhanced visible–light –driven photocatalytic activity for the decomposition of CH_3CHO, which was almost eight times higher than that of the Bi_2WO_6 sample, and the decomposition rate by the bare TiO_2 could be neglected under visible light irradiation. This high photocatalytic activity was ascribed to the reduced probability of electron–hole recombination and the promoted migration of photogenerated carriers. Similarly, Wang et al. [352] fabricated SnO_2–TiO_2 heterostructured photocatalysts based on TiO_2 nanofibers by combining the electrospinning technique with the hydrothermal method (Fig 30C and D). This SnO_2–TiO_2 composite possessed a high photocatalytic activity for the degradation of rhodamine B (RhB) dye under UV light irradiation, which was almost 2.5 times higher than that of the bare TiO_2. The enhanced photocatalytic efficiency was attributed to the improvement of the separation of photogenerated electrons and holes. Wang's group [352] prepared a graphene–Bi_2WO_6 composite via an in situ hydrothermal reaction (Fig. 31(A) and (B)). This graphene–Bi_2WO_6 photocatalyst showed significantly enhanced photocatalytic activity for the degradation of RhB under visible light ($\lambda > 420$ nm), which was three times greater than that of the pure Bi_2WO_6. The enhanced photocatalytic activity could be attributed to the negative shift in the Fermi level of graphene–Bi_2WO_6 and the high migration efficiency of photoinduced electrons; these electrons may not only be effectively involved in the oxygen reduction reaction but also suppress the charge recombination. Kudo et al. [353] reported the composite of reduced graphene oxide (RGO) with

$BiVO_4$, where an significantly improved PCE activity of a near 10–fold enhancement was observe compared with pure $BiVO_4$ under visible-light irradiation. The longer photoexcited electron lifetime of $BiVO_4$ is mainly responsible for this improvement as the electrons are injected to RGO instantly at the site of generation, leading to a significant reduction in charge recombination.

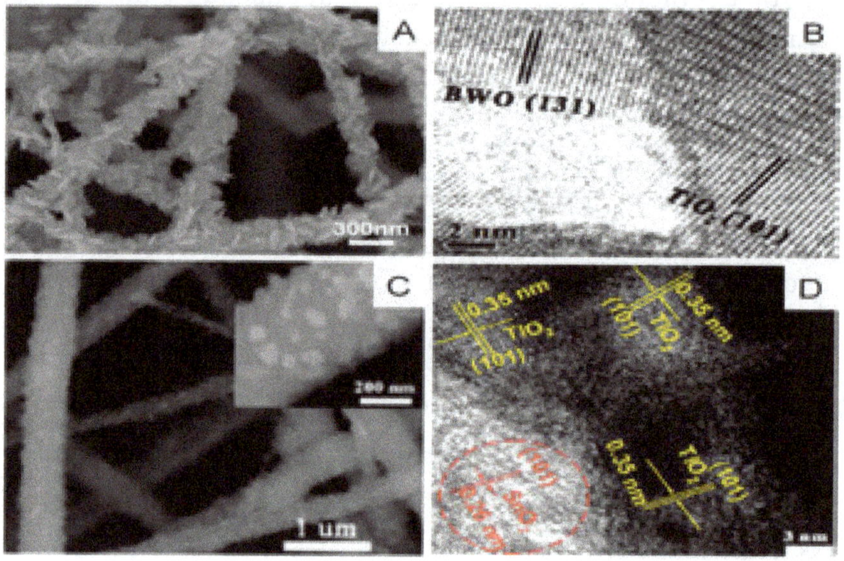

Figure 30. SEM (A) and HRTEM (B) images of Bi_2WO_6/TiO_2 [343]; SEM (C) and HRTEM (D) images of SnO_2/TiO_2 heterostructures [351].

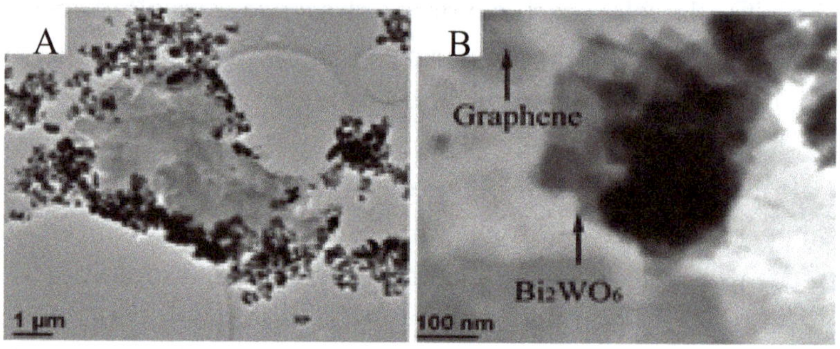

Figure 31. (A and B) TEM images of graphene decorated with Bi_2WO_6 composite [352].

A simple soft-chemical method was used to synthesize the BiOI/TiO_2 heterostructures with different Bi to Ti molar ratios at low temperature of 80°C. The degradation of methyl orange under visible-light irradiation (λ > 420 nm) of the material revealed that the BiOI/TiO_2 heterostructures exhibited much higher photocatalytic activities than pure BiOI or TiO_2, where 50%BiOI/TiO_2 showed the best activity among all these heterostructured photocatalysts [348]. BiOBr–Bi_2WO_6 mesoporous nanosheet composite enhanced photocatalytic activity is attributed to well-matched band edge positions of BiOBr and Bi_2WO_6 and the large specific surface area of the mesoporous nanosheet composites in view of the incorporation of mesopores and the highly exposed BiOBr (001) facet, compared with pure BiOBr and Bi_2WO_6 under exposure to a 3–W LED light [354].

During the past few years, another promising carbon material, graphene, which possess many unique properties, has been used for corporation with BiOX (X = Cl, Br and I) and significantly improved photocatalytic efficiencies was achieved [356-360]. Ai and coworkers [357] have developed a facile solvothermal route tosynthesize BiOBr/graphene hybrids using graphene oxide (GO), bismuth nitrite, and CTAB as the precursors. As shown in Fig. 32, BiOBr nanoplates with hundreds nanometers in size are dispersed randomly on the 2D graphene sheet surface. Evaluated by the removal of gaseous NO under visible-light irradiation, the as-prepared BiOBr/graphene hybrid displays a two times higher removal rate than that of pure BiOBr. It is evidenced that the strong chemical bonding between BiOBr and graphene is mainly responsible for the fast photogenerated electrons transfer from BiOBr to graphene, which further inhibities the unwanted recombination and leading to its enhanced photocatalytic activity.

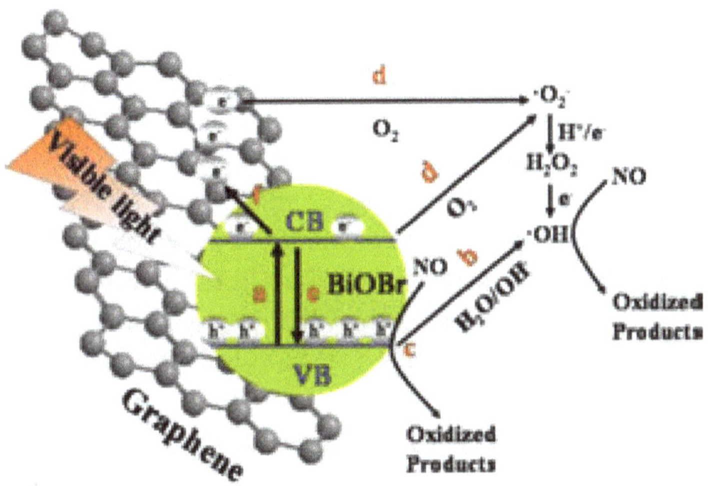

Figure 32. Schematic illustration of the visible–light photocatalytic enhancement of BiOBr/graphene nanocomposites [357].

4.2. Ternary composite

To date, although a variety of approaches have been developed to prepare many kinds of visible–light–driven semiconductor heterojunction photocatalysts, many shortage is still needed to be overcomed, for example, the limited region of visible-light photo–response. To solve these problems, multi–component heterojunction systems have been developed [361,362], in which two or more visible-light active components and an electron-transfer system are spatially integrated as shown in Fig. 33 [363].

As demonstrated in Figure 33, since both semiconductor A (S-A) and semiconductor B (S-B) can be excited by UV/visible light and have different photoabsorption ranges, the conjunction of the two materials can overlap and broaden the range of UV/visible–light photoresponse. At the same time, it is well-known that the photocatalytic reaction is initiated by the incident UV/

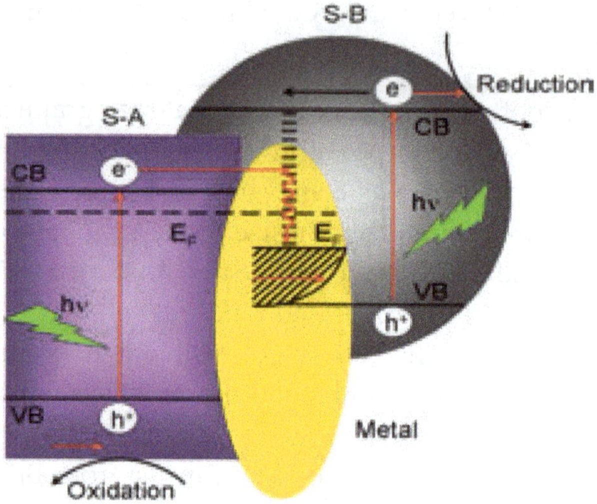

Figure 33. Schematic structure of multicomponent heterojunction systems [363].

visible photons with energy equal or higher than the band-gap in both S–A and S-B, which lead to the creation of photogenerated holes in their VB and electrons in their CB. On the one hand, the electrons in the CB of S–A easily flow into metal (electron transfer I: S–A metal) through the Schottky barrier because the CB (or the Fermi level) of S–A is higher than that of the loaded metal, which is consistent with the previous study on electron transfer from the semiconductor (such as TiO_2) to metal (such as Ag and Au) [361,364]. This process of electron transfer I is faster than the electron-hole recombination between the VB and the CB of S–A. Thus, plenty of electrons in the CB of S–A can be stored in the metal component. As a result, more holes with a strong oxidation power in the VB of S–A escape from the pair recombination and are available to oxidize the pollutants or OH-. On the other hand, since the energy level of metal is above the VB of S–B, holes in the VB of S–B also easily flow into metal (electron transfer II: metal S–B, see Fig. 33), which is faster than the electron-hole recombination between the VB and CB of S–B. More electrons with a strong reduction power in the CB of S–B can escape from the pair recombination and are available to reduce some absorbed compounds (such as O_2 and H^+). Therefore, simultaneous electron transfer I and II (i.e., vectorial electron transfer of S–A metal to S–B in Fig. 33) can occur as a result of UV/visible-light excitation of both S–A and S–B. In these vectorial electron–transfer processes, metal in multicomponent heterojunction systems acts as a storage and/or a recombination center for electrons in the CB of S–A and holes in the VB of S–B, and contributes to enhancing interfacial charge transfer and realizing the complete separation of holes in the VB of S–A and electrons in the CB of S–B. Therefore, the multi-component heterojunction systems can simultaneously and efficiently generate holes with a strong oxidation power in the VB of S–A and electrons with a strong reduction power in the CB of S–B, resulting in greatly enhanced photocatalytic activity, compared with the single semiconductor or semiconductor heterojunctions mentioned above.

In 2006, by using a facile photo–chemical technique, Tada et al. [361] developed a CdS–Au –TiO_2 ternary component nanojunction system (Fig. 34 (A) and (B)). This CdS–Au–TiO_2 triple nanojunction shows significantly improved photocatalytic activity, which was far higher than that of either the single-component or two–components systems. For this photocatalytic CdS

$-Au-TiO_2$ nanojunction system, 52.2% of methylviologen (MV^{2+}) have been reduced in 100 min, which are 1.6, 1.8 and 2.3 times higher than that of Au/TiO_2, CdS/TiO_2 and TiO_2 [361].

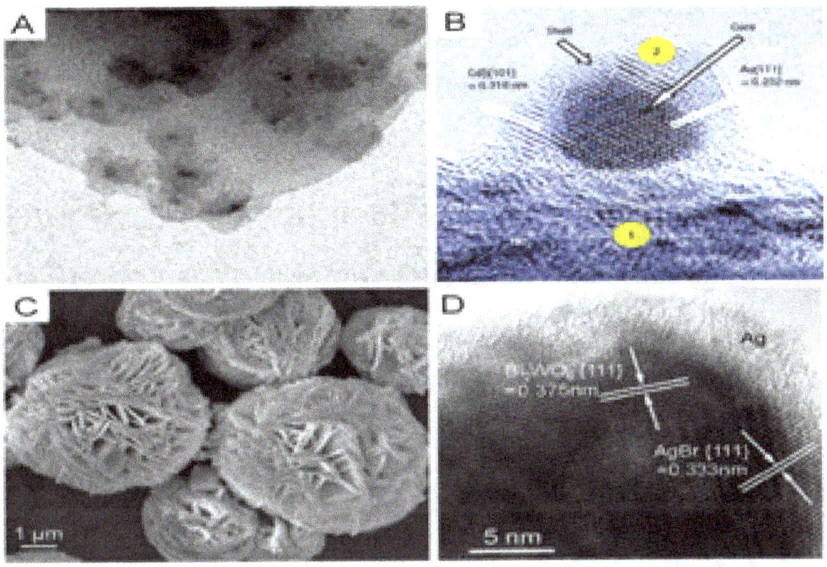

Figure 34. TEM (A) and HRTEM (B) images of Au@CdS–TiO_2 [361]; SEM(C) and HRTEM (D) images of the AgBr–Ag –Bi_2WO_6 nanojunction system [364].

Subsequently, an AgBr–Ag–Bi_2WO_6 nanojunction system was developed by a facile deposition –precipitation method (Fig. 34 (C) and (D)) [364]. This AgBr–Ag–Bi_2WO_6 nanojunction system shows much higher visible–light–driven photocatalytic activity than a photocatalyst with single visible-light response components, such as Bi_2WO_6 nanostructures, Ag–Bi_2WO_6 and AgBr–Ag–TiO_2. For example, with the AgBr–Ag–Bi_2WO_6 nanojunction system as the photocatalyst, the MX–5B could be photocatalytically degraded (42.8 mg L^{-1}) within 60 min under visible-light irradiation, which is higher than that of Bi_2WO_6 nanostructures (2.0 mg L^{-1}), Ag-Bi_2WO_6 (2.9 mg L^{-1}) and AgBr–Ag–TiO_2 (34.1 mg L^{-1}). Furthermore, 65% of pentachlorophenol could be mineralized within 4 h by AgBr–Ag–Bi_2WO_6, which is much higher than that (34.5%) of the AgBr–Ag–TiO_2 composite. This excellent visible–light–driven photocatalytic performance was mainly attributed to the vectorial interparticle electron transfer driven by the two-step excitation of both visible-light-driven components (AgBr and Bi_2WO_6).

A one-step low–temperature chemical bath method was developed to synthesize the flower –like Ag/AgCl/BiOCl composite [365]. The as–prepared Ag/AgCl/BiOCl composite exhibited enhanced visible–light photocatalytic activity on photodegradation of rhodamine B, which was greatly improved in comparison with either pure Ag/AgCl or BiOCl. It is evidenced that the superoxide radical, chlorine radical and the hole play a critical role in the photocatalytic degradation of RhB over the Ag/AgCl/BiOCl. Next, Ag/AgX/BiOX (X = Cl, Br) three-component visible-light-driven photocatalysts were synthesized by a low–temperature chemical bath method (Fig. 35) [366]. The Ag/AgX/BiOX composites showed enhanced visible–light–driven photocatalytic activity for the degradation of rhodamine B, which was much higher than Ag/AgX and BiOX. The photocatalytic mechanisms were analyzed by active species trapping and superoxide radical quantification experiments. The role of metallic Ag in Ag/AgCl/BiOCl and Ag/AgBr/BiOBr were analyzed, and we found that the role of metallic Ag was a surface plasmon resonance and the Z-scheme bridge for Ag/AgCl/BiOCl and Ag/AgBr/BiOBr, respectively. This results suggests that no matter in narrow band gap photocatalysts (Eg < 3.1 eV) or wide band gap photocatalysts (Eg > 3.1 eV), metallic Ag can enhance visible-light-driven photocatalytic activity though the different roles.

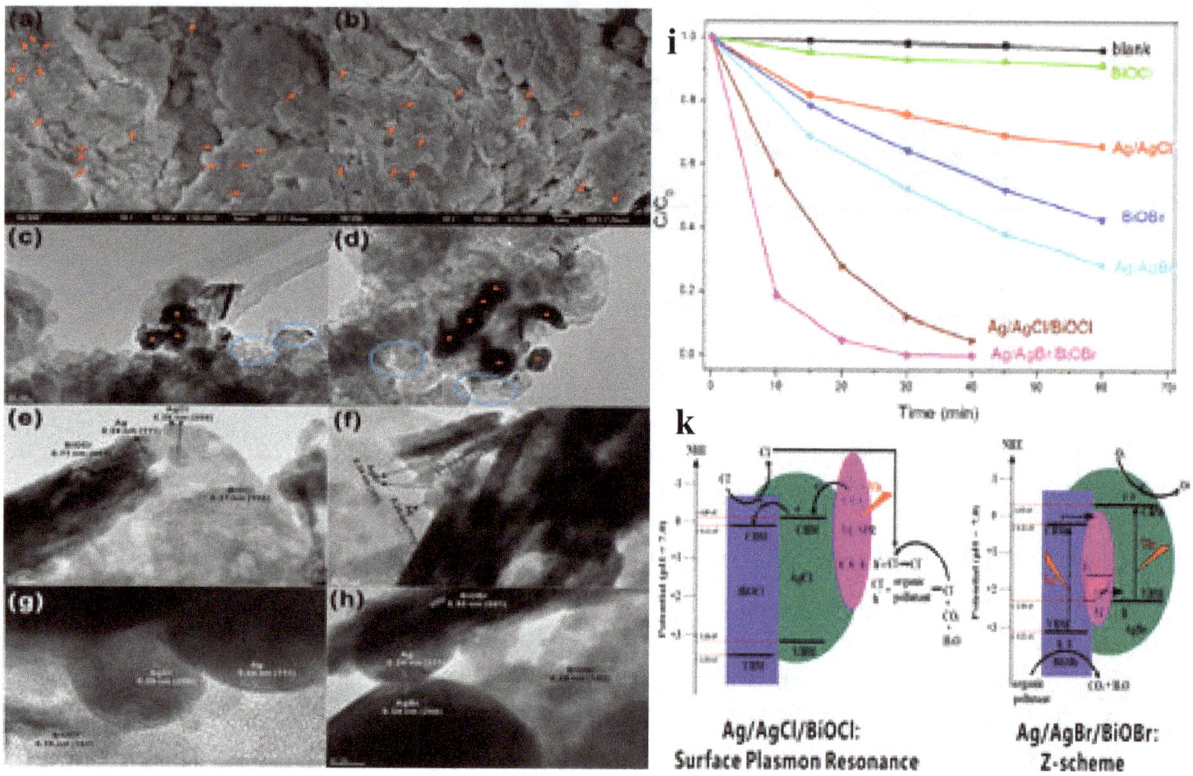

Figure 35. FESEM images of Ag/AgCl/BiOCl (a) and Ag/AgBr/BiOBr (b): the red arrows pointing out the Ag/AgX. TEM images of Ag/AgCl/BiOCl (c) and Ag/AgBr/BiOBr (d): blue rings show the small Ag/AgX, and the red dots point out the large Ag/AgX. HRTEM images of Ag/AgCl/BiOCl (e) and Ag/AgBr/BiOBr (f) with small Ag/AgX and HRTEM images of Ag/AgCl/BiOCl (g) and Ag/AgBr/BiOBr (h) with large Ag/AgX (i)the photocatalytic degradation percentage of RhB under visible-light irradiation ($\lambda \geq 400$ nm) and (k) schematic structure of multicomponent heterojunction systems [366].

Recently, our group has demonstrated a simple and efficient one-pot approach to prepare Ag/r-GO/TiO_2 composites using solvothermal method under atmospheric pressures (Fig. 36) [367], where N,N-dimethylacetamide serves as the reducing agent for Ag and GO reduction. On account of the experimental result, we concluded that the introduction of Ag into classical graphene/TiO_2 system (i) availably expands the absorption range, (ii) improves the photogenerated electron separation and (iii)increases the photocatalysis reaction active sites. The optimized composite sample exhibits outstanding photocatalysis activity compared with pure TiO_2 under simulated sunlight. We further proposed that besides the above three advantages of Ag, different sizes of Ag nanoparticles are also responsible for the improved photocatalysis ability, where small-sized Ag nanoparticles (2~5 nm) could store photoexcited electrons that generated from TiO_2, while large-sized Ag nanoparticles could utilize visible light due to their localized surface plasmon resonance (LSPR) absorption. Our work gives a new insight into the photocatalysis mechanism of noble metal/r–GO/TiO_2 composites and provides a new pathway into the design of TiO_2–based photocatalysts and promote their practical application in various environmental and energy issues.

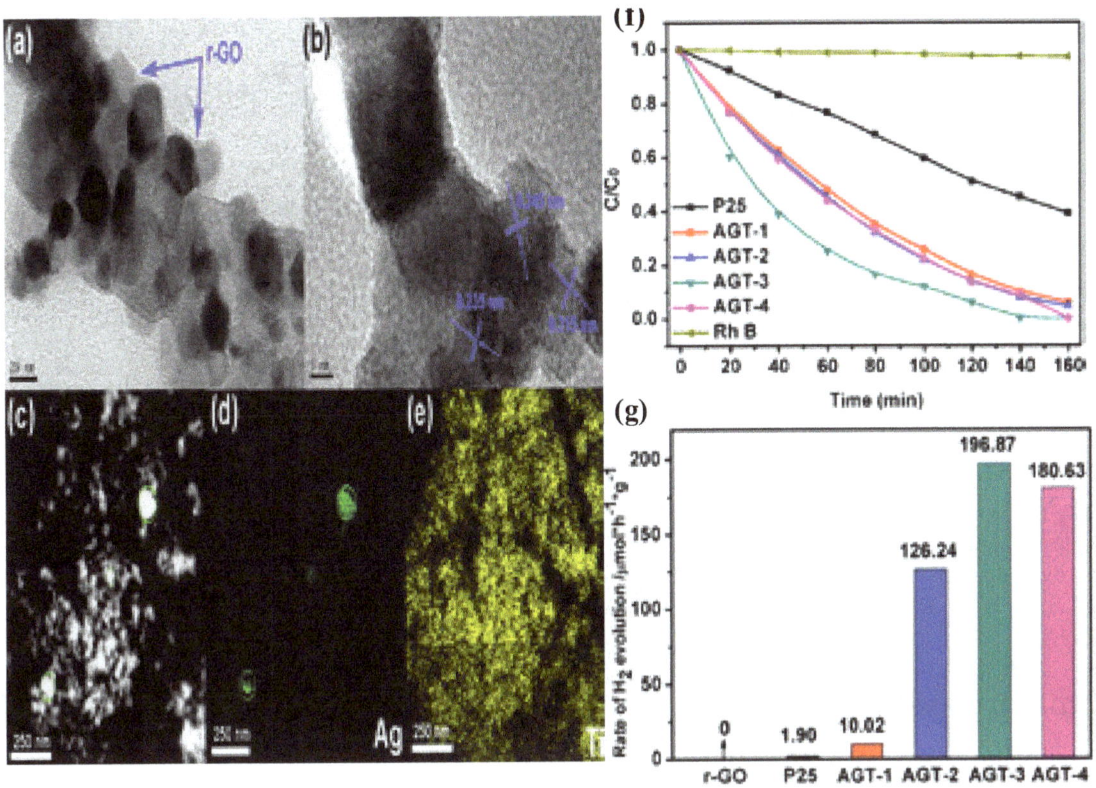

Figure 36. (a) TEM and (b) HRTEM images of sample AGT. (c) STEM model of Ag/r–GO/TiO_2. Elemental mapping of (d) Ag and (e) Ti in the same area in (c). (f)Photocatalytic degradation of Rh B under simulated sunlight irradiation over P25, Ag/r–GO/TiO_2 composites with different $AgNO_3$ contents. (g) Comparison of the photocatalytic activity of r –GO, P25 and Ag/r–GO/TiO_2 composites with different $AgNO_3$ contents for the photocatalytic H_2 production under simulated sunlight irradiation [367].

Next, we have replaced the Ag with MoS_2 quantum dots (QDs) and demonstrated a simple and an efficient one–pot approach to prepare MoS_2 quantum dots–graphene–TiO_2 (MGT) composites using a solvothermal method under obtained atmospheric pressures and at low temperatures (Fig. 37) [368]. The shape of MoS_2 obtained using this method is quantum dot instead of a layered sheet because of the interaction between functional groups on GO sheets and Mo precursors in a suitable solvent environment. In addition, it shows significantly increased photodegradation performance even without a noble-metal cocatalyst, which is due to the increased charge separation, visible-light absorbance, specific surface area and reaction sites upon the introduction of MoS_2 QDs. Besides, the enhancement mainly came from holes left in the TiO_2 crystals rather than electrons transferring to reduced graphene oxide (RGO).

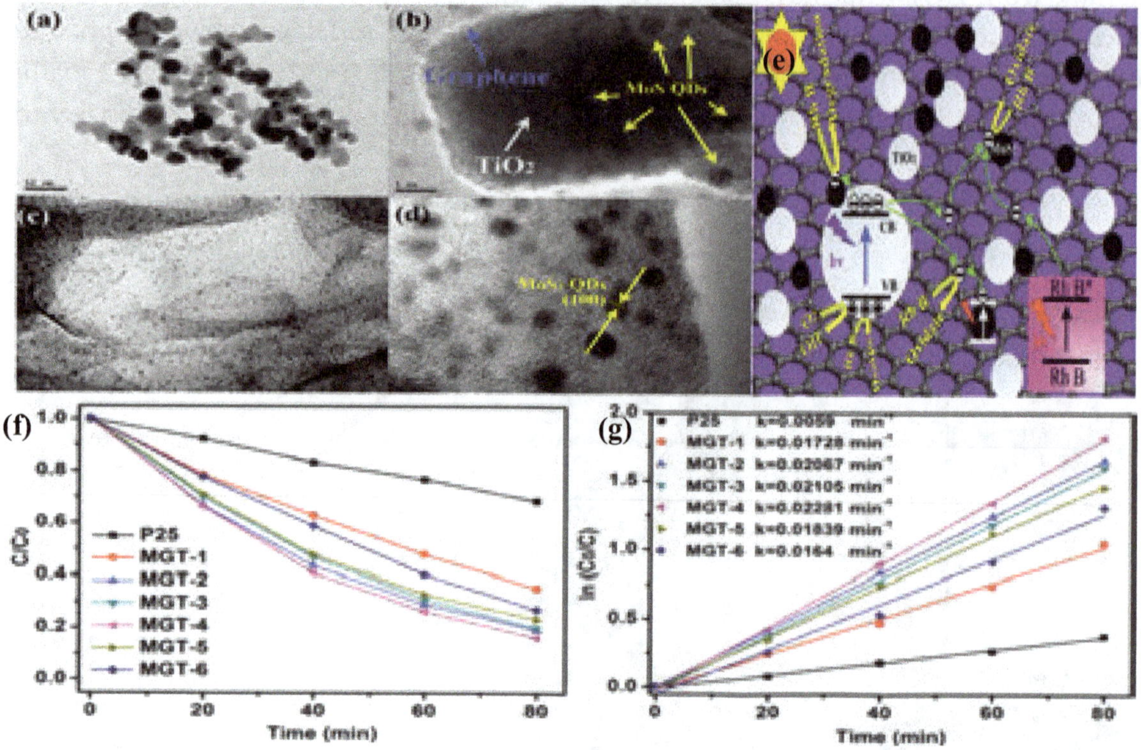

Figure 37. TEM and HRTEM images of the sample (a), (b) MGT–4 and (c), (d)MoS_2–graphene. (e) Proposed mechanism for the photodegradation of RhB by MGT under simulated sunlight irradiation (f) Photocatalytic degradation and (g) photocatalytic degradation reaction of RhB under simulated sunlight irradiation over P25, MGT composites with different MoS_2 contents [368].

5. Summary and outlook

Photocatalysis appears to be a promising avenue to solve environmental and energy issues in the future. Although the photocatalytic processes involve a complicated sequence of multiple synergistic or competing steps, the efficient utilization of solar energy (especial visible-light energy) and improvement in separation and transportation of charge carriers are the main challenges and current trend to design highly effective photocatalysts. Finally, we conclude that this chapter, after discussing with various materials and its composites for photocatalytic process, may be useful for further applications in the area of energy and environment. In summary, we have discussed the general strategies and recent progress in photocatalysis for developing highly efficient and stable photocatalysts, including: (1) Titania (TiO_2), iron oxides (α-Fe_2O_3); (2) ternary oxide photocatalytic matericals, such as Bi systems photocatalytic materials and (3) semiconducting materials and its composites. The achieved progress in photocatalysis indicates a promising route to enhance the photocatalytic efficiencies of photocatalytic semiconductors.

To date, in addition to different kinds of semiconductor materials and its composites significant advances have been reported to improve the photocatalytic efficiencies that range from environmental remediation to clean-energy harvesting by enhancing the utilization of sunlight or improving the separation/transportation of the electron–hole pairs some examples are highlighted in this chapter. Extending light-response to the visible– or even infrared regions, decreasing the amount of recombination of electrons and holes, and increasing the light–harvesting efficiency have been the major tools that have led to such advances. Although great advancements have been made in investigation of heterostructured photocatalytsts, it is still challenging to design more challenges in high efficiency of photocatalytic systems. First, there is no detailed understanding of the charge generation, separation and transportation across nanoscale interfaces of heterostructured photocatalysts, which are critical for the design and optimization of highly more-efficient and more-reliable photocatalysts. Second, while most available photocatalysts so far can only function in the UV or near–UV regime, the highly effective utilization of visible light is another challenge of heterostructured photocatalysts. Third, photostability of heterostructured photocatalyst is and will still be a major challenge for practical applications. Finally, elucidating and understanding the mechanisms that are involved in various photocatalytic reactions. Therefore, the deepening knowledge of the photocatalytic mechanism and exploration of new materials are indispensible to make substantial breakthroughs for practical application of photocatalysts.

Author details

Le Li and Minqiang Wang*

*Address all correspondence to: mqwang@mail.xjtu.edu.cn

Electronic Materials Research Laboratory (EMRL), Key Laboratory of Education Ministry, International Center for Dielectric Research, Xi'an Jiaotong University, Xi'an, China

References

[1] M. R. Hoffmann, S. T. Martin, W. Y. Choi, D. W. Bahnemann, Chem Rev. 1995, 95, 69–96.

[2] A. Kudo, Y. Miseki, Chem. Soc. Rev. 2009, 38, 253–278.

[3] X. Li, J. Q. Wen, J. X. Low, Y. P. Fang, J. G. Yu, Sci. China Mater. 2014, 57,70–100.

[4] Y. Q. Qu, X. F. Duan, Chem. Soc. Rev. 2013, 42, 2568–2580.

[5] H. Tong, S. Ouyang, Y. Bi, N. Umezawa, M. Oshikiri, J. Ye, Adv. Mater. 2012, 24, 229–251.

[6] M. G. Kibria, H. P. T. Nguyen, K. Cui, S. Zhao, D. Liu, H. Guo, M. L.Trudeau, S. Paradis, A. R. Hakima, Z. Mi, ACS Nano. 2013, 7,7886–7893.

[7] Q. J. Xiang, J. G. Yu, M. Jaroniec, J. Am. Chem. Soc. 2012, 134, 6575–6578.

[8] J. A. Christians, R. C. M. Fung, P. V. Kamat, J. Am. Chem. Soc. 2014,136, 758–764.

[9] Q. Zhang, E. Uchaker, S. L. Candelaria, G. Cao, Chem. Soc. Rev. 2013, 42, 3127–3171.

[10] J. G. Yu, X. X. Yu, Environ. Sci. Technol. 2008, 42, 4902–4907.

[11] J. Kim, J. Lee, W. Choi, Chem. Commun. 2008, 756–761.

[12] Y. Bi, S. Ouyang, N. Umezawa, J. Cao, J. Ye, J. Am. Chem. Soc. 2011,133, 6490–6492.

[13] E. V. Kondratenko, G. Mul, J. Baltrusaitis, G. O. Larrazábal, J. Pérez-Ramírez, Energy Environ. Sci. 2013, 6, 3112–3135.

[14] S. C. Roy, O. K. Varghese, M. Paulose, C. A. Grimes, ACS Nano 2010, 4, 1259–1278.

[15] A. Dhakshinamoorthy, S. Navalon, A. Corma, H. Garcia, Energy Environ. Sci. 2012, 5, 9217–9233.

[16] P Zhou, J G Yu, M Jaroniec, Adv. Mater. 2014, 26, 4920–4935.

[17] E. Fujita, J. T. Muckerman, K. Domen, ChemSusChem. 2011, 4, 155–157.

[18] S. Linic, P. Christopher, D. B. Ingram, Nat. Mater. 2011, 10, 911–921.

[19] C. Li, F. Wang, J. C. Yu, Energy Environ. Sci. 2011, 4, 100–113.

[20] H. Zhou, Y. Qu, T. Zeid, X. Duan, Energy Environ. Sci. 2012, 5, 6732–6743.

[21] Q. Li, X. Li, S. Wageh, A. Al-Ghamdi, J. G. Yu, Adv. Energy. Mater. 2015, 17, 10010–10039.

[22] J. M. Elward, A. Chakraborty, J. Chem. Theory Comput. 2013, 9, 4351–4357.

[23] F. L. Formal, S. R. Pendlebury, M. Cornuz, S. D. Tilley, M. Grä tzel, J. R. Durrant, J. Am. Chem. Soc. 2014, 136, 2564–2574.

[24] P. Zhou, J. G. Yu, M. Jaroniec, Adv. Mater. 2014, 26, 4920–4935.

[25] L. L. Tan, S. P. Chai, A. R. Mohamed, Chemsuschem. 2012, 5,1868–1882.

[26] L. Sun, Z. L. Zhao, Y. C. Zhou, L. Liu, Nanoscale. 2012, 4, 613–620.

[27] Q. J. Xiang, J. G. Yu, M. Jaroniec, Chem. Soc. Rev. 2012, 41, 782–796.

[28] M. R. Hoffmann, S. T. Martin, W. Y. Choi, D. W. Bahnemann, Chem. Rev. 1995, 95, 69–96.

[29] O. Carp, C. L. Huisman, A. Reller, Prog. Solid State Chem. 2004, 32, 33–177.

[30] D. Dambournet, I. Belharouak, K. Amine, Chem. Mater. 2010, 22, 1173–1179.

[31] U. Diebold, Surf. Sci. Rep. 2003, 48, 53–229.

[32] O. Carp, C. L. Huisman, A. Reller, Prog. Solid State Chem. 2004, 32, 33–177.

[33] M. A. Fox, M. T. Dulay, Chem. Rev. 1993, 93, 341–357.

[34] A. Fujishima, T. N. Rao, D. A. Tryk, J. Photochem. Photobiol. C. 2000, 1, 1–21.

[35] U. Diebold, N. Ruzycki, G. S. Herman, A. Selloni, Catal.Today. 2003, 85, 93–100.

[36] A. S. Barnard, H. Xu, ACS Nano. 2008, 2, 2237–2242.

[37] J. K. Burdett, T. Hughbands, J. M. Gordon, J. W. Richarson, J. Smith, J. Am. Chem. Soc. 1987, 109, 3639–3646.

[38] R. Asahi, Y. Taga, Y. Mannstadt, A. Freeman, J. Phys. Rev. B. 2000, 61, 7459–7467.

[39] C. Sousa, F. Illas, J. Chem. Phys. 1994, 50, 13974–13983.

[40] A. Kubacka, M. Fernández-García, G Colón, Chem. Rev. 2012, 112, 1555–1614.

[41] A. Fujishima, K. Honda, Nature. 1972, 238, 37–41.

[42] A. L. Linsebigler, G. Lu, J. T. Yates, Chem. Rev. 1995, 95, 735–758.

[43] A. Fujishima, T. N. Rao, D. A. Tryk, J. Photochem. Photobiol. C: Photochem. Rev. 2000, 1, 1–21.

[44] K. Hashimoto, H. Irie, A. Fujishima, Jpn. J. Appl. Phys. 2005, 44, 8269–8285.

[45] U. I. Gaya, A. H. Abdullah, J. Photochem. Photobiol. C: Photochem. Rev. 2008, 9, 1–12.

[46] D. P. Macwan, P. N. Dave, S. Chaturvedi, J. Mater. Sci. 2011, 46, 3669–3686.

[47] X. Guan, J. Du, X. Meng, Y. Sun, B. Sun, Q. Hu, J. Hazard. Mater. 2012, 1, 215–216.

[48] M. D. Hernández-Alonso, F. Fresno, S. Suárez, J. M. Coronado, Energy Environ. Sci. 2009, 2, 1231–1257.

[49] W. Zhou, H. Liu, R. I. Boughton, G. Du, J. Lin, J. Wang, D. Liu, J. Mater. Chem. 2010, 20, 5993–6008.

[50] G. Liu, L. Wang, H. G. Yang, H. M. Cheng, G. Q. Lu, J. Mater. Chem. 2010, 20, 831–843.

[51] S. G. Kumar, L. G. Devi, J. Phys. Chem. A. 2011, 115, 13211–13241.

[52] A. A. Ismail, D. W. Bahnemann, J. Mater. Chem. 2011, 21, 11686–11707.

[53] R. Leary, A. Westwood, Carbon. 2011, 49, 741–772.

[54] A. Primo, A. Corma, H. García, Phys. Chem. Chem. Phys. 2011, 13, 886–910.

[55] A. S. Weber, A. M. Grady, R. T. Koodali, Catal. Sci. Technol. 2012, 2, 683–693.

[56] R. Asahi, T. Morikawa, T. Ohwaki, K. Aoki, Y. Taga, Science. 2001, 293, 269–271.

[57] C. Burda, Y. B. Lou, X. B. Chen, A. C. S. Samia, J. Stout, J. L. Gole, Nano Lett. 2003, 3,1049–1051.

[58] S. U. M. Khan, M. Al. Shahry, W. B. Ingler, Science, 2002, 297, 2243–2245.

[59] C. Di Valentin, G. Pacchioni, A. Selloni, Chem. Mater. 2005,17, 6656–6665.

[60] S. Sakthivel, H. Kisch, Angew. Chem. Int. Ed. 2003, 42, 4908–4911.

[61] T. Umebayashi, T. Yamaki, H. Itoh, K. Asai, Appl. Phys. Lett. 2002, 81, 454–456.

[62] W. K. Ho, J. C. Yu, S. C. Lee, J. Solid State Chem. 2006,179, 1171–1176.

[63] W. Zhao, W. H. Ma, C. C. Chen, J. C. Zhao, Z. G. Shuai, J. Am.Chem. Soc. 2004, 126, 4782–4783.

[64] G. Liu, Y. N. Zhao, C. H. Sun, F. Li, G. Q. Lu, H. M. Cheng, Angew. Chem. Int. Ed. 2008, 47, 4516–4520.

[65] S. In, A. Orlov, R. Berg, F. Garcia, S. Pedrosa-Jimenez, M. S. Tikhov, D. S. Wright, R. M. Lambert, J. Am. Chem. Soc. 2007, 129, 13790–13791.

[66] J. G. Yu, W. K. Ho, Z. T. Jiang, L. Z. Zhang, Chem. Mater. 2002, 14, 3808–3816.

[67] D. Li, H. Haneda, N. K. Labhsetwar, S. Hishita, N. Ohashi, Chem. Phys. Lett. 2005, 401, 579–584.

[68] C. Di Valentin, E. Finazzi, G. Pacchioni, A. Selloni, S. Livraghi, A. M. Czoska, M. C. Paganini, E. Giamello, Chem. Mater. 2008, 20, 3706–3714.

[69] H. M. Luo, T. Takata, Y. G. Lee, J. F. Zhao, K. Domen, Y. S. Yan, Chem. Mater. 2004, 16, 846–849.

[70] X. T. Hong, Z. P. Wang, W. M. Cai, F. Lu, J. Zhang, Y. Z. Yang, N. Ma, Y. J. Liu, Chem. Mater. 2005, 17, 1548–1552.

[71] G. Liu, Z. G. Chen, C. L. Dong, Y. N. Zhao, F. Li, G. Q. Lu, H. M. Cheng, J. Phys. Chem. B. 2006, 110, 20823–20828.

[72] W. Y. Su, Y. F. Zhang, Z. H. Li, L. Wu, X. X. Wang, J. Q. Li, X. Z. Fu, Langmuir. 2008, 24, 3422–3428.

[73] S. Tojo, T. Tachikawa, M. Fujitsuka, T. Majima, J. Phys. Chem. C. 2008, 112, 14948–14954.

[74] L. Lin, W. Lin, Y. X. Zhu, B. Y. Zhao, Y. C. Xie, Chem. Lett. 2005, 34, 284–285.

[75] G. Liu, L. Z. Wang, H. G. Yang, H. M. Cheng, G. Q. Lu, J. Mater. Chem. 2010, 20, 831–843.

[76] M. Lazzeri, A. Vittadini, A. Selloni, Phys. Rev. B: Condens. Matter. 2001, 63, 155409–155412.

[77] M. Lazzeri, A. Vittadini, A. Selloni, Phys. Rev. B: Condens. Matter. 2002, 65, 119901–119908.

[78] M. Ramamoorthy, D. Vanderbilt, R. D. Kingsmith, Phys. Rev. B. 1994, 49, 1672–1679.

[79] X. Q. Gong, A. Selloni, Phys. Rev. B: Condens. Matter Mater. Phys. 2007, 76, 235307–235313.

[80] G. Liu, J. C. Yu, G. Q. Lu, H. M. Cheng, Chem. Commun. 2011, 47, 6763–6783.

[81] J. Jiang, K. Zhao, X. Xiao, L. Zhang, J. Am. Chem. Soc. 2012, 134, 4473–4476.

[82] H. G. Yang, C. H. Sun, S. Z. Qiao, J. Zou, G. Liu, S. C. Smith, H. M. Cheng, G. Q. Lu, Nature, 2008, 453, 638–641.

[83] M. D'Arienzo, J. Carbajo, A. Bahamonde, M. Crippa, S. Polizzi, R. Scotti, L. Wahba, F. Morazzoni, J. Am. Chem. Soc. 2011, 133, 17652–17661.

[84] X. G. Han, Q. Kuang, M. S. Jin, Z. X. Xie, L. S. Zheng, J. Am. Chem. Soc. 2009, 131, 3152–3153.

[85] H. G. Yang, G. Liu, S. Z. Qiao, C. H. Sun, Y. G. Jin, S. C. Smith, J. Zou, H. M. Cheng, G. Q. Lu, J. Am. Chem. Soc. 2009, 131, 4078–4083.

[86] D. Zhang, G. Li, X. Yang, J. C. Yu, Chem. Commun. 2009, 27, 4381–4383.

[87] J. Zhu, S. H. Wang, Z. F. Bian, S. H. Xie, C. L. Cai, J. G. Wang, H. G. Yang, H. X. Li, CrystEngComm. 2010, 12, 2219–2224.

[88] D. Q. Zhang, G. S. Li, H. B. Wang, K. M. Chan, J. C. Yu, Cryst. Growth Des. 2010, 10, 1130–1137.

[89] F. Amano, O. O. Prieto-Mahaney, Y. Terada, T. Yasumoto, T. Shibayama, B. Ohtani, Chem. Mater. 2009, 21, 2601–2603.

[90] J. S. Chen, Y. L. Tan, C. M. Li, Y. L. Cheah, D. Y. Luan, S. Madhavi, F. Y. C. Boey, L. A. Archer, X. W. Lou, J. Am. Chem. Soc. 2010, 132, 6124–6130.

[91] M. Liu, L. Y. Piao, W. M. Lu, S. T. Ju, L. Zhao, C. L. Zhou, H. L. Li, W. J. Wang, Nanoscale. 2010, 2, 1115–1117.

[92] X. H. Yang, Z. Li, G. Liu, J. Xing, C. H. Sun, H. G. Yang, C. Z. Li, CrystEngComm. 2011, 13, 1378–1383.

[93] X. N. Wang, B. B. Huang, Z. Y. Wang, X. Y. Qin, X. Y. Zhang, Y. Dai, M. H. Whangbo, Chem. Eur. J. 2010, 16, 7106–7109.

[94] X. Y. Ma, Z. G. Chen, S. B. Hartono, H. B. Jiang, J. Zou, S. Z. Qiao, H. G. Yang, Chem. Commun. 2010, 46, 6608–6610.

[95] S. W. Liu, J. G. Yu, M. Jaroniec, J. Am. Chem. Soc. 2010, 132,11914–11916.

[96] M. Liu, L. Piao, L. Zhao, S. Ju, Z. Yan, T. He, C. Zhou, W. Wang, Chem. Commun. 2010, 46, 1664–1666.

[97] G. Liu, C. H. Sun, H. G. Yang, S. C. Smith, L. Z. Wang, G. Q. Lu, H. M. Cheng, Chem. Commun. 2010, 46, 755–757.

[98] Z. K. Zheng, B. B. Huang, X. Y. Qin, X. Y. Zhang, Y. Dai, M. H. Jiang, P. Wang, M. H. Whangbo, Chem. Eur. J. 2009, 15, 12576–12579.

[99] Y. Q. Dai, C. M. Cobley, J. Zeng, Y. M. Sun, Y. N. Xia, Nano Lett. 2009, 9, 2455–2459.

[100] J. M. Li, D. S. Xu, Chem. Commun. 2010, 46, 2301–2303.

[101] M. D'Arienzo, J. Carbajo, A. Bahamonde, M. Crippa, S. Polizzi, R. Scotti, L. Wahba, F. Morazzoni, J. Am. Chem. Soc. 2011, 133, 17652–17661.

[102] T. Tachikawa, S. Yamashita, T. Majima, J. Am. Chem. Soc. 2011, 133, 7197–7204.

[103] J. Pan, G. Liu, G. Q. Lu, H. M. Cheng, Angew. Chem. Int. Ed. 2011, 50, 2133–2137.

[104] T. R. Gordon, M. Cargnello, T. Paik, F. Mangolini, R. T. Weber, P. Fornasiero, C. B. Murray, J. Am. Chem.Soc. 2012, 134, 6751–6761.

[105] H. Xu, P. Reunchan, S. Ouyang, H. Tong, N. Umezawa, T. Kako, J. Ye, Chem. Mater. 2013, 25, 405–411.

[106] J. Yu, J. Low, W. Xiao, P. Zhou, M. Jaroniec, J. Am. Chem. Soc. 2014, 136, 8839–8842.

[107] S. Liu, J. Yu, M. Jaroniec, Chem. Mater. 2011, 23, 4085–4093.

[108] T. Tachikawa, S. Yamashita, T. Majima, J. Am. Chem. Soc. 2011,133, 7197–7204.

[109] H. B. Jiang, Q. A. Cuan, C. Z. Wen, J. Xing, D. Wu, X. Q. Gong, C. Z. Li, H. G. Yang, Angew. Chem. 2011, 123, 3848–3852.

[110] J. Zhang, Q. Xu, M. J. Li, Z. C. Feng, C. Li, J. Phys. Chem. C. 2009, 113, 1698–1704.

[111] G. Liu, F. Li, Z. G. Chen, G. Q. Lu, H. M. Cheng, J. Solid State Chem. 2006, 179, 331–335.

[112] X. Chen, S. S. Mao, Chem. Rev. 2007, 107, 2891–2959.

[113] J. M. Wu, H. C. Shih, W. T. W, Nanotechnology, 2006, 17,105–113.

[114] C. R. Xiong, X. Y. Deng, J. B. Li, Appl. Catal. B. 2010, 94,234–240.

[115] D. Q. Zhang, G. S. Li, F. Wang, J. C. Yu, CrystEngComm, 2010, 12, 1759-1763.

[116] K. Kakiuchi, E. Hosono, H. Imai, T. Kimura, S. Fujihara, J. Cryst. Growth, 2006, 293, 541–545.

[117] X. F. Yang, J. Chen, L. Gong, M. M. Wu, J. C. Yu, J. Am. Chem. Soc. 2009, 131, 12048–12049.

[118] Y. W. Wang, L. Z. Zhang, K. J. Deng, X. Y. Chen, Z. G. Zou, J. Phys. Chem. C, 2007, 111, 2709–2714.

[119] H. M. Cheng, J. M. Ma, Z. G. Zhao, L. M. Qi, Chem. Mater. 1995, 7, 663–671.

[120] Y. Sang, B. Y. Geng, J. Yang, Nanoscale. 2010, 2, 2109–2113.

[121] B. Liu, E. S. Aydil, J. Am. Chem. Soc. 2009, 131, 3985–3390.

[122] T. Y. Ke, C. W. Peng, C. Y. Lee, H. T. Chiu, H. S. Sheu, CrystEngComm, 2009, 11, 1691–1695.

[123] E. Bae, T. Ohno, Appl. Catal. B: Environ. 2009, 91, 634–639.

[124] J. Yu, Y. Chen, A. M. Glushenkov, Cryst. Growth Des. 2009, 9, 1240–1244.

[125] F. Y. Wei, H. L. Zeng, P. Cui, S. C. Peng, T. H. Cheng, Chem. Eng. J. 2008, 144, 119–123.

[126] H. Kaper, F. Endres, I. Djerdj, M. Antonietti, B. M. Smarsly, J. Maier, Y. S. Hu, Small. 2007, 3, 1753–1763.

[127] H. Xu, F. L. Jia, Z. H. Ai, L. Z. Zhang, Cryst. Growth Des. 2007, 7, 1216–1219.

[128] A. Dessombz, D. Chiche, P. Davidson, P. Panine, C. Chaneac, J. P. Jolivet, J. Am. Chem. Soc. 2007, 129, 5904–5909.

[129] M. N. Tahir, P. Theato, P. Oberle, G. Melnyk, S. Faiss, U. Kolb, A. Janshoff, M. Stepputat, W. Tremel, Langmuir. 2006, 22, 5209–5912.

[130] B. Zhao, F. Chen, Q. W. Huang, J. L. Zhang, Chem. Commun. 2009, 34, 5115–5117.

[131] M. Addamo, V. Augugliaro, M. Bellardita, A. Di Paola, V. Loddo, G. Palmisano, L. Palmisano, S. Yurdakal, Catal. Lett. 2008, 126, 58–62.

[132] F. Iskandar, A. B. D. Nandiyanto, K. M. Yun, C. J. Hogan, K. Okuyama, P. Biswas, Adv. Mater. 2007, 19, 1408–1412.

[133] J. G. Li, C. C. Tang, D. Li, H. Haneda, T. Ishigaki, J. Am. Ceram. Soc. 2004, 87, 1358–1361.

[134] R. Buonsanti, V. Grillo, E. Carlino, C. Giannini, T. Kipp, R. Cingolani, P. D. Cozzoli, J. Am. Chem. Soc. 2008, 130, 11223–11233.

[135] R. M. Cornell, U. Schwertmann, The Iron Oxides: Structure, Properties, Reactions, Occurrences, and Uses, 2nd Ed. Wiley-VCH, Weinheim, 2003.

[136] K Sivula, F Le Formal, M Grätzel, ChemSusChem 2011, 4, 432–449.

[137] K. Sivula, R. Zboril, F. Le Formal, R. Robert, A. Weidenkaff, J. Tucek, J. Frydrych, M. Gra°tzel, J. Am. Chem. Soc. 2010, 132, 7436–7444.

[138] A.I. Galuza, A.B. Beznosov, V.V. Eremenko, Low Temp. Phys. 1998, 24, 726–729.

[139] J.H. Kennedy, K.W. Frese, J. Electrochem. Soc. 1978, 125, 709–714.

[140] L.A. Marusak, R. Messier, W.B. White, J. Phys. Chem. Solids. 1980, 41, 981–984.

[141] N.C. Debnath, A.B. Anderson, J. Electrochem. Soc. 1982, 129, 2169–2174.

[142] B. Wang, J.S. Chen, H.B. Wu, Z.Y. Wang, X.W. Liu, J. Am. Chem. Soc. 2011, 133, 17146–17148.

[143] J. Chen, L. Xu, W. Li, X. Gou, Adv. Mater. 2005, 17, 582–586.

[144] Z. Sun, H. Yuan, Z. Liu, B. Han, X. Zhang, Adv. Mater. 2005, 17, 2993–2997.

[145] X. Hu, J.C. Yu, J. Gong, Q. Li, G. Li, Adv. Mater. 2007, 19, 2324–2329.

[146] J. Yu, X. Yu, B. Huang, X. Zhang, Y. Dai, Cryst. Growth Res. 2009, 9, 1474–1480.

[147] G. Liu, Q. Deng, H. Wang, D.H.L. Ng, M. Kong, W. Cai, G. Wang, J. Mater. Chem. 2012, 22, 9704–9713.

[148] C.Y. Cao, J. Qu, W.S. Yan, J.F. Zhu, Z.Y. Wu, W.G. Song, Langmuir 2012, 28, 4573–4579.

[149] K. Sivula, F. Le Formal, M. Gratzel, Chem. Sus. Chem. 2011, 4, 432–449.

[150] Z. Chen, T.F. Jaramillo, T.G. Deutsch, A. Kleiman-Shwarsctein, A.J. Forman, N.Gaillard, R. Garland, K. Takanabe, C. Heske, M. Sunkara, E.W. McFarland, K.Domen, E.L. Miller, J.A. Turner, H.N. Dinh, J. Mater. Res. 2010, 25, 3–16.

[151] J. Brillet, M. Cornuz, F. Le Formal, J.-H. Yum, M. Grätzel, K. Sivula, J. Mater. Res. 2010, 25, 17–24.

[152] C.J. Sartoretti, M. Ulmann, B.D. Alexander, J. Augustynski, A. Weidenkaff, Chem. Phys. Lett. 2003, 376, 194–200.

[153] J.H. Kennedy, J.K.W. Frese, J. Electrochem. Soc. 1978, 125, 709–714.

[154] F.J. Morin, Phys. Rev. 1951, 83, 1005–1010.

[155] F.J. Morin, Phys. Rev. 1954, 93, 1195–1199.

[156] M.P. Dare-Edwards, J.B. Goodenough, A. Hamnett, P.R. Trevellick, J. Chem. Soc.Faraday Trans. 1983, 79, 2027–2041.

[157] K. Sivula, F.L. Formal, M. Grätzel, Chem. Sus. Chem. 2011, 4, 432–449.

[158] M.J. Katz, S.C. Riha, N.C. Jeong, A.B.F. Martinson, O.K. Farha, J.T. Hupp, Coord.Chem. Rev. 2012, 256, 2521–2529.

[159] T.K. Townsend, E.M. Sabio, N.D. Browning, F.E. Osterloh, Energy Environ. Sci. 2011, 4, 4270–4275.

[160] S.N. Dang, S.X. Lu, W.G. Xu, J. Sa, J. Non-Cryst. Solids 2008, 354, 5018–5021.

[161] M.A. Valenzuela, P. Bosch, J. Jiménez-Becerrill, O. Quiroz, A.I. Páez, J. Pho-tochem. Photobiol. A 2002, 148, 177–182.

[162] G. Zhang, Y. Feng, Y. Xu, D. Gao, Y. Sun, Mater. Res. Bull. 2012, 47, 625–630.

[163] L. Li, Y. Chu, Y. Liu, L. Dong, J. Phys. Chem. C. 2007, 111, 2123–2127.

[164] X. Li, X. Yu, J. He, Z. Xu, J. Phys. Chem. C. 2009, 113, 2837–2845.

[165] S. Cao, Y. Zhu, Nanoscale Res. Lett. 2011, 6, 27–36.

[166] J. Xu, Y. Zhu, Cryst. Eng. Comm. 2011, 13, 5162–5169.

[167] L. Xu, J. Xia, K. Wang, L. Wang, H. Li, H. Xu, L. Huang, M. He, Dalton Trans. 2013, 42, 6468–6477.

[168] S.K. Maji, N. Mukherjee, A. Mondal, B. Adhikary, Polyhedron 2012, 33,145–149.

[169] S.K. Apte, S.D. Naik, R.S. Sonawane, B.B. Kale, J. Am. Ceram. Soc. 2007, 90, 412–414.

[170] S. Bharathi, D. Nataraj, K. Senthil, Y. Masuda, J. Nanopart. Res. 2013, 15, 1346–1352.

[171] A. Hosseinian, H. Rezaei, A.R. Mahjoub, World Acad. Sci. Eng. Technol. 2011, 52, 4–21.

[172] C.T. Seip, C. Connor, Nanostruct. Mater. 1999, 12, 183–188.

[173] S. Yang, Y. Xu, Y. Sun, G. Zhang, D. Gao, Cryst. Eng. Comm. 2012, 14, 7915–7921.

[174] X. Zhou, H. Yang, C. Wang, X. Mao, Y. Wang, Y. Yang, G. Liu, J. Phys. Chem. C. 2010, 114, 17051–17061.

[175] M.J. Pawar, A.D. Khajone, M.D. Gaoner, P.S. Chandel, Int. J. Adv. Sci. Res. Tech-nol. 2012, 2, 471–476.

[176] Y. Liu, H. Yu, S. Zhan, Y. Li, Z. Lv, X. Yang, Y. Yu, J. Sol-Gel Sci. Technol. 2011, 58, 716–723.

[177] J. Sundaramurthy, P.S. Kumar, M. Kalaivani, V. Thavasi, S.G. Mhaisalkar, RSC Adv. 2012, 2, 8201–8208.

[178] B. Geng, B. Tao, X. Li, W. Wei, Nanoscale 2012, 4, 1671–1676.

[179] Y. Xu, G. Zhang, G. Du, Y. Sun, D. Gao, Mater. Lett. 2013, 92, 321–324.

[180] H. Zhou, S.S. Wong, ACS Nano 2008, 2, 944–958.

[181] X. Cheng, J. Jiang, M. Hu, G. Mao, F. Bu, C. Lin, Y. Zeng, Q. Zhang, Cryst. Eng.Comm. 2012, 14, 7701–7708.

[182] Y. Mao, S.S. Wong, J. Am. Chem. Soc. 2006, 128, 8217–8226.

[183] J. Bandara, U. Klehm, J. Kiwi, Appl. Catal. B. 2007, 76, 73–81.

[184] S. Cao, Y. Zh, G. Cheng, Y. Huang, J. Phys. Chem. Solids. 2010, 71, 1680–1683.

[185] W. Wu, R. Hao, F. Liu, X. Su, Y. Hou, J. Mater. Chem. A. 2013, 1, 6888–6894.

[186] W. Du, Q. Sun, X. Lv, Y. Xu, Catal. Commun. 2009, 10, 1854–1858.

[187] B. Lv, Z. Liu, H. Tian, Y. Xu, D. Wu, Y. Sun, Adv. Funct. Mater. 2010, 20, 3987–3996.

[188] A.B. Isaev, Z.M. Aliev, N.K. Adamadzieva, N.A. Alieva, G.A. Magomedova, Nanotechnol. Russia. 2009, 4, 475–479.

[189] Zhou X, Lan J, Liu G, Deng K, Yang Y, Nie G, Yu J, Zhi L, Angew Chem Int Ed. 2012, 51, 178–182.

[190] J. D. Bierlein, A. W. Sleight, Solid State Commun. 1975,16, 69–76.

[191] R. S. Roth, J. L. Waring, Am. Mineral. 1963, 48, 1348–1356.

[192] Y Park, KJ McDonald, KS Choi, Chem. Soc. Rev. 2013, 42, 2321–2337.

[193] A. W. Sleight, H.-y. Chen, A. Ferretti, D. E. Cox, Mater. Res. Bull. 1979, 14, 1571–1577.

[194] G. Dreyer, E. Tillmanns, Neues Jahrb. Mineral. Monatsh.e, 1981, 81, 151–156.

[195] Y Park, K J. McDonald, Kyoung-Shin Choi, Chem. Soc. Rev. 2013, 42, 2321–2337.

[196] A. K. Bhattacharya, K. K. Mallick, A. Hartridge, Mater. Lett. 1997, 30, 7-10.

[197] A. Kudo, K. Omori, H. Kato, J. Am. Chem. Soc. 1999, 121, 11459–11461.

[198] S. Tokunaga, H. Kato, A. Kudo, Chem. Mater. 2001, 13, 4624–4628.

[199] H. Tong, S. Ouyang, Y. Bi, N. Umezawa, M. Oshikiri, J. Ye, Adv. Mater. 2012, 24, 229–251.

[200] H. L. Zhou, Y. Q. Qu, T. Zeid, X. F. Duan, Energy Environ. Sci. 2012, 5, 6732–6743.

[201] Y. Ma, X. Wang, Y. Jia, X. Chen, H. Han, C. Li, Chem. Rev. 2014, 114, 9987–10043.

[202] A. Kubacka, M. Fernandez-Garcia, G. Colon, Chem. Rev. 2011, 112, 1555–1614.

[203] S.-i. Eda, M. Fujishima, H. Tada, Appl. Catal. B. 2012, 125, 288–293.

[204] H. He, S. P. Berglund, A. J. E. Rettie, W. D. Chemelewski, P. Xiao, Y. Zhang, C. B. Mullins, J. Mater. Chem. A, 2014, 2, 9371–9379.

[205] W. Wang, Y. Yu, T. An, G. Li, H. Y. Yip, J. C. Yu, P. K. Wong, Environ. Sci. Technol. 2012, 46, 4599–4606.

[206] Y. Sun, Y. Xie, C. Wu, S. Zhang, S. Jiang, Nano Res. 2010, 3, 620–631.

[207] X. Chen, J. Liu, H. Wang, Y. Ding, Y. Sun, H. Yan, J. Mater. Chem. A. 2013, 1, 877–887.

[208] J. Sun, G. Chen, J. Wu, H. Dong, G. Xiong, Appl. Catal. B. 2013, 132, 304–314.

[209] Y. Sun, C. Wu, R. Long, Y. Cui, S. Zhang, Y. Xie, Chem. Commun. 2009, 30, 4542–4544.

[210] Y. Zhao, Y. Xie, X. Zhu, S. Yan, S. Wang, Chem. Eur. J. 2008, 14, 1601–1606.

[211] G. Li, D. Zhang, J. C. Yu, Chem. Mater. 2008, 20, 3983–3992.

[212] Y. Park, K. J. McDonald, K.-S. Choi, Chem. Soc. Rev. 2013, 42, 2321–2337.

[213] F. F. Abdi, N. Firet, R. van de Krol, ChemCatChem, 2013, 5, 490–496.

[214] K. J. McDonald, K.-S. Choi, Energy Environ. Sci. 2012, 5, 8553–8557.

[215] M. Zhou, H. B. Wu, J. Bao, L. Liang, X. W. Lou, Y. Xie, Angew. Chem. Int. Ed. 2013, 52, 8579–8583.

[216] E. S. Kwak, W. Lee, N.-G. Park, J. Kim, H. Lee, Adv. Funct. Mater. 2009, 19, 1093–1100.

[217] G. Liu, J. C. Yu, G. Q. Lu, H. M. Cheng, Chem. Commun. 2011, 47, 6763–6783.

[218] G. Xi, J. Ye, Chem. Commun. 2010, 46, 1893–1895.

[219] D. Wang, H. Jiang, X. Zong, Q. Xu, Y. Ma, G. Li, C. Li, Chem. Eur. J. 2011, 17, 1275–1282.

[220] J. Yang, D. Wang, X. Zhou, C. Li, Chem. Eur. J. 2013, 19, 1320–1326.

[221] J. Pan, G. Liu, G. Q. Lu, H. M. Cheng, Angew. Chem. Int. Ed. 2011, 50, 2133–2137.

[222] R. Li, F. Zhang, D. Wang, J. Yang, M. Li, J. Zhu, X. Zhou, H. Han, C. Li, Nat. Commun. 2013, 4, 1432–1437.

[223] R. Li, H. Han, F. Zhang, D. Wang, C. Li, Energy Environ. Sci. 2014, 7, 1369–1376.

[224] Y.Y. Li, J.P. Liu, X.T. Huang, Nanoscale Res. Lett. 2007, 3, 365–371.

[225] N. Kim, R.N. Vannier, C.P. Grey, Chem. Mater. 2005, 17, 1952–1958.

[226] L. Zhang, H. Wang, Z. Chen, P. K. Wong, J. Liu, Appl. Catal. B. 2011, 106, 1–13.

[227] S. H. Chen, Z. Yin, S. L. Luo, X. J. Li, L. X. Yang, F. Deng, Appl. Surf. Sci. 2012, 259, 7–12.

[228] A. Kudo, S. Hijii, Chem. Lett. 1999, 10, 1103–1104.

[229] J. Tang, Z. Zou, J. Ye, Catal. Lett. 2004, 92, 53–56.

[230] C. Zhang, Y. Zhu, Chem. Mater. 2005, 17, 3537–3545.

[231] N. Zhang, R. Ciriminna, M. Pagliaro, Y. J. Xu, Chem Soc Rev. 2014, 43, 5276–5287.

[232] A. Kudo, S. Hijii, Chem. Lett. 1999, 10, 1103–1104.

[233] G. K. Zhang, F. Lü, M. Li, J. L. Yang, X. Y. Zhang, B. B. Huang, J. Phys. Chem. Solids. 2010, 71, 579–582.

[234] Z. J. Zhang, W. Z. Wang, M. Shang, W. Z. Yin, J. Hazard. Mater. 2010, 177, 1013–1018.

[235] L. Zhou, W. Z. Wang, L. S. Zhang, J. Mol. Catal. 2007, 268, 195–200.

[236] S. O. Alfaro, A. Martínez-de la Cruz, Appl. Catal. A. 2010, 383, 128–133.

[237] S. C. Zhang, C. Zhang, Y. Man, Y. F. Zhu, J. Solid State Chem. 2005, 178, 3823–3830.

[238] L. S. Zhang, W. Z. Wang, Z. G. Chen, L. Zhou, H. L. Xu, W. Zhu, J. Mater. Chem. 2007, 17, 2526–2532.

[239] L. S. Zhang, W. Z. Wang, L. Zhou, H. L. Xu, Small. 2007, 3, 1618–1625.

[240] M. Shang, W. Z. Wang, S. M. Sun, L. Zhou, L. Zhang, J. Phys. Chem. C. 2008, 112, 10407–10411.

[241] J. X. Xia, H. M. Li, Z. J. Luo, H. Xu, K. Wang, S. Yin, Y. S. Yan, Mater. Chem. Phys. 2010, 121, 6–9.

[242] C. X. Xu, X. Wei, Y. M. Guo, H. Q. Wu, Z. H. Ren, G. Xu, G. Shen, G. R. Han, Mater. Res. Bull. 2009, 44, 1635–1641.

[243] C. Y. Wang, H. Zhang, F. Li, L. Y. Zhu, Environ. Sci. Technol. 2010, 44, 6843–6848.

[244] H. B. Fu, W. Q. Yao, L. W. Zhang, Y. F. Zhu, Mater. Res. Bull. 2008, 43, 2617–2625.

[245] H. B. Fu, L. W. Zhang, W. Q. Yao, Y. F. Zhu, Appl. Catal. B. 2006, 66, 100–110.

[246] C. Zhang, Y. F. Zhu, Chem. Mater. 2005, 17, 3537–3545.

[247] H. B. Fu, C. S. Pan, W. Q. Yao, Y. F. Zhu, J. Phys. Chem. B. 2005, 109, 22432–22439.

[248] Y. Tian, G. M. Hua, W. Xu, N. Li, M. Fang, L. D. Zhang, J. Alloys Compd. 2011, 509, 724–730.

[249] F. Amano, K. Nogami, R. Abe, B. Ohtani, Chem. Lett. 2007, 36, 1314–1315.

[250] F. Amano, K. Nogami, R. Abe, B. Ohtani, J. Phys. Chem. C. 2008, 112, 9320–9326.

[251] F. Amano, K. Nogami, R. Abe, B. Ohtani, J. Phys. Chem. C. 2009, 113, 1536–1542.

[252] F. Amano, K. Nogami, R. Abe, M. Tanaka, B. Ohtani, Langmuir 2010, 26, 7174–7180.

[253] S. W. Liu, J. G. Yu, J. Solid State Chem. 2008, 181,1048–1055.

[254] L. Z. Zhang, J. C. Yu, Chem. Commun. 2003, 16, 2078–2079.

[255] D. R. Rolison, Science. 2003, 299, 1698–1701.

[256] F. Amano, A. Yamakata, K. Nogami, M. Osawa, B. Ohtani, J. Am. Chem. Soc. 2008, 130, 17650–17651.

[257] Y. Zhou, Z. Tian, Z. Zhao, Q. Liu, J. Kou, X. Chen, J. Gao, S. Yan, Z. Zou, ACS Appl. Mater. Interfaces. 2011, 3, 3594–3601.

[258] J. Rotmensch, J. L. Whitlock, M. L. Dietz, J. J. Hines, R. C. Reba, E. P. Horwitz, P. V. Harper, Abstr. Pap. Am. Chem. Soc. 1998, 216, U926.

[259] G. G. Briand, N. Burford, Chem. Rev. 1999, 99, 2601–2657.

[260] S. K. Poznyak, A. I. Kulak, Electrochim. Acta. 1990, 35, 1941-1948.

[261] K. L. Zhang, C. M. Liu, F. Q. Huang, C. Zheng, W. D. Wang, Appl. Catal. B. 2006, 68, 125–129.

[262] L. J. Zhao, X. C. Zhang, C. M. Fan, Z. H. Liang, P. D. Han, Phys. B. 2012, 407, 3364–3370.

[263] X. C. Zhang, L. J. Zhao, C. M. Fan, Z. H. Liang, P. D. Han, Comput. Mater. Sci. 2012, 61, 180–184.

[264] H. J. Zhang, L. Liu, Z. Zhou, RSC Adv. 2012, 2, 9224–9229.

[265] W. C. Wang, W. J. Yang, R. Chen, X. B. Duan, Y. L. Tian, D. W. Zeng, B. Shan, Phys. Chem. Chem. Phys. 2012, 14, 2450–2454.

[266] H. J. Zhang, L. Liu, Z. Zhou, Phys. Chem. Chem. Phys. 2012, 14, 1286–1292.

[267] M. L. Guan, C. Xiao, J. Zhang, S. J. Fan, R. An, Q. M. Cheng, J. F. Xie, M. Zhou, B. J. Ye, Y. Xie, J. Am. Chem. Soc. 2013, 135, 10411–10417.

[268] Z. H. Ai, W. K. Ho, S. C. Lee, L. Z. Zhang, Environ. Sci. Technol. 2009, 43, 4143–4150.

[269] K. Li, Y. L. Xu, Y. He, C. Yang, Y. L. Wang, J. P. Jia, Environ. Sci. Technol. 2013, 47, 3490–3497.

[270] Y. F. Fang, Y. P. Huang, J. Yang, P. Wang, G. W. Cheng, Environ. Sci. Technol. 2011, 45, 1593–1600.

[271] H. T. Tian, J. W. Li, M. Ge, Y. P Zhao, L. Liu, Catal. Sci. Technol. 2012, 2, 2351–2355.

[272] F. Chen, H. Q. Liu, S. Bagwasi, X. X. Shen, J. L. Zhang, J. Photochem. Photobiol. A. 2010, 215, 76–80.

[273] J. Xu, W. Meng, Y. Zhang, L. Li, C. S. Guo, Appl. Catal. B. 2011, 107, 355–362.

[274] Z. Y. Yu, B. Detlef, D. Ralf, L. Song, L. Q. Lu, J. Mol. Catal. A: Chem. 2012, 365, 1–7.

[275] L. Zhang, W. Z. Wang, S. M. Sun, Y. Y. Sun, E. P. Gao, J. Xu, Appl. Catal. B. 2013, 132, 315–320.

[276] R. S. Yuan, S. L. Fan, H. X. Zhou, Z. X. Ding, S. Lin, Z. H. Li, Z. Z. Zhang, C. Xu, L. Wu, X. X. Wang, X. Z. Fu, Angew. Chem. Int. Ed. 2013, 52, 1035–1039.

[277] N. Kijima, K. Matano, M. Saito, T. Oikawa, T. Konishi, H. Yasuda, T. Sato, Y. Yoshimura, Appl. Catal. A. 2001, 206, 237–244.

[278] Kubacka A, Fernández-García M, Colón G. Chem. Rev. 2012, 112, 1555–1614.

[279] Y. Xia, P. Yang, Y. Sun, Y. Wu, B. Mayers, B. Gates, Y. Yin, F. Kim, H. Yan, Adv. Mater. 2003, 15, 353–389.

[280] R. S. Devan, R. A. Patil, J. H. Lin, Y. R. Ma, Adv. Funct. Mater. 2012, 22, 3326–3370.

[281] C. H. Wang, C. L. Shao, Y. C. Liu, L. N. Zhang, Scr. Mater. 2008, 59, 332–335.

[282] R. S. Yuan, C. Lin, B. C. Wu, X. Z. Fu, Eur. J. Inorg. Chem. 2009, 24, 3537–3540.

[283] S. J. Wu, C. Wang, Y. F. Cui, T. M. Wang, B. B. Huang, X. Y. Zhang, X. Y. Qin, P. Brault, Mater. Lett. 2010, 64, 115–118.

[284] Q. H. Wang, K. Kalantar-Zadeh, A. Kis, J. N. Coleman, M. S. Strano, Nat. Nanotechnol. 2012, 7, 699–712.

[285] M. Chhowalla, H. S. Shin, G. Eda, L. Li, K. P. Loh, H. Zhang, Nat. Chem. 2013, 5, 263–275.

[286] M. Q. Zhao, Q. Zhang, J. Q. Huang, F. Wei, Adv. Funct. Mater. 2012, 22, 675–694.

[287] J. H. Han, S. Lee, J. Cheon, Chem. Soc. Rev. 2013, 42, 2581–2591.

[288] K. L. Zhang, C. M. Liu, F. Q. Huang, C. Zheng, W. D. Wang, Appl. Catal. B. 2006, 68, 125–129.

[289] L. Q. Ye, L. Zan, L. Tian, T. Peng, J. Zhang, Chem. Commun. 2011, 47, 6951-6953.

[290] D. Zhang, J. Li, Q. G. Wang, Q. S. Wu, J. Mater. Chem. A. 2013, 1, 8622–8629.

[291] X. F. Chang, J. Huang, C. Cheng, Q. Sui, W. Sha, G. B. Ji, S. B. Deng, G. Yu, Catal. Commun. 2010, 11, 460–464.

[292] X. Chang, M. A. Gondal, A. A. Al-Saadi, M. A. Ali, H. Shen, Q. Zhou, J. Zhang, M. Du, Y. Liu, G. Ji, J. Colloid Interface Sci. 2012, 377, 291–298.

[293] J. Jiang, K. Zhao, X. Y. Xiao, L. Z. Zhang, J. Am. Chem. Soc. 2012, 134, 4473–4476.

[294] J. Y. Xiong, G. Cheng, G. F. Li, F. Qin, R. Chen, RSC Adv. 2011, 1, 1542–1553.

[295] M. Shang, W. Z. Wang, L. Zhang, J. Hazard. Mater. 2009, 167, 803–809.

[296] L. Q. Ye, L. H. Tian, T. Y. Peng, L. Zan, J. Mater. Chem. 2011, 21, 12479–12484.

[297] H. J. Zhang, L. Liu, Z. Zhou, RSC Adv. 2012, 2, 9224–9229.

[298] G. M. Whitesides, B. Grzybowski, Science, 2002, 295, 2418–2421.

[299] M. A. Snyder, M. Tsapatsis, Angew. Chem. Int. Ed. 2007, 46, 7560–7573.

[300] D. R. Rolison, J. W. Long, J. C. Lythe, A. F. Fischer, C. P. Rhodes, T. M. McEvoy, M. E. Bourg, A. M. Lubers, Chem. Soc. Rev. 2009, 38, 226–252.

[301] Y. Li, Z. Y. Fu, B. L. Su, Adv. Funct. Mater. 2012, 22, 4634–4667.

[302] Q. F. Zhang, E. Uchaker, S. L. Candelaria, G. Z. Cao, Chem. Soc. Rev. 2013, 42, 3127–3171.

[303] X. Zhang, Z. H. Ai, F. L. Jia, L. Z. Zhang, J. Phys. Chem. C. 2008, 112, 747–753.

[304] X. Y. Qin, H. F. Cheng, W. J. Wang, B. B. Huang, X. Y. Zhang, Y. Dai, Mater. Lett. 2013, 100, 285–288.

[305] J. X. Xia, J. Zhang, S. Yin, H. M. Li, H. Xu, L. Xu, Q. Zhang, J. Phys. Chem. Solids. 2013, 74, 298–304.

[306] J. M. Song, C. J. Mao, H. L. Niu, Y. H. Shen, S. Y. Zhang, CrystEngComm, 2010, 12, 3875–3881.

[307] L. P. Zhu, G. H. Liao, N. C. Bing, L. L. Wang, Y. Yang, H. Y. Xie, CrystEngComm, 2010, 12, 3791–3796.

[308] S. J. Peng, L. L. Li, P. N. Zhu, Y. Z. Wu, M. Srinivasan, S. G. Mhaisalkar, S. Ramakrishna, Q. Y. Yan, Chem. Asian J. 2013, 8, 258–268.

[309] D. H. Wang, G. Q. Gao, Y. W. Zhang, L. S. Zhou, A. W. Xu, W. Chen, Nanoscale. 2012, 4, 7780–7785.

[310] J. Zhang, F. Shi, J. Lin, D. Chen, J. Gao, Z. Huang, X. Ding, C. Tang, Chem. Mater. 2008, 20, 2937–2941.

[311] Z. H. Ai, W. Ho, S. Lee, L. Z. Zhang, Environ. Sci. Technol. 2009, 43, 4143–4150.

[312] J. Xu, W. Meng, Y. Zhang, L. Li, C. S. Guo, Appl. Catal. B. 2011, 107, 355–362.

[313] H. F. Cheng, B. B. Huang, Z. Y. Wang, X. Y. Qin, X. Y. Zhang, Y. Dai, Chem. Eur. J. 2011, 17, 8039–8043.

[314] Y. J. Chen, M. Wen, Q. S. Wu, CrystEngComm, 2011, 13, 3035–3039.

[315] J. X. Xia, S. Yin, H. M. Li, H. Xu, L. Xu, Y. G. Xu, Dalton Trans. 2011, 40, 5249–5258.

[316] L. Zhang, X. F. Cao, X. T. Chen, Z. L. Xue, J. Colloid Interface Sci. 2011, 354, 630–636.

[317] D. Q. Zhang, M. C. Wen, B. Jiang, G. S. Li, J. C. Yu, J. Hazard. Mater. 2012, 211, 104–111.

[318] Y. C. Feng, L. Li, J. W. Li, J. F. Wang, L. Liu, J. Hazard. Mater. 2011, 192, 538–544.

[319] Y. N. Huo, J. Zhang, M. Miao, Y. Jin, Appl. Catal. B. 2012, 111, 334–341.

[320] H. T. Tian, J. W. Li, M. Ge, Y. P. Zhao, L. Liu, Catal. Sci. Technol. 2012, 2, 2351–2355.

[321] Y. Y. Li, J. S. Wang, H. C. Yao, L. Y. Dang, Z. J. Li, J. Mol. Catal. A: Chem. 2011, 334, 116–122.

[322] J. X. Xia, S. Yin, H. M. Li, H. Xu, Y. S. Yan, Q. Zhang, Langmuir. 2011, 27, 1200–1206.

[323] L. Chen, S. F. Yin, R. Huang, Y. Zhou, S. L. Luo, C. T. Au, Catal. Commun. 2012, 23, 54–57.

[324] K. Zhang, J. Liang, S. Wang, J. Liu, K. X. Ren, X. Zheng, H. Luo, Y. J. Peng, X. Zou, X. Bo, J. H. Li, X. B. Yu, Cryst. Growth Des. 2012, 12, 793–803.

[325] Y. Q. Lei, G. H. Wang, S. Y. Song, W. Q. Fan, M. Pang, J. K. Tang, H. J. Zhang, Dalton Trans. 2010, 39, 3273–3278.

[326] R. Hao, X. Xiao, X. X. Zuo, J. M. Nan, W. D. Zhang, J. Hazard. Mater. 2012, 209, 137–145.

[327] Y. Q. Lei, G. H. Wang, S. Y. Song, W. Q. Fan, H. J. Zhang, CrystEngComm, 2009, 11, 1857–1862.

[328] C. H. Deng, H. M. Guan, Mater. Lett. 2013, 107, 119–122.

[329] G. Cheng, J. Y. Xiong, F. J. Stadler, New J. Chem. 2013, 37, 3207–3213.

[330] X. Xiao, W. D. Zhang, J. Mater. Chem. 2010, 20, 5866–5870.

[331] J. Y. Xiong, Z. B. Jiao, G. X. Lu, W. Ren, J. H. Ye, Y. P. Bi, Chem. Eur. J. 2013, 19, 9472–9475.

[332] J. Jiang, K. Zhao, X. Y. Xiao, L. Z. Zhang, J. Am. Chem. Soc. 2012, 134, 4473–4476.

[333] L. Q. Ye, L. Zan, L. H. Tian, T. Y. Peng, J. J. Zhang, Chem. Commun. 2011, 47, 6951–6953.

[334] L. Q. Ye, K. J. Deng, F. Xu, L. H. Tian, T. Y. Peng, L. Zan, Phys. Chem. Chem. Phys. 2012, 14, 82–85.

[335] J. Jiang, L. Z. Zhang, H. Li, W. W. He, J. J. Yin, Nanoscale. 2013, 5, 10573–10581.

[336] A. Kudo, Y. Miseki, Chem. Soc. Rev. 2009, 38, 253–278.

[337] X. B. Chen, S. S. Mao, Chem. Rev. 2007, 107, 2891–2959.

[338] H. F. Cheng, B. B. Huang, Y Dai, Nanoscale, 2014, 6, 2009–2026

[339] S. Xuan, W. Jiang, X. Gong, Y. Hu, Z. Chen, J. Phys. Chem. C, 2008, 113, 553–558.

[340] Q. Yuan, N. Li, W. C. Geng, Y. Chi, X. T. Li, Mater. Res. Bull. 2012, 47, 2396–2402.

[341] Y. Z. Wang, X. B. Fan, S. L. Wang, G. L. Zhang, F. B. Zhang, Mater. Res. Bull. 2013, 48, 785–789.

[342] B. Cui, H. X. Peng, H. Q. Xia, X. H. Guo, H. L. Guo, Sep. Purif. Technol. 2013, 103, 251–257.

[343] Z. Wang, L. Shen, S. Zhu, Int. J. Photoenergy. 2012, 202, 519–524.

[344] R. Chalasani, S. Vasudevan, ACS Nano, 2013, 7, 4093–4104.

[345] H. Liu, L. Gao, J. Am. Ceram. Soc. 2006, 89, 370–373.

[346] W. Zhou, H. Fu, K. Pan, C. Tian, Y. Qu, P. Lu, C. C. Sun, J. Phys. Chem. C. 2008, 112, 19584-19589.

[347] F. X. Chen, W. Q. Fan, T. Y. Zhou, W. H. Huang, Acta Phys. Chim. Sin. 2013, 29, 167–175.

[348] W. S. Tung, W. A. Daoud, ACS Appl. Mater. Interfaces. 2009, 1, 2453–2461.

[349] B. Palanisamy, C. M. Babu, B. Sundaravel, S. Anandan, V. Murugesan, J. Hazard. Mater. 2013, 23, 233–242.

[350] M. Shang, W. Wang, L. Zhang, S. Sun, L. Wang, L. Zhou, J. Phys. Chem. C, 2009, 113, 14727–14731.

[351] C. Wang, C. Shao, X. Zhang, Y. Liu, Inorg. Chem. 2009, 48, 7261–7268.

[352] E. Gao, W. Wang, M. Shang, J. Xu, Phys. Chem. Chem. Phys. 2011, 13, 2887–2893.

[353] Y. H. Ng, A. Iwase, A. Kudo, R. Amal, J Phys Chem Lett, 2010, 1, 2607–2612.

[354] X. Zhang, L. Z. Zhang, T. F. Xie, D. J. Wang, J. Phys. Chem. C. 2009, 113, 7371–7378.

[355] Y. Li, Y. Liu, E. Uchaker, Q. Zhang, S. Sun, J. Wang, Y. Huang, J. Li, G. Cao, J. Mater. Chem. A. 2013, 1, 7949–7956.

[356] F. D. Gao, D. W. Zeng, Q. W. Huang, S. Q. Tian, C. S. Xie, Phys. Chem. Chem. Phys. 2012, 14, 10572–10578.

[357] Z. H. Ai, W. Ho, S. Lee, J. Phys. Chem. C. 2011, 115, 25330–25337.

[358] X. M. Tu, S. L. Luo, G. X. Chen, J. H. Li, Chem. Eur. J. 2012, 18, 14359–14366.

[359] S. Y. Song, W. Gao, X. Wang, X. Y. Li, D. P. Liu, Y. Xing, H. J. Zhang, Dalton Trans. 2012, 41, 10472–10476.

[360] H. Liu, W. R. Cao, Y. Su, Z. Chen, Y. Wang, J. Colloid Interface Sci. 2013, 398, 161–167.

[361] H. Tada, T. Mitsui, T. Kiyonaga, T. Akita, K. Tanaka, Nat. Mater. 2006, 5, 782–786.

[362] M. R. Elahifard, S. Rahimnejad, S. Haghighi, M. R. Gholami, J. Am. Chem. Soc. 2007, 129, 9552–9553.

[363] H. L. Wang, L. S. Zhang, Z. G. Chen, J. Q. Hu, S. J. Li, Z. H. Wang, J. S. Liu, X. C. Wang, Chem. Soc. Rev. 2014, 43, 5234–5244.

[364] L. S. Zhang, K. H. Wong, Z. G. Chen, J. C. Yu, J. C. Zhao, C. Hu, C. Y. Chan, P. K. Wong, Appl. Catal. A. 2009, 363,221–229.

[365] W. Xiong, Q. D Zhao, X. Y. Li, D. K. Zhang, Catalysis Communications. 2011, 16, 229–233.

[366] L. Q. Ye, J. Y. Liu, C. Q. Gong, L. H. Tian, T. Y. Peng, L. Zan. ACS Catal. 2012, 2, 1677–1683.

Metal/Semiconductor Hybrid Nanocrystals and Synergistic Photocatalysis Applications

Jiatao Zhang, Muwei Ji, Jiajia Liu and Meng Xu

Abstract

This review focuses on recent research efforts to synthesize metal/semiconductor hybrid nanocrystals, understand and control the photocatalytic applications. First, we summarize the synthesis methods and recent presented metal/seminconductor morphologies, including heterodimer, core/shell, and yolk/shell etc. The metal clusters and nanocrystals deposition on semiconductor micro/nano substrates with well-defined crystal face exposure will be clarified into heterodimer part. The outline of this synthesis part will be the large lattice mismatch directed interface, contact and morphologies evolution. For detailed instructions on each synthesis, the readers are referred to the corresponding literature. Secondly, the recent upcoming photocatalysis applications and research progress of these hybrid nanocrystals will be reviewed, including the photocatalytic hydrogen evolution (water splitting), photo-reduction of CO_2 and other newly emerging potential photosynthesis applications of metal/semiconductor hybrid nanocrystals. Finally, we summarize and outlook the future of this topic. From this review, we try to facilitate the understanding and further improvement of current and practical metal/semiconductor hybrid nanocrystals and photocatalysis applications.

Keywords: Metal/semiconductor, hybrid nanocrystals, photocatalysis, sythesis, interface, lattice mismatch, Plasmon, exciton, photooxidation, photoreduction

1. Introduction

The shape and size effect, the controllable doping, heterocomposite, and interface are the prerequisite of colloidal nanocrystals for exploring their optoelectronic properties, such as fluorescence, plasmon–exciton coupling, efficient electron/hole separation, and enhanced photoelectric conversion [1–4]. The use of photoexcited electrons and holes in semiconductor nanocrystals as reduction and oxidation reagents is an intriguing way of harvesting photon energy to drive chemical reactions since increasing energy demand and environmental pollution create a pressing need for clean and sustainable energy solution. The high efficient separation and collection of photoexcited electrons (e-) and holes (h+) are the key points to get high efficient photocatalysis applications. Hybrid nanocrystals composed of semiconductor and metal components are receiving extensive attention in recent years due to their high efficient separation of photoexcited electrons and holes and potentional photocatalysis applications [1,4–5]. Furthermore, hybrid nanocrystals composed of semiconductor and plasmonic metal components are receiving extensive attention. The simultaneous existence and coupling of localized surface plasmon resonance induced plasmon and excitons in semiconductors, as well as the synergistic interactions between the two components [1, 6–7].

This review focuses on recent research efforts to synthesize metal/semiconductor hybrid nanocrystals to understand and control the photocatalytic applications. First, we summarize the synthesis methods and recent presented metal/semiconductor morphologies, including heterodimer, core/shell, and yolk/shell. The metal clusters and nanocrystals deposition on semiconductor micro/nanosubstrates with well-defined crystal face exposure will be clarified into heterodimer part. The outline of this synthesis part will be the large lattice mismatch-directed interface, contact, and morphology evolution. For detailed instructions on each synthesis, the readers are referred to the corresponding literature.

Second, the recent upcoming photocatalysis applications and research progress of these hybrid nanocrystals will be reviewed, including the photocatalytic hydrogen evolution (water splitting), photoreduction of CO_2, and other newly emerging potential photosynthesis applications of metal/semiconductor hybrid nanocrystals. Finally, we provide a summary and outlook on the future of this topic. From this review, we try to facilitate the understanding and further improvement of current and practical metal/semiconductor hybrid nanocrystals and photocatalysis applications.

2. Metal/semiconductor hybrid nanocrystal synthesis

According to the classical hetero epitaxial growth theory [8], when a secondary material (referred to as "2") has to be deposited over a preexisting seed substrate of a different material (denoted as "1"), total Gibbs free surface energy change function ΔG_s can be described as follows: $\Delta G_s = \gamma_1 - \gamma_2 + \gamma_{1,2}$, where γ_1 and γ_2 are the surface energies of seed substrate and the deposited materials, respectively, the solid/solution interfacial energies of colloidal nanostructure in the liquid medium, and $\gamma_{1,2}$ is strain-related solid/solid interfacial energy that depends on the bonding strength and degree of crystallographic compatibility between materials 1 and 2. As shown in Figure 1, if material 2 exposes lower energy surfaces ($\gamma_2 < \gamma_1$) and/or attains good lattice matching with the material 1 ($\gamma_{1,2}$ is small), then it is easy to get continuous and uniform core/shell nanostructure ($\Delta G_s > 0$, Frank—van der Merwe (FM) growth mode). On the contrary, if material 2 is featured by higher energy surfaces ($\gamma_2 > \gamma_1$) and/or $\gamma_{1,2}$ is high owing to the large lattice mismatch, then it will tend to deposit as a discontinuous island-like domain to be heterodimer structure (ΔG_s <0, Volmer—Weber (VW)growth mode) or even separate from each other. Therefore, the metal/semiconductor hybrid nanocrystal synthesis should follow the mechanism of Figure 1. However, in colloidal phase, γ_1 and γ_2 would be strongly influenced by adhesion of solvent, capping ligands and precursors, thus changing $\gamma_{1,2}$ too [9,10]. Considering of the significant impact of binding of organic stabilizers or other solution species to the surface energy terms and therefore altering the ultimate ΔG_s, the colloidal phase synthesis will be potential for the control of large lattice mismatch-directed shape evolution and morphology control when choosing different solvent, capping ligands, and reactant precursors.

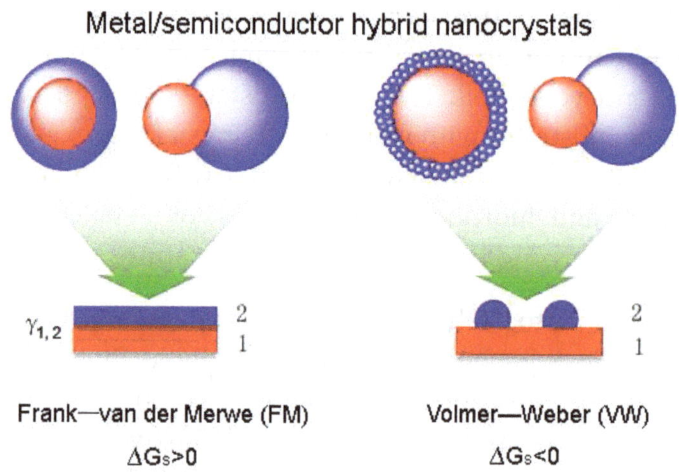

Figure 1. General metal/semiconductor hybrid nanocrystal synthesis sketches by traditional heterogeneous epitaxy growth mechanism: (A) Franck–van der Merwe (FM) mode; (B) Volmer–Weber (VM) mode.

2.1. Metal/semiconductor heterodimers by direct heterogeneous deposition

A broad family of metal/semiconductor hybrid nanocrystals has been obtained by accomplishing direct heterogeneous nucleation and growth of one or more secondary material layers onto preformed metal NC seeds serving as starting "cores" [9,10]. Synthetic strategies aim, on one side, at inhibiting homogeneous self-nucleation of isolated NCs made of the shell material and, on the other side, at achieving size- and shape-mediated deposition of the shell beyond limitations imposed by misfit strain at the metal/semiconductor interface. Practical techniques to realize these objectives rely on the colloidal synthesis, the "active" surface of "core" and the slow-nucleation of another material to deposit on "core" homogeneously, depending on the inherent chemical accessibility of the "core" or "seeds" as well on the reactivity of the shell molecular precursors (generally nonaqueous) solution media. The regulation of the thermodynamics and kinetics of reactions, namely, the temperature and selection of suitable surface-adhering organic ligands or surfactants and the optimal reactant injection rate all critically influence on the temporal evolution of the supersaturation degree.

To date, disparate combinations of metals, semiconductors, and oxides have been addressed by manipulating direct heterogeneous deposition pathways [11–25]. Here, we typically review most recent achievement in the field, the metal/semiconductor heterodimers (as scheme in Figure 1). Representative transmission electron microscopy (TEM) and high-resolution TEM examples of core–shell NCs can be found in Figure 2, respectively.

The one-pot synthesis of metal/semiconductor heterodimers mostly focus on the bifunctional heterodimers of nanoparticles, such as the conjugate of quantum dots (QDs) and magnetic nanoparticles reported by Gu et al. [20], the noble metal/metal oxide heterodimers by Wang et al. [21] and Wu et al. [22], and the bimagnetic FePt–iron oxide heterodimer nanocrystals by Figuerola et al. [23] and Cozzoli et al. [24] (see also Figure 2).

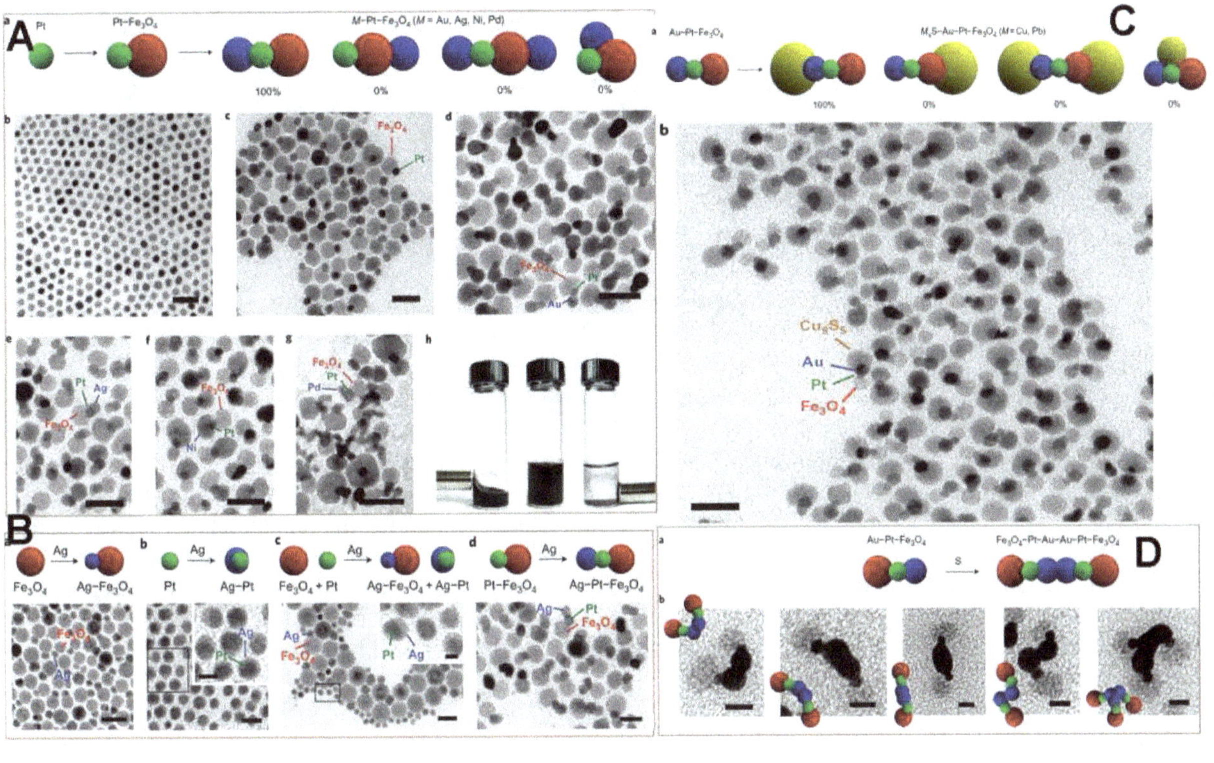

Figure 2. (A) FePt/CdS heterodimers by one-pot synthesis: (a) the scheme of synthesis process, (b) a TEM image, and (c) a high-resolution TEM image of FePt/CdS heterodimers. Adapted from Gu et al. [20] with permission; copyright American Chemical Society. (B) Seed-mediated high temperature growth of M/metal oxide heterodimers: (a) the scheme of synthesis, (b) 5–12 nm Au–Fe_3O_4, (c) 5–12 nm Ag-Fe_3O_4, and (d) 5–12 nm Pt–Fe_3O_4. Adapted from Wang et al. [21] with permission; copyright American Chemical Society. FePt/In_2O_3 heterodimers were adapted from Wu et al. [22] with permission; copyright American Chemical Society. (C) One-pot, two-step colloidal strategy to prepare FePt/ iron oxide heterodimer nanocrystals: (a) the scheme of synthesis method; (b) the TEM image of FePt/ Iron Oxide heterodimers; and (c) the HRTEM image of FePt/ Iron Oxide heterodimer. Adapted from Figuerola et al. [23] and Cozzoli et al. [24] with permission; copyright American Chemical Society.

The Au/ CdSe or Au/CdS heterodimer nanocrystals, because of the noble metal and II-VI QDs, are potential for photocatalysis applications. Banin et al. used surface nucleation and growth of a second phase to get Au/CdSe nanorod (NR) heterodimers, the first time to realize site-selective growth of Au NPs on CdSe NRs tips (see also Figure 3A) [13, 25, 26]. Du et al. used similar way to directly deposit the Au NPs on CdS NRs to get heterodimers (see also Figure 3B) [27]. The as-obtained Au/CdS and Au/CdSe NR heterodimers promote the electron/hole separation and exhibit promising photocatalytic activity for the water-splitting reaction in photoelectrochemical cells and photodegradation applications.

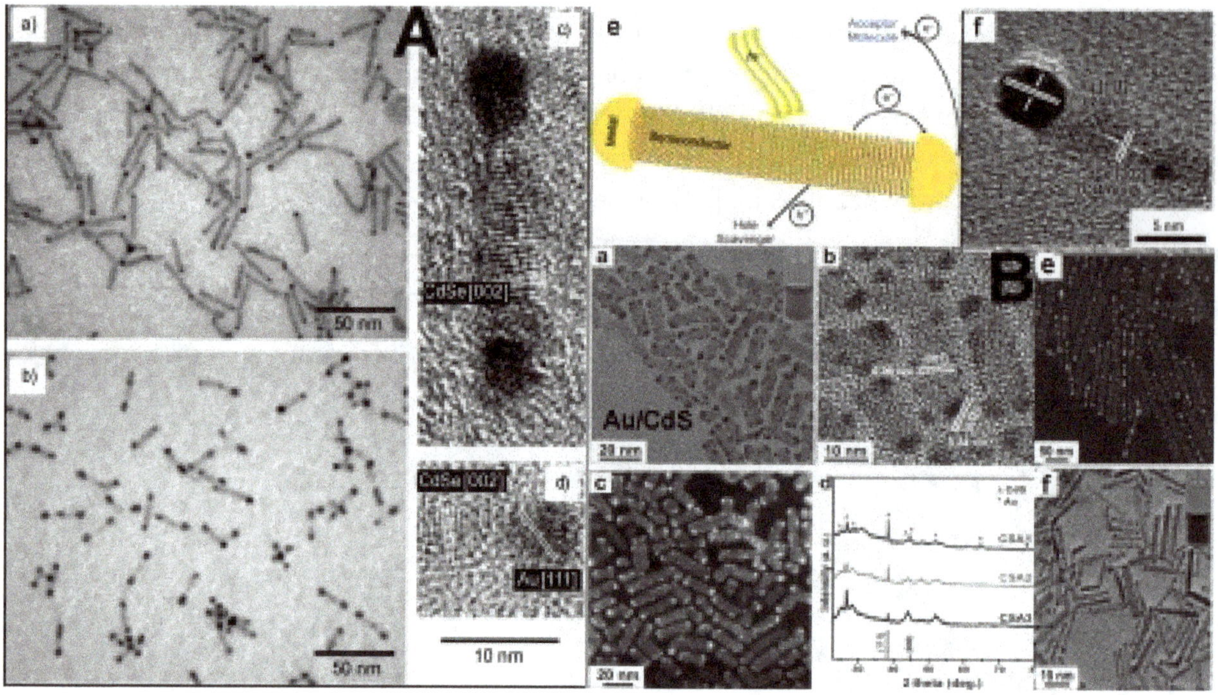

Figure 3. (A) TEM images showing controlled growth of gold onto the tips of CdSe nanorods. (a, b) The size of the gold tips can be controlled by varying the amount of gold precursor added during growth. (c, d) HRTEM images of a single nanodumbbell (c) and a nanodumbbell tip (d); the CdSe lattice for the rod in the center and gold tips at the rod edges can be identified. (From [25, 26], reprinted with permission from the AAAS.) (B) (a) TEM image, (b) HRTEM image, and (c) HAADF-STEM image of Au/CdS NRs hybrid nanocrystals. (d) XRD patterns of hybrid nanocrystals of Au/CdS NRs. (e) HAADF-STEM images in which the size of Au NPs is ~8 nm. (f) TEM images of Au/CdS hybrid nanocrystals in which the size of Au NPs is ~3 nm. Adapted from Du et al. [27] with permission; copyright Wiley-VCH Verlag GmbH & Co. KGaA.

Colloidal metal/semiconductor hybrid nanoparticles contain multiple nanoscale domains fused together by solid-state interfaces. They represent an emerging class of multifunctional lab-on-a-particle architectures that underpin future advances in solar energy conversion, photocatalysis. Buck et al. reported that the known direct heterogeneous deposition could be applied in a predictable and stepwise manner to build complex hybrid nanoparticle architectures that include M–Pt–Fe_3O_4 (M=Au, Ag, Ni, Pd) heterotrimers, M_xS–Au–Pt–Fe_3O_4 (M=Pb, Cu) heterotetramers, and higher-order oligomers based on the heterotrimeric Au–Pt–Fe_3O_4 building block [18]. This synthetic framework conceptually mimics the total synthesis approach used by chemists to construct complex organic molecules (see Figure 4).

In other words, the advantage of direct heterogeneous deposition colloidal method is it could get heterodimers with high monodispersity. The disadvantages are as follows: (1) The metal part or semiconductor part could not be mediated in more larger size range, such as tens of nanometers to hundreds of nanometers due to the Lar Mer growth mechanism of high

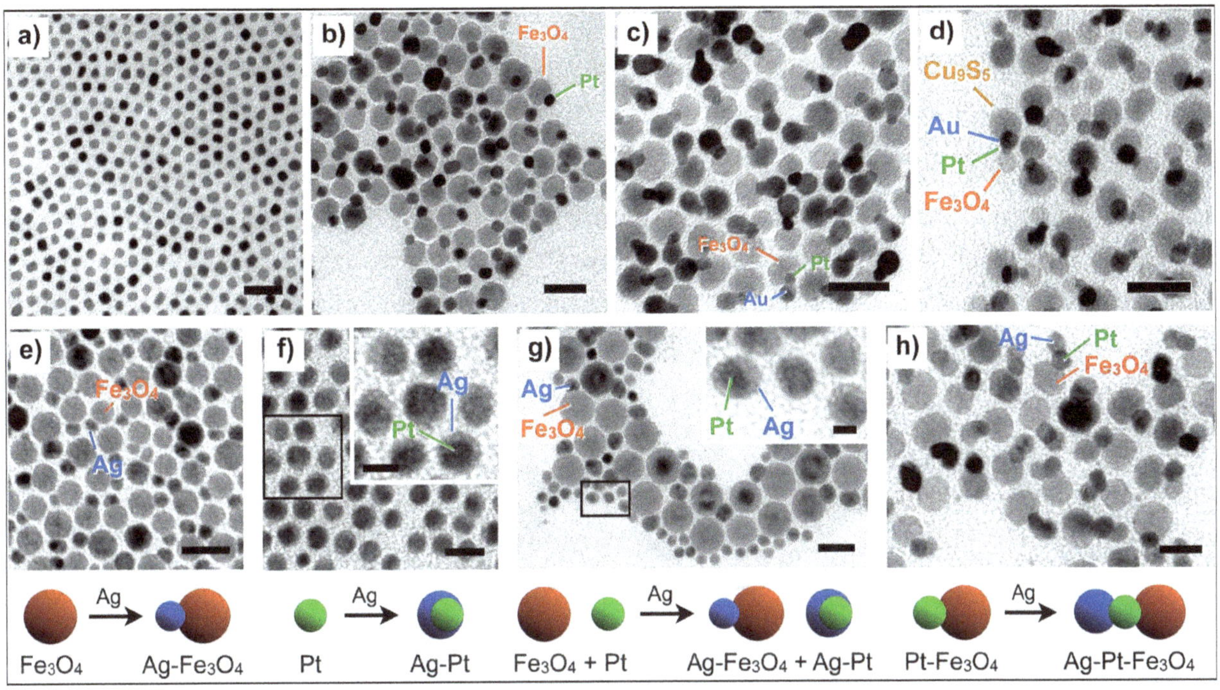

Figure 4. (A–D) TEM images showing the stepwise direct heterogeneous deposition of linear nanoparticle heterotetramers: (A) Pt nanoparticles, (B) Pt–Fe_3O_4 heterodimers, (C) Au–Pt–Fe_3O_4 heterotrimers, and (D) Cu_9S_5–Au–Pt–Fe_3O_4 heterotetramers. (E–H) TEM images and schematic representations showing a series of metal/semiconductor heterodimers (same synthetic conditions for all steps) aimed at understanding the site-selective deposition of Ag onto Pt–Fe_3O_4 heterodimers: (E) Ag grown off Fe_3O_4 nanoparticles, (F) Ag grown off Pt nanoparticles, (G) Ag grown indiscriminately off both Fe_3O_4 and Pt nanoparticles when both are present as a physical mixture, and (H) Ag grown exclusively off the Pt domain when Pt and Fe_3O_4 are directly attached as heterodimers. Reprinted from Buck et al. [18] with permission (Copyright of Nature group 2011, Macmillan Publishers Limited).

temperature organic phase synthesis. (2) The interface between metal and semiconductor is still not controlled well to be clear enough because of the size of two parts mostly still limited at <10 nm range. The lattice mismatch-induced stain energy here is usually too large to get more defects on the interface, which is not helpful for the highly efficient photoinduced electron/hole separation and the following photocatalysis applications.

2.2. Large lattice mismatch-directed shape evolution and morphology control

The growth of monocrystalline semiconductor-based metal/semiconductor hybrid nanostructures with modulated composition, morphology, and interface strain are the prerequisite for exploring their plasmon–exciton coupling, efficient electron/hole separation, and enhanced photocatalysis properties. As the schematic process in Figure 1, different from generally used one-pot epitaxial growth on performed metal nanoparticle seeds, Ouyang et al. and Zhang et al. unprecedentedly took two steps of facile chemical thermodynamics processes to maximally adjust the surface energies γ_2, $\gamma_{1,2}$, and then ΔG_s to change their overgrowth modes from FM mode to VW mode gradually. Furthermore, Ouyang et al. and Zhang et al. unprecedentedly used the cation exchange initiated the topotactic in situ conversion from amorphous to be single-crystalline structure [28–33]. In this case, the more flexible size range (from smaller than

10 nm to tens of nanometers, to hundreds of nanometers, and to micrometer scale), morphology (isotropic and anisotropic), interface, and single-crystallinity of metal and semiconductor part could be tailored synergistically.

By controlling soft acid–base coordination reactions between molecular complexes and colloidal nanostructures, Zhang et al. and Ouyang et al. showed that chemical thermodynamics could drive nanoscale monocrystalline growth of the semiconductor shell on metal nanosubstrates and then enhanced light–matter-spin interactions in these judiciously engineered nanostructures could be achieved [28, 3].

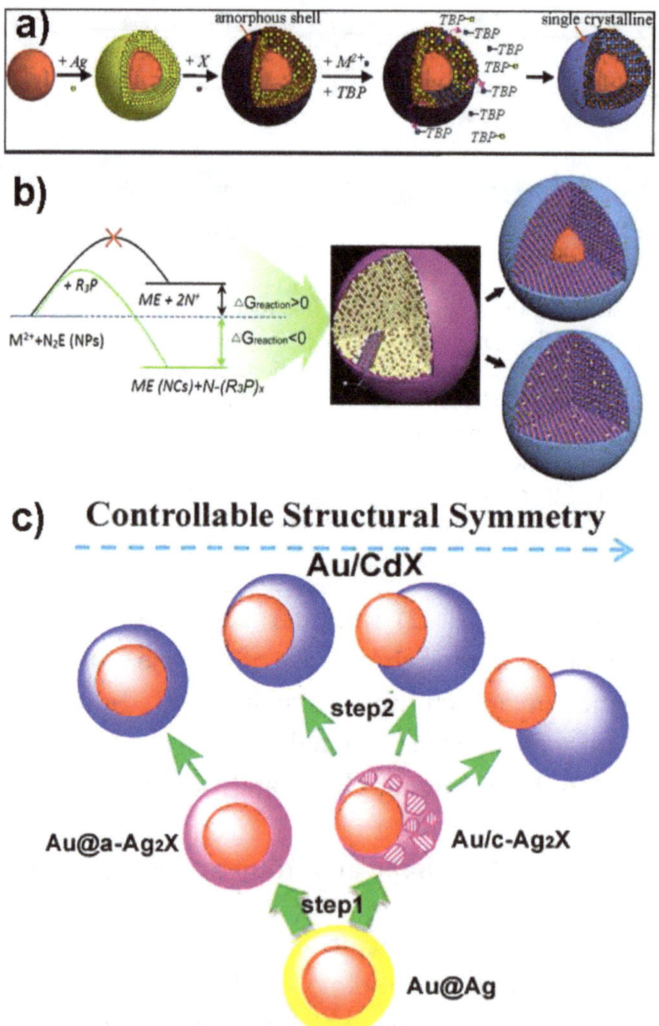

Figure 5. Schemes of large lattice mismatch-directed shape evolution and morphology control by Ouyang et al. and Zhang et al.: (A) Nonepitaxial growth process and mechanism of hybrid core–shell nanostructures with substantial lattice mismatches. From Zhang et al. [28], reprinted with permission from the AAAS. (B) The cation exchange reaction initiated by different phosphines or phosphites (R3P) and phosphine-initiated cation exchange for precisely tailoring composition and properties of metal/semiconductor nanostructures. (C) The schematic process of controllable structural symmetries in Au/CdX (X = S, Se, and Te) hybrid structures with large lattice mismatch by two steps of in situ chemical conversion. The a-Ag_2X means amorphous Ag_2X, c-Ag_2X means crystalline Ag_2X. Adapted from Gui et al. [31] and Zhao et al. [32] with permission; copyright Wiley-VCH Verlag GmbH & Co. KGaA.

First, as shown in Figure 5A, based on the Lewis acid–base reaction mechanism, where the entire nanostructure is spatially confined by an amorphous matrix, the monocrystalline growth of the semiconductor shell is fully directed by chemical thermodynamic properties of reactions within the matrix; the shell's lattice structure can be independent of that of the core NPs, thus circumventing the limitations imposed by epitaxial strategies. Starting from the core NPs, an overlayer of metal with soft Lewis acidity is grown onto the core. For all hybrid core–shell structures, we choose an Ag metal overlayer. The silver shells can be modified to form silver compound shells (Ag_2X) with an amorphous structure, providing a crucial platform for the next chemical transformation stage, ultimately leading to monocrystalline growth. It has been demonstrated that nanoscale chemical transformations, such as cation exchange, represent a versatile route for converting one crystalline solid to another [34, 35]. They show that this process can be harnessed to drive the single-crystal growth by carefully controlling the thermodynamic properties of reaction (1).

Tributylphosphine (TBP) was selected because it is a soft base and can behave as a phase-transfer agent to transport metal ions (Mn^+) to the surface of the core NPs by binding to free cations in solution. The high acid softness of Ag^+ favors the exchange process between Ag^+ in the amorphous matrix and Mn^+ in solution as long as the softness of Mn^+ is small enough to result in a positive ΔG.

Second, based on above research achievements, as shown in Figure 5B, Zhang et al. further studied the cation exchange reaction here. It has been explored that different phosphines could modulate the thermodynamic and kinetic parameters of the cation exchange reaction to synthesize complex semiconductor nanostructures [31]. Here, we take the examples of cation exchange between N^+ in amorphous N_2E (E means chalcogen) nanoparticles (NPs) and M^{2+} (such as Cd^{2+}) ions in solution, as shown in reaction (2) and Figure 1. Besides TBP, many other phosphine choices have been studied to mediate the thermodynamics and kinetics of reaction (2). Initiated by trace phosphine (R3P), the crystallization, the morphology, and the composition of metal/ME core/shell NCs have been tailored well.

The thermodynamics and kinetics of different phosphine and phosphite agents to synthesize semiconductor shell have been studied. The prerequisite to forward the cation exchange is the lone-pair electrons of P atom in phosphines or phosphites. Based on this, different π-accepting and σ-donating capabilities of them to M and N ions make reaction (2) to be exothermic (ΔG reaction < 0). Different phosphine coordinating to Ag^+ and Cd^{2+} and their coordination abilities study further prove reaction (2) is exothermic in principle. Therefore, thermodynamics and kinetics of reaction (2) could be mediated. The stronger coordination ability of R3P to N^+ than to M^{2+} ions enables in situ conversion of amorphous N_2E nanoparticles to be ME NCs. Although many kinds of phosphine or phosphite are applicable in reaction (2) thermodynamically, the steric effect derived from carbonyl ligands, such as alkyl and aryl ligands, would influence the kinetics of reaction (2), and the crystallization and composition of produced ME NCs distinctly. They took the example of cation exchange from amorphous Ag_2S nanostructure to single-crystalline CdS nanostructure. The kinetic activity order here is $PEt_3 > P(MeO)_3 > P(EtO)_3 > P(n\text{-}Bu)_3 > PPh_3 > P(n\text{-}Oct)_3 > P(PhMe)_3 > P(PhOMe)_3$. This is almost consistent with their σ donation ability order and contrary to their π back-bonding order. In particular, the steric hindrance from carbonyl ligands would influence σ donation ability and then chemical kinetics distinctly. That is the reason why $P(PhOMe)_3$ phosphines have low activities to reaction (2). These findings are concordant with Chad Tolman's classic phosphine ligands ordering in terms of their electron-donating ability and steric bulk. Besides preserving the original shape and size, phosphine-initiated cation exchange reactions show potential to precisely tune the crystallinity and composition of metal/semiconductor hybrid nanocrystals.

Third, as illustrated in Figure 5C, Zhang et al. further showed that the structural symmetry of such metal/semiconductor hybrid heterostructures can be finely tuned with controllable separation between metal and semiconductor components by taking advantage of chemical thermodynamics-directed colloidal strain tuning, from symmetric core–shell to asymmetric heterodimer gradually by in situ conversion of amorphous/crystalline Ag_2X shell to be single crystalline CdX. Typically, nanoparticle/solution interfacial energies and heterointerfacial energy (due to the lattice mismatch) can be strongly influenced by adhesion of solvent, capping ligands and precursors in colloidal phase [9]. Here, Zhang et al. took a facile two-step approach to maximize these effects to precisely control gradual shape evolution from symmetric core/shell to asymmetric heterodimer nanostructures, as shown in Figure 5C [32]. This approach is fundamentally different from generally used one-pot epitaxial overgrowth with metal nanoparticle seeds [11-25]. First, they started with concentric Au–Ag core/shell. The fact of close lattice feature between Au and Ag makes it possible to achieve precise control of core size, shell thickness, and monodispersity based on Frank-van der Merwe growth mode [9]. Moreover, the higher reactivity of silver metal nanostructures than gold enables the in situ conversion of Ag shell to silver chalcogenide (Ag_2X) without modification of the Au metal core. Depending on reaction condition (such as temperature and reactants), the crystallinity of Ag_2X can be controlled from amorphous to partial crystalline, thus leading to different lattice mismatch between Au core and Ag_2X shell. As a result, different interfacial lattice strain between Au and Ag_2X can be induced to initiate phase separation to a certain extent. Different from the long time aging-induced elemental Au diffusion in Au–Ag_2X core/shell nanocrystals via Ostwald ripening, the Au–Ag_2X here was used as intermediate precursor to carry out next step quickly to get Au/CdX heteronanocrystals. Second, they utilized cation exchange process based on the theory of hard-soft acid and base to realize the in situ conversion of Ag_2X to monocrystalline CdX [28]. The cation exchange-induced cation rearrangement and crystallography texture transformation provide further impetus to shift CdX shell gradually because of the larger strain energy between Au and monocrystalline CdX. In general, the higher crystallization of Ag_2X shell as well as the higher reaction temperature of cation exchange reaction leads to larger phase separation between Au and CdX to reduce interfacial and grain boundary energies. These tunable relocations of gold to CdX in quantum size region would enable maximum degree of tunability of their optoelectronic properties coupling.

2.3. Metal/semiconductor core/shell nanocrystals

The growth of single-crystal semiconductor-based heterostructures with modulated composition is a prerequisite for exploring fundamental nanoscale semiconductor physics [36] and can offer technological devices with optimum characteristics, including enhanced optical properties with high quantum yields [37], engineered electronic band gaps [38–40], and various solid-state optoelectronic properties [41–43]. Unintentional crystalline imperfections (such as polycrystallinity, dislocations, and other structural defects) lead to performance degradation or even premature failure of devices. For example, although the optical quality of semiconductor CdSe nanoparticles (NPs) could be improved by an overlayer of epitaxially grown CdS or ZnS, problems appear once the shell thickness becomes larger than the critical layer thickness (about two monolayers) due to the existence of strain-induced defects [38, 44, 45].

Current methods that achieve high-quality monocrystalline heterostructures are all based on epitaxial growth, as shown in Figure 1, which requires moderate lattice mismatches (<2%) between the two different materials. This lattice-matching constraint is a severe obstacle, particularly for growth of core–shell nanostructures with (quasi)spherical core NPs with highly curved surfaces that present many different crystallographic facets [46]. In addition to such lattice-matching requirements, the issues related to differences in crystal structure, bonding, and other properties have been found to inhibit epitaxial growth of dissimilar hybrid materials such as monocrystalline semiconductors on metals [47]. Attempts to use epitaxy to achieve hybrid metal core–semiconductor shell nanostructures have been unsuccessful, resulting in either polycrystalline semiconductor shells or anisotropic structures with segregation of the core and shell. This is only because as Figure 1 schemed, under large lattice mismatch (>40%), the semiconductor nanostructures would grow as small island-shaped NPs on metal core surface to decrease the surface strain energy. In this case, the Volmer—Weber (VW) mode growth would happen. Thus, based on these direct heterogeneous deposition methods, such as Figures 2–4 demonstrated, the polycrystalline semiconductor shell formed finally. As Figure 6 demonstrated, the Klimov, Talapin, and Wang groups have tried these kinds of method to prepare Au–PbS, Co–CdSe, Au–CdS, and Au–ZnS core/shell hybrid nanocrystals. Although they are highly monodispersed and could self-assemble into superlattice, the polycrystalline shell and too many defects at the interface limited their usefulness, especially in photocatalysis applications [48–53].

The photocatalysis application in the integration of semiconductor nanocrystals with noble metals uses the localized surface plasmon resonance (LSPR) effect of the metal component to enhance the light absorption, charge separation, and facilitate the absorbed light energy transfer from the metal to the semiconductor component for technologically important light-involved applications. Abundant studies have been devoted to the controllable productions of various hybrid nanocrystals. Heterostructure preparation is the basis for any application of semiconductor/noble metal hybrid nanocrystals. The crystalline imperfections of each component and the defects on the interface lead to the performance degradation or lost. Currently, high-quality monocrystalline heterostructures are usually produced by thermal decomposition epitaxial growth or through cation-exchange processes. The reported studies of monocrystalline hybrid nanocrystal synthesis are mainly based on epitaxial growth, which requires a moderate lattice mismatches between the different components. The lattice-mismatching constraint seriously limits the application of this method, particularly for growth of core/shell hybrid nanocrystals with highly curved surfaces. Therefore, it is a great challenge to synthesize the large lattice-mismatching semiconductor/metal hybrid nanocrystals with monocrystalline compounds and clear interface. Cation exchange reaction is a successful

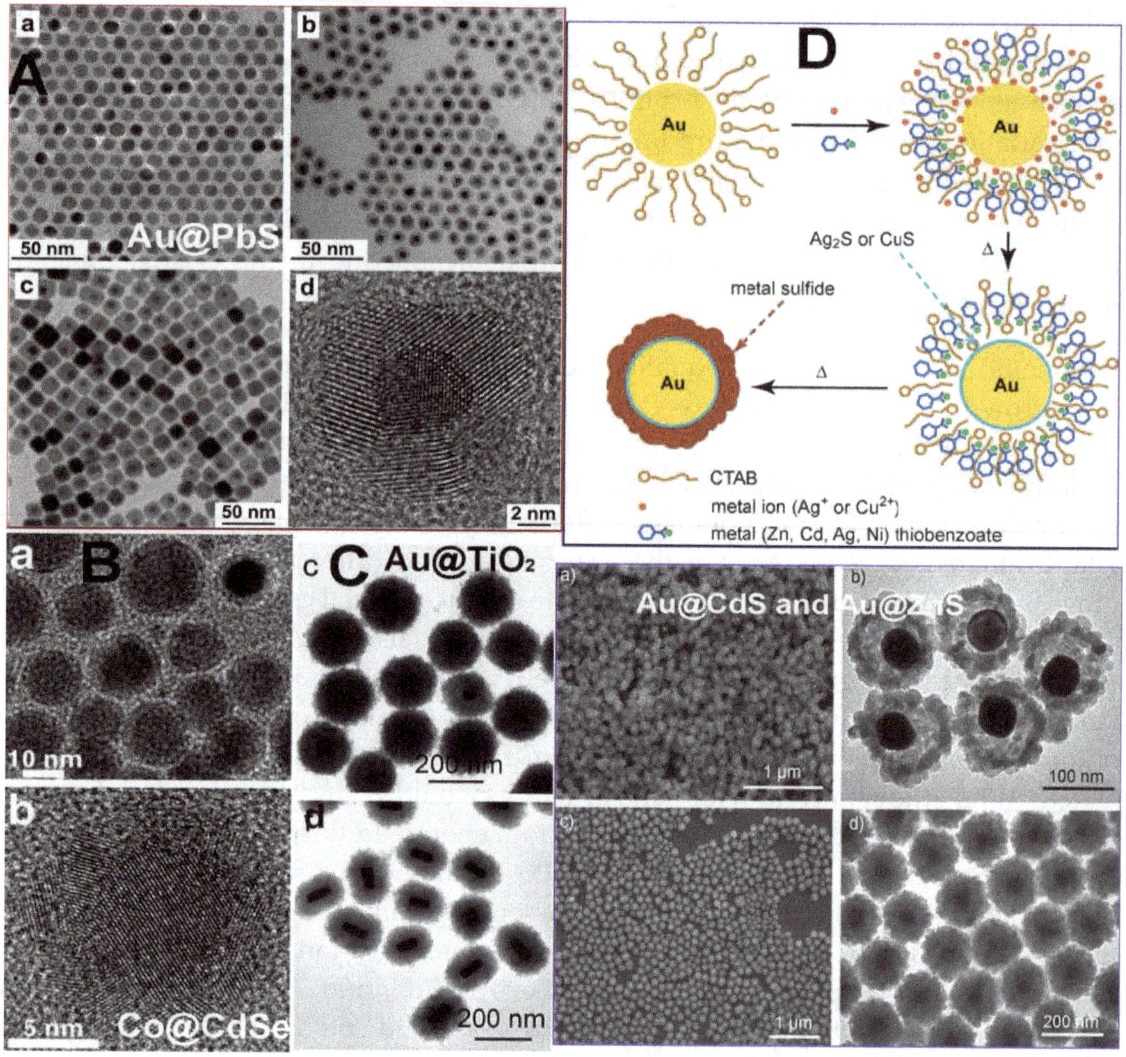

Figure 6. The metal/semiconductor core/shell NCs with polycrystalline semiconductor shell by direct heterogeneous deposition method. (A) Au–PbS core/shell NCs. (B) Co–CdSe core/shell NCs. Adapted from Lee et al. [48] and Kim et al. [49], respectively, with permission; copyright American Chemical Society. (C) Au–TiO_2 core/shell NCs. Adapted from Fang et al. [53] with permission; copyright Royal Society of Chemistry. (D) Au–CdS and Au–ZnS core/shell NCs. Adapted from Sun et al. [54] with permission; copyright Wiley-VCH Verlag GmbH & Co. KGaA.

synthetic method to preparation hybrid large lattice-mismatching heterodimer and core/shell nanocrystals with monocrystalline compounds and clear interface, which cannot be obtained by conventional epitaxial techniques [48–54]. Based on the Lewis acid–base reaction between molecular complexes and colloidal nanocrystals, the amorphous nanostructures can be transferred into monocrystalline compounds. Hence, by combining the sulfidation, selenization, or hyperoxidation of silver with cation exchange, other monocrystalline metal chalcogenide, selenide, or oxide nanostructures can been nonepitaxially grown on the large lattice-mismatching metal surface [28–33]. For the catalytic applications of semiconductor/noble metal hybrid nanocrystals, the key challenges are to obtain a clear semiconductor–metal interface and precise control over the size and shape of the heterostructures. Via cation exchange reaction, a series of semiconductor/metal hybrid nanocrystals can be obtained from concentric core–shell to nonconcentric heterodimer with precisely controlled separation and clear interface between the semiconductor and noble metal compounds. The symmetry evolution of semiconductor/metal hybrid nanocrystal has led to novel control of light absorption and photocatalytic activity, which indicates the advantage of cation exchange nonepitaxial growth and the importance of nanoscale interface control. For the phosphine-initiated cation exchange, different phosphines have been used to modulate the thermodynamic and kinetic process of the cation exchange reaction in semiconductor nanocrystal synthesis. By using different phosphines, the crystallinity, composition, morphology, and related properties of semiconductors can be precisely controlled.

Different from the reports in Figure 6, Ouyang et al. and Zhang et al. used the nonepitaxial growth scheme (Figure 5A) to get metal–semiconductor core/shell NCs with single crystalline semiconductor shell. The size of metal, semiconductor shell, and the composition of semiconductor shell could be precisely controlled (Figures 7–9) [28].

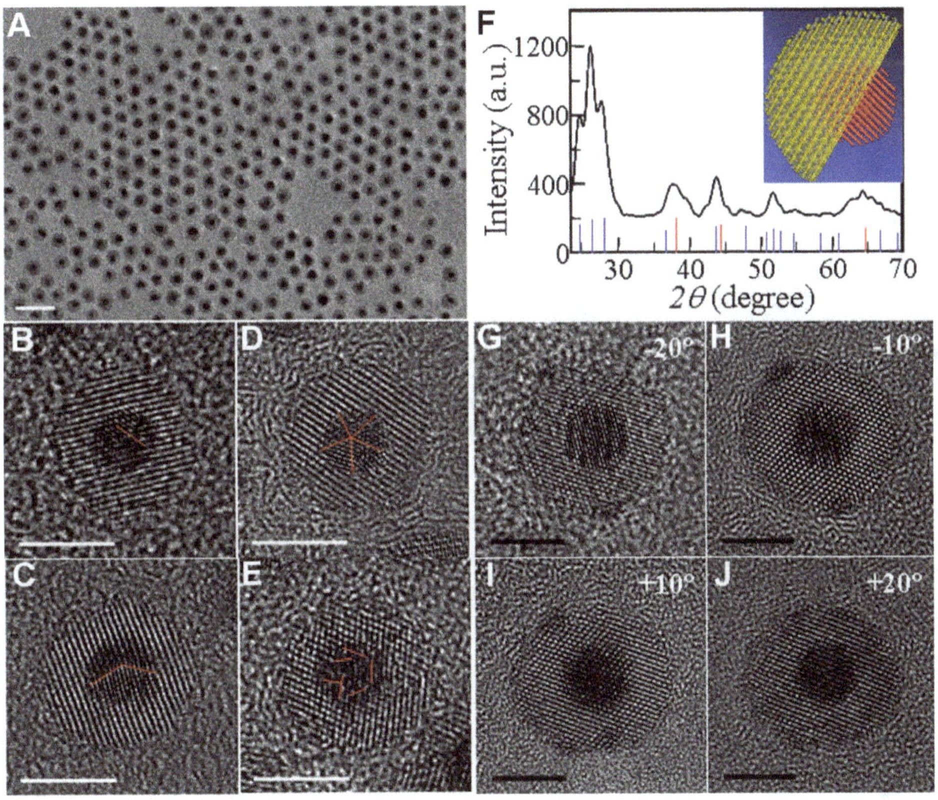

Figure 7. Au–CdS core–shell nanostructures with monocrystalline shell. (A) Typical TEM image showing uniform core–shell nanostructures. Scale bar, 20 nm. (B–E) High-resolution TEM images of core–shell nanostructures from (A). Whereas Au core NPs can manifest monocrystalline (B), single-fold twin (C), fivefold twin (D), and multiple-twin (E) lattice structures, all CdS shells are monocrystalline. The red lines highlight the lattice orientations within the Au core NPs. Scale bar, 5 nm. (F) XRD pattern of Au–CdS sample shown in panel A. Bulk Au [red solid lines, Joint Committee on Powder Diffraction Standards (JCPDS) #04-0784] and wurtzite CdS (blue solid lines, JCPDS #41-1049) are also provided for reference and comparison. (Inset) A ball-and-stick molecular model of Au–CdS, illustrating a cubic core and wurtzite shell. (G–J) Angle-dependent high-resolution TEM characterization. The sample depicted has a larger shell thickness than the one in panel A to emphasize the extremely high-quality crystallinity of the shell. The CdS shell shows perfect monocrystalline features without detectable structural defects under a different viewing angle. Scale bar, 5 nm. From Zhang et al. [28], reprinted with permission from the AAAS.

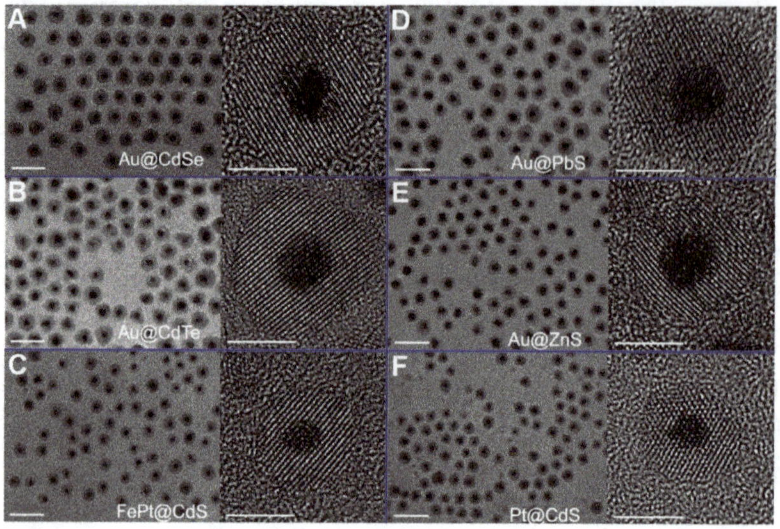

Figure 8. Large-scale (left) and high-resolution (right) TEM images of different hybrid core–shell nanostructures with various combinations of the core and shell components. All semiconductor shells show monocrystalline features. Scale bars for large-scale and high-resolution TEM images are 20 and 5 nm, respectively. (A) Au–CdSe; (B) Au–CdTe; (C) FePt–CdS; (D) Au–PbS; (E) Au–ZnS; and (F) Pt–CdS. From Zhang et al. [28], reprinted with permission from the AAAS.

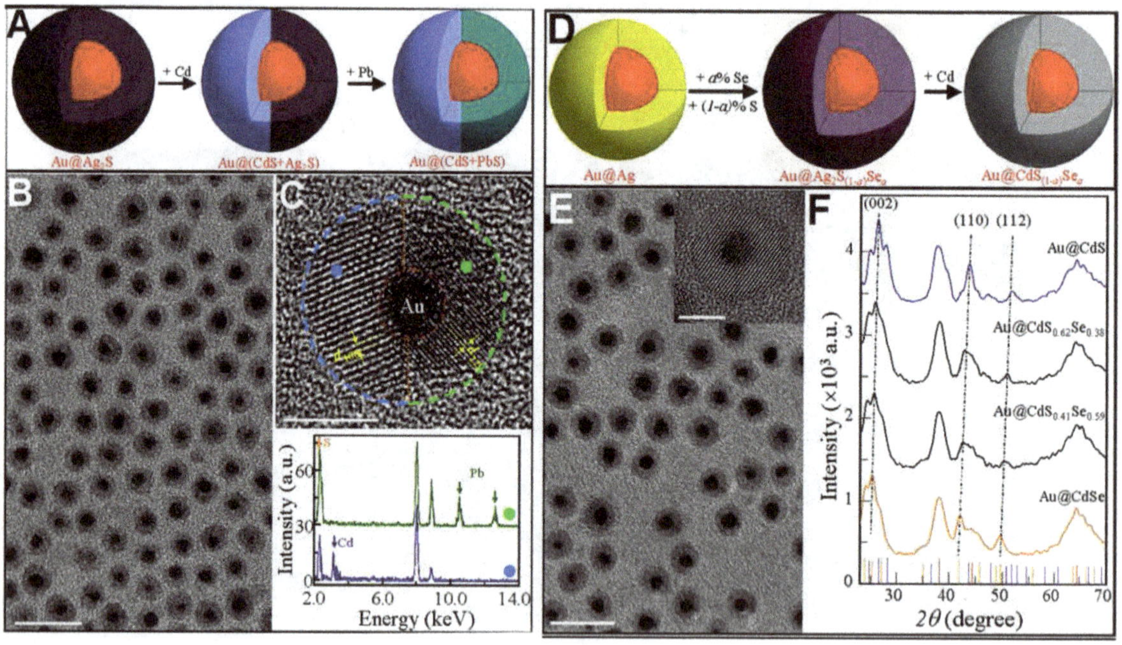

Figure 9. Growth of complex hybrid core–shell nanostructures with tailored structures and compositions of the monocrystalline shells. (A–C) Control of the monocrystalline cation species within the shell: the case of Au–(CdS+PbS). (A) Schematic of the growth procedure. (B) Large-scale TEM image. Scale bar, 20 nm. (C) (Top) High-resolution TEM image. Blue and green dashed arc curves highlight the monocrystalline CdS and PbS regimes, respectively. CdS and PbS manifest distinct lattice planes that can be assigned to (100) and (220), respectively. Scale bar, 5 nm. (Bottom) Single-particle EDS measurements in the CdS and PbS regimes. Peaks from Cd, Pb, and S elements are highlighted. (D–F) Control of the monocrystalline anion species within the shell: the case of Au–CdS1–aSea. (D) Schematic growth procedure. (E) Large-scale TEM image. Scale bar, 20 nm. (Inset) High-resolution TEM image showing the monocrystalline alloy shell. Scale bar, 5 nm. (F) XRD patterns highlighting lattice evolution from CdSe to CdS with different ratio a. From Zhang et al. [28], reprinted with permission from the AAAS.

Based on the strategy in Figure 5B, Zhang et al. use air-stable PPh_3 for the first time to initiate cation exchange reaction (2). Figure 10 demonstrates the Au–CdS, Pt–CdS core/shell NCs preparation by PPh_3 initiated reaction (2). Following the nonepitaxial growth process we published before [28] after cation exchange from amorphous Ag_2S shell, Au–CdS and Pt–CdS NCs could preserve thick monocrystalline CdS shell. LRTEM and HRTEM images in Figure 10A–D confirmed their good crystallization. Especially, as shown in Figure 10D, despite of anisotropic shape of Pt nanocube, the thick CdS shell has good single-crystallinity. PPh_3 further facilitates the monocrystalline engineering to break through critical layer thickness limit of heteroepitaxy. PPh_3 is "green" choice than TBP because it is air-stable enough to enable reaction (2) under more flexibly conditions, such as higher temperature and longer time to facilitate versatile ME NCs crystallization. Moreover, besides crystallization tailoring, PPh_3 could initiate cation exchange to get more complex heterostructures with precise compositional tailoring. Figure 11A and B showed the LRTEM and HRTEM images of as-prepared Au–$CdS_{1-x}Se_x$ core/shell nanocrystals with ternary single-crystal alloys shell. The S-to-Se ratio could be tailored ($CdS_{0.58}Se_{0.42}$ and $CdS_{0.45}Se_{0.55}$, obtained by EDS elemental analysis) to engineer their band gaps. The consistent shift of their powder XRD peaks from $CdS_{1-x}Se_x$ shell (Figure 11C) and the distinct colloid color changing (insert in Figure 3D) confirmed the homogeneous composition modulation. The strong visible light absorption (550–700 nm) (Figure 11D) due to the surface plasmon resonance (SPR) and exciton coupling in as-prepared Au–$CdS_{1-x}Se_x$ NCs indicated their potential photocatalysis and photovoltaic applications [55].

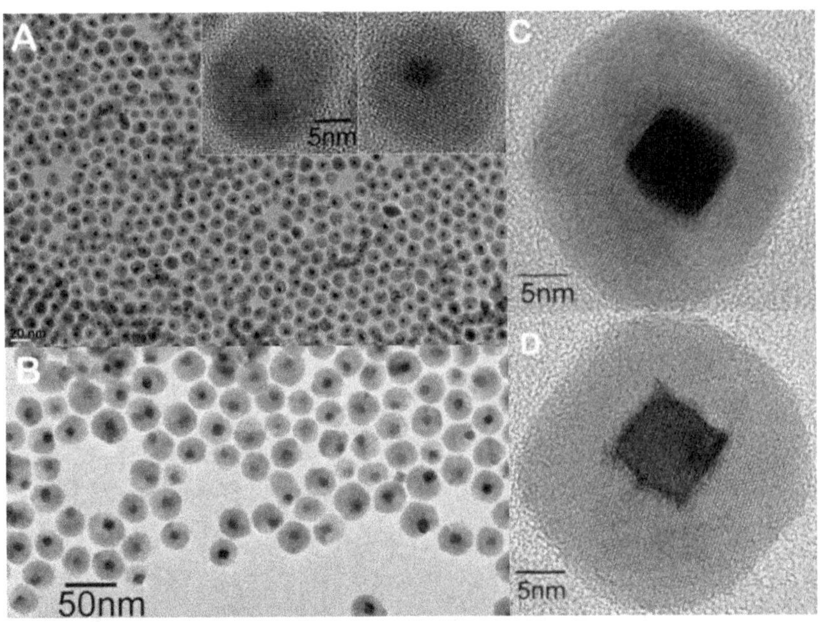

Figure 10. Core/shell metal–semiconductor NCs with thick monocrystalline CdS shell synthesized by PPh_3 initialization under large lattice mismatches: LRTEM and HRTEM images of Au–CdS NCs (A) and Pt–CdS NCs with Pt nanocube core (B-D). Adapted from Gui et al. [31] with permission; copyright Wiley-VCH Verlag GmbH & Co. KGaA)

Recently, through the strategy in Figure 5C, Zhang et al. have demonstrated evolution of relative position of Au and CdX in Au–CdX (X means S, Se, and Te) from symmetric to asymmetric configuration (Figures 12 and 13) [32].

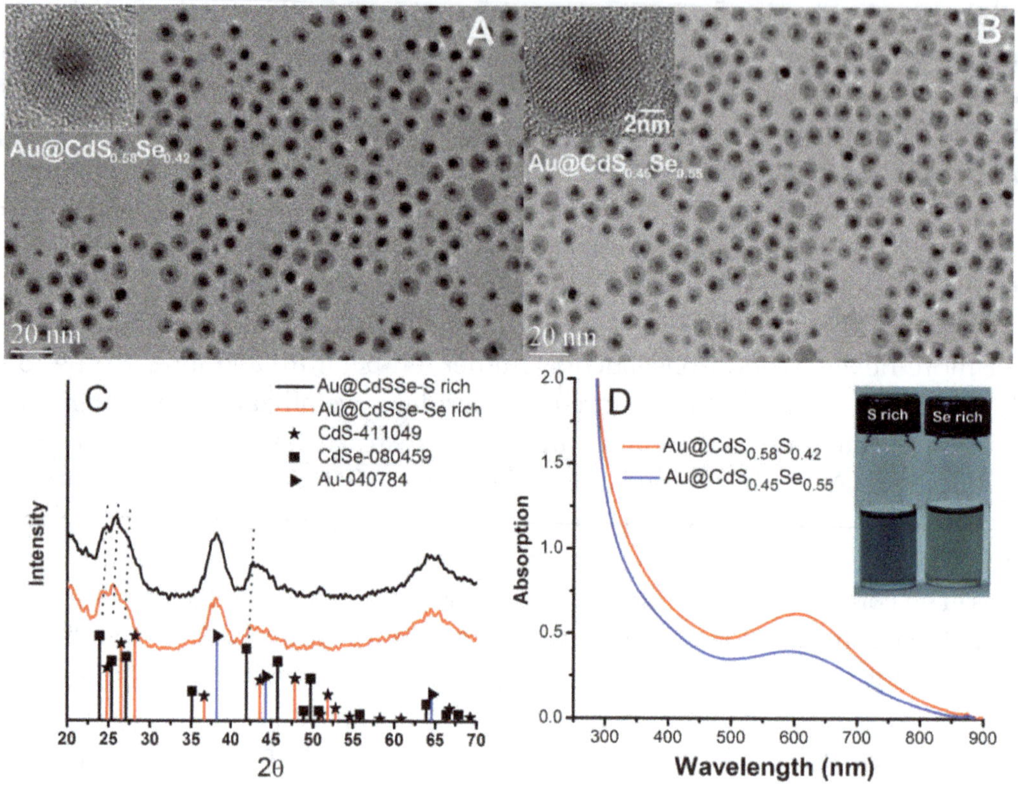

Figure 11. Au–$CdS_{1-x}Se_x$ NCs synthesized by PPh_3 initialization. (A, B) LRTEM and HRTEM images of prepared Au–$CdS_{0.58}Se_{0.42}$ NCs and Au–$CdS_{0.45}Se_{0.55}$ NCs. (C) XRD patterns comparison of them. (D) UV-Vis extinction spectra comparison of them. Inserted pictures showed their colloid color. Adapted from Gui et al. [31] with permission; copyright Wiley-VCH Verlag GmbH & Co. KGaA)

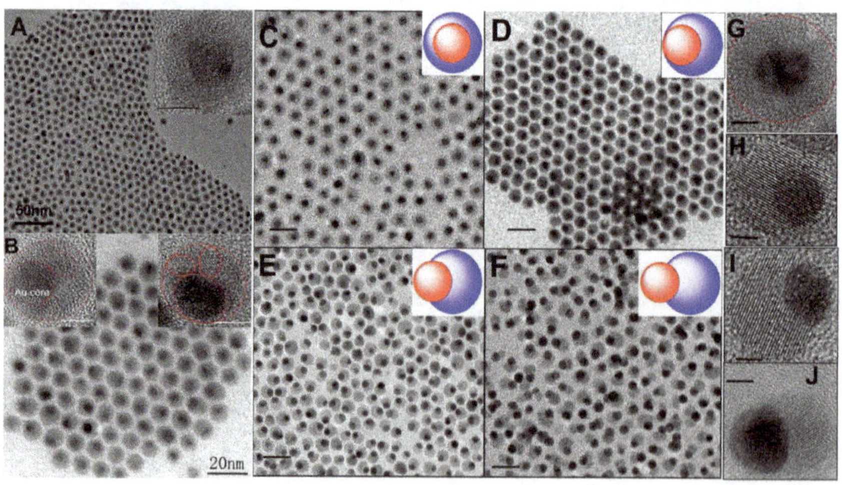

Figure 12. TEM characterizations on controllable structural symmetry from core/shell to heterodimers. (A, B) Au–Ag_2S with amorphous (A) and crystalline Ag_2S (B) shells. The inserts are HRTEM images of a-Ag_2S (A) and c-Ag_2S shell (B); scale bar, 5 nm. Red solid lines are guides for the eye, distinguishing the Au core and c-Ag_2S shell boundaries, respectively. (C–F) Au/CdS hybrid nanostructure with controllable structural symmetry. (C) Concentric core/shell. (D) Non-concentric core/shell. (E, F) Heterodimers. The inset diagrams highlight the phase separation-induced Au/CdS morphologies. Scale bar: 20 nm. (G–J) HRTEM images highlight the shape separations of panels C–F, respectively. Scale bar: 2.5 nm. Adapted from Zhao et al. [32] with permission; copyright Wiley-VCH Verlag GmbH & Co. KGaA.

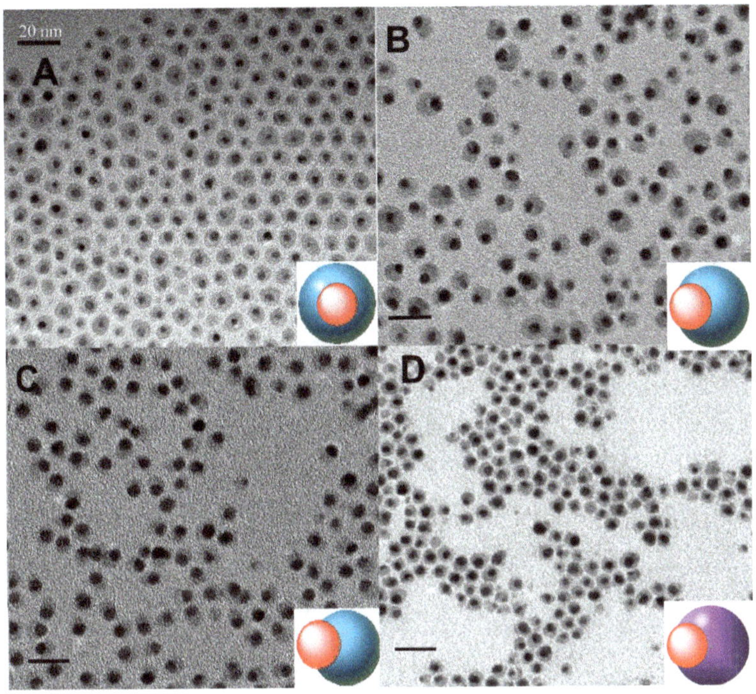

Figure 13. Large-scale TEM images of shape evolutions in Au/CdSe and Au/CdTe hybrid nanostructures with ~5 nm sized Au. (A) Concentric core/shell of Au/CdSe; (B, C) heterodimer of Au/ CdSe; (D) heterodimer of Au/CdTe. The inset diagrams highlight the phase separation-induced Au/CdSe and Au/CdTe morphologies. Scale bar: 20 nm. Adapted from Zhao et al. [32] with permission; copyright Wiley-VCH Verlag GmbH & Co. KGaA.

Based on Figure 5, the metal/semiconductor core/shell NCs can further lead to fine tuning of plasmon–exciton coupling, different hydrogen photocatalytic performance, and enhanced photovoltaic, electrical properties.

2.4. Metal/semiconductor yolk/shell nanocrystals

With a unique structure, metal-semiconductor yolk/shell nanocrystals play an important role in photocatalysis and other fields [56–63]. As photocatalysts, metal-semiconductor yolk/shell nanocrystals process some advantages, such as stronger absorption, higher catalysts selectivity, and higher quantum yield. Nowadays, different kinds of metal-semiconductor yolk/shell nanoarchitecture have been designed to enhance their photocatalysis properties. Following it are several methods for preparing metal-semiconductor yolk/shell nanocrystal.

2.4.1. Template Free for formation of metal/semiconductor yolk–shell nanocrystals

Following this method, metal/semiconductor yolk–shell nanocrystals are synthesized directly. During the process, no any nanostructure is used as template or core–shell nanocrystals forming during the process are taken as self-template [56, 64, 65]. For example, Pt–CeO_2 yolk–shell nanoparticles can be synthesized by hydrothermal method, although the mechanism may involve several factors [64]. Pt nanoparticles and CeO_2 precursors are put into the autoclave and heated at 90ºC in a period. The authors illustrate that the ratio of $CeCl_3$ concentration and

Pt colloid solution is the main factor to formation of yolk–shell nanoparticles. Only can Pt–CeO_2 core–shell nanoparticles be obtained when a larger $CeCl_3$/Pt ratio is employed. With a smaller ratio, the looser shell of Pt–CeO_2 yolk–shell nanoparticles can be obtained. The yolk–shell nanoparticles enhance the properties of photocatalysis of hydride of Pt and CeO_2.

Au–Cu_2O yolk–shell nanocrystals are synthesized by prolonging the reaction time after formation of Au–Cu_2O core–shell nanocrystals [65]. Cu_2O shell grows on the surface of Au nanoparticles to form Au–Cu_2O core–shell nanoparticles. As the reaction time increasing, Au core and Cu_2O shell spare forming yolk–shell nanostructure. The longer the reaction time, the more obvious the yolk–shell nanostructure. Furthermore, the thickness of Cu_2O shell can be tuned with Au colloid/Cu^{2+} ratio. As the authors showed in the articles, the distance between Au core and Cu_2O shell and the thickness of Cu_2O shell have a strong influence on the optical properties of Au–Cu_2O yolk–shell nanoparticles (Figure 14). It means that the method for preparation of Au–Cu_2O with tunable yolk–shell nanoparticle provides a route to tune the optical properties which is of importance for photocatalysis.

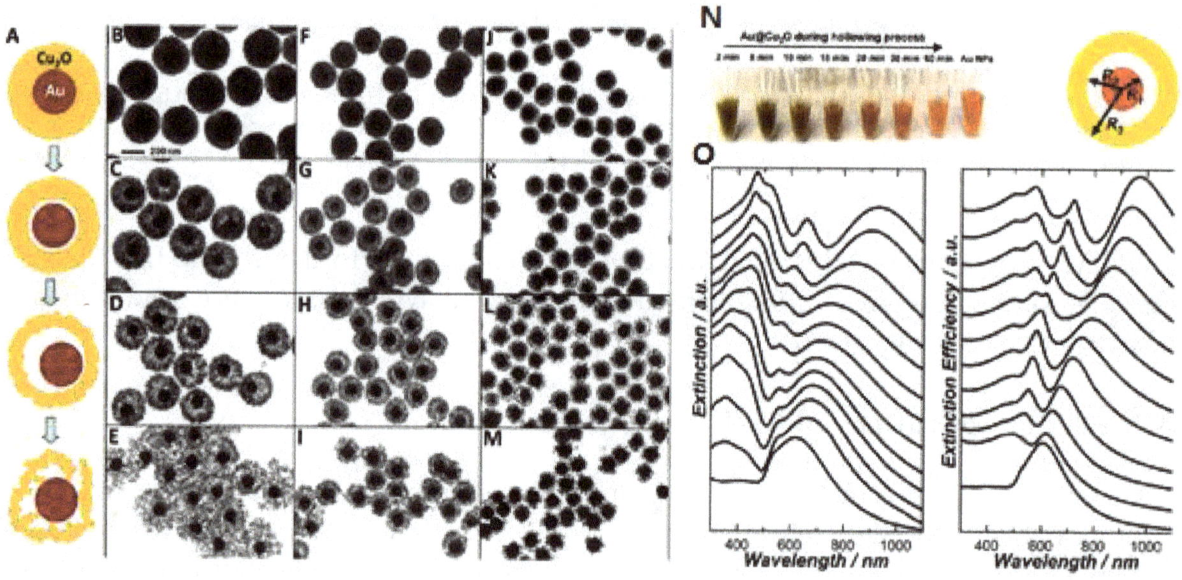

Figure 14. (A) Schematic illustration of formation of Au–Cu_2O yolk–shell nanocrystals. (B–M) TEM images of prepared Au–Cu_2O yolk–shell nanoparticles. (N) Photograph of Au–Cu_2O during hollow process. (O) Their optical properties collected by experiment and simulation.

Co–SiO_2 yolk–shell nanoparticles are obtained by reduction of CoO–SiO_2 core–shell nanoparticles. Park et al. coat silica on CoO nanoparticles to form CoO–SiO_2 nanoparticles then reduce CoO–SiO_2 nanoparticles with hydrogen to form Co–SiO_2 yolk–shell nanoparticles [66].

With template free method, some metal cation is introduced into hollow nanocrystals then reduced forming metal nanoparticles in the cavities of hollow structure [67, 68]. Cu–SiO_2 yolk–shell nanocrystal is synthesized by reducing Cu^{2+} cation in cavity of SiO_2 hollow particle and the size of Cu core can be tuned by multiple reduction cycles (Figure 15A–D) [67]. Besides that, yolk–shell nanocrystals can be employed because they also have cavities for reduction of metal cation. By this method, multiplied cores or cores with different materials can be synthesized. Au–SiO_2 yolk–shell with tunable core is prepared by reducing HAuCl4 in the cavity of SiO_2–SiO_2 yolk–shell nanostructure (Figure 15F–I) [68]. On account of that, metal cation can be introduced into the cavities of hollow nanoparticles or yolk–shell nanoparticles, and other kinds of metal cation can be introduced into the cavities to react with metal core forming a new metal core or alloy core. For instance, Ag–SiO_2 yolk–shell nanoparticles can be obtained by displacing Cu with Ag^+ from Cu–SiO_2 yolk–shell nanoparticles (Figure 15E) [67]. Moreover, because the cavity of yolk–shell can be taken as reactor, the synthesis way can be applied to form other yolk–shell nanostructures.

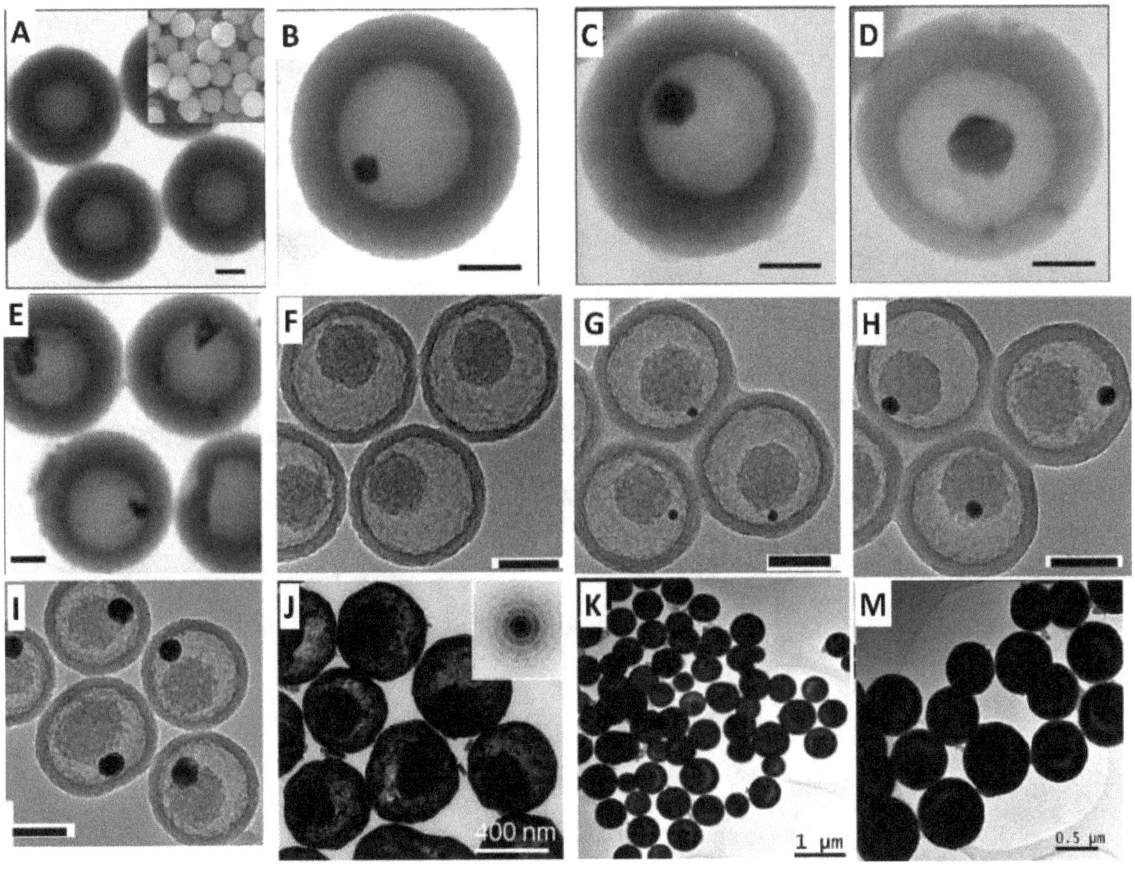

Figure 15. TEM images of various yolk–shell nanoparticles prepared with different way. (A) SiO_2 hollow nanosphere, (B–D) Cu-SiO_2 yolk–shell nanoparticles, and (E) Ag–SiO_2 yolk–shell prepared by replacing the Cu core of Cu–SiO_2 yolk–shell nanoparticles with Ag+; scale bar: 100 nm [67]. Copyright: Royal Society of Chemistry, 2004. (F) SiO_2–SiO_2 yolk–shell nanoparticles, (G-I) Au–SiO_2 yolk–shell nanoparticles with different sizes Au core; scale bar: 50 nm [68]. Copyright: Wiley-VCH, 2010. (J) Au–CeO_2 yolk–shell nanoparticles [69]. Copyright: RSC, 2012. (K-M) Pt–CeO_2 yolk–shell nanoparticles [64]. Copyright: RSC, 2011.

Although some of mechanism of template free of yolk–shell nanostructure formation is unclear, it provides the method to design and prepare metal/semiconductor yolk–shell nanocrystals.

2.4.2. Using template for metal-semiconductor yolk–shell nanocrystals

Template employed for synthesizing metal-semiconductor yolk–shell nanocrystals can be conclude into two kinds of templates: soft template and hard template.

Using soft template for metal-semiconductor yolk–shell nanocrystals: By soft template protocol, several kinds of chemical are usually employed as soft template, such as biomolecular, polymers, surfactants, or microemulsions [56, 70]. In most of case, the large molecular would form microemulsions in solution and precursors react on the boundary of microemulsions. Therefore, sizes or morphologies of yolk–shell nanoparticles could be tuned by tuning cell of microemulsions [71–73]. For metal semiconductor yolk–shell nanoparticles, metal core is reduced in the cavities of microemulsions then semiconductor shell grows on the boundary of microemulsions. For instance, Priebe and Fromm synthesize the Ag–SiO_2 yolk–shell nanoparticles in cyclohexane and Igepal CO-520 system [72]. Ag^+ is introduced into cavity of microemulsions and reduced to nanoparticles. Subsequently, silica precursor is introduced into the system and form the shell on the boundary of microemulsions (Figure 16). Besides, because Ag^+ is reduced in the cavities, the size of Ag core can be tuned by using the series of concentration of $AgNO_4$ solution.

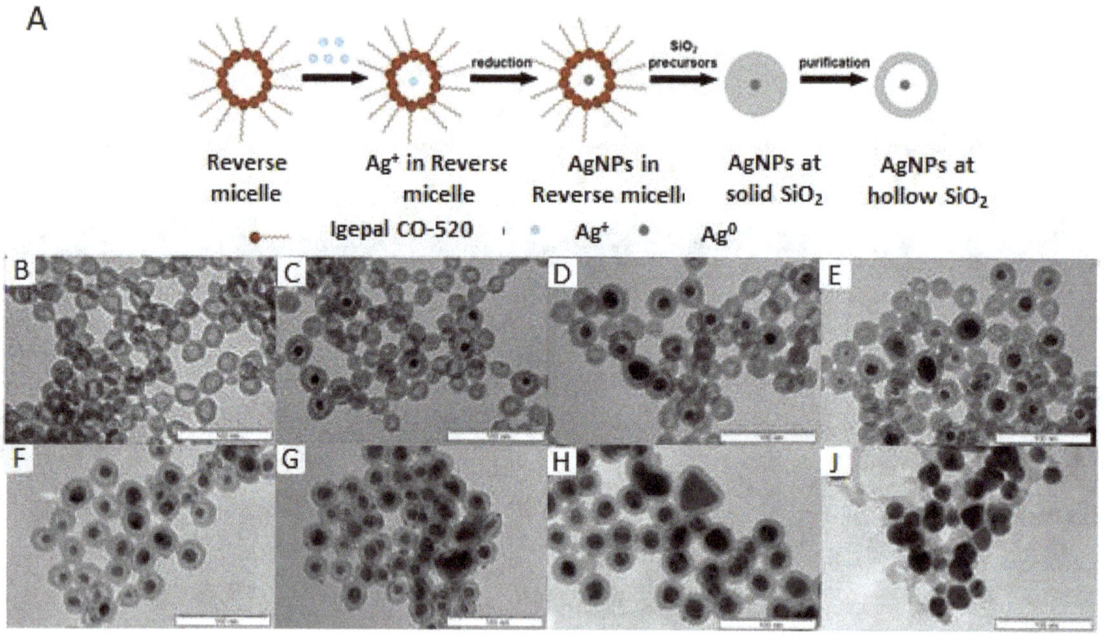

Figure 16. (A) Schematic for formation of Ag–SiO_2 yolk–shell nanoparticles and (B–I) TEM images of Ag–SiO_2 yolk–shell nanoparticles with various Ag core size; scale bar: 500 nm, Copyright: Wiley-VCH, 2014.

Metal cores can also be introduced the cavities of microemulsions and then the shell grows to form metal/semiconductor yolk–shell nanoparticles directly [73]. Hyunjoon Song and his coworkers introduce Pd nanoparticles into the system of hexanol and igepal CO-630, followed by adding tetraethyl orthosilicate (TMOS) and octadecyltrimethoxysilane (C18TMS) for shell growth (Figure 17). Their protocol provides a possible method for preparing metal/semiconductor yolk–shell nanoparticles with other metal core and semiconductor shell.

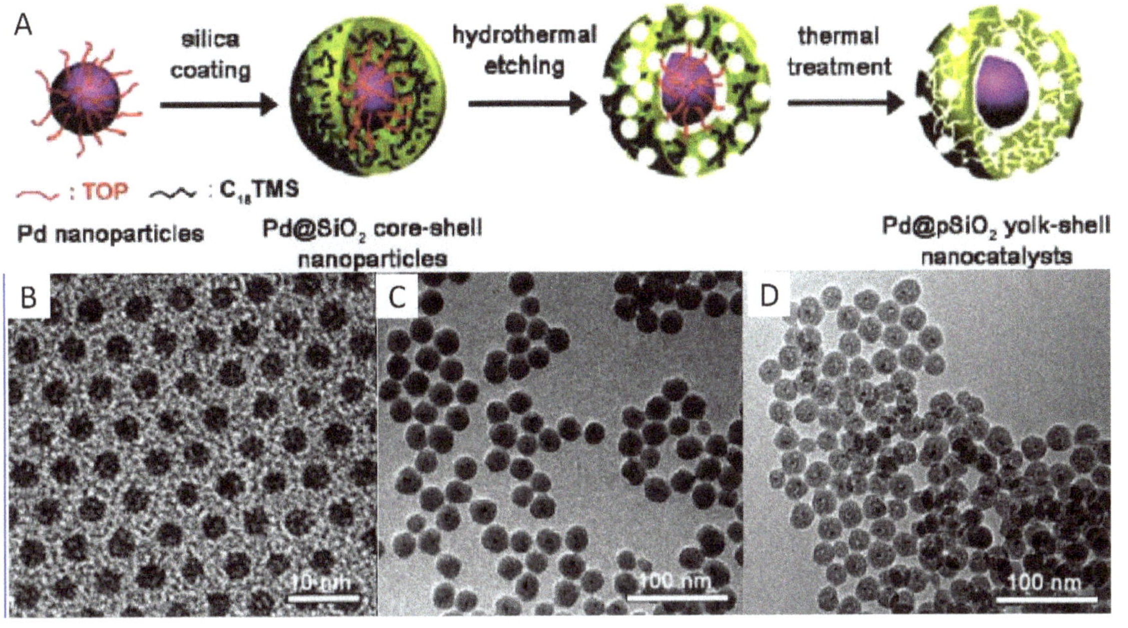

Figure 17. (A)Schematic for preparing Pb–SiO_2 yolk–shell nanoparticles and TEM images of Pb cores (B), Pb–SiO_2 core–shell nanoparticles (C), and Pb–SiO_2 yolk–shell nanoparticles(D). Copyright: ACS, 2011.

Using hard template for metal/semiconductor yolk–shell nanostructure: By hard template method, shell coating on core is synthesized as middle layer and subsequently removed or etched. The material of middle shell includes SiO_2, carbon, polymer, and so on. For example, Au–TiO_2, Au–ZrO_2, Au–SnO_2, and Au–SiO_2 yolk–shell nanoparticles are synthesized by coating semiconductor shell on Au–oxides and subsequently etching oxides layer (Figure 18) [74–77]. In most of case, SiO_2 is employed as hard template to form the middle layer and etched by NaOH, HF, or other reagents.

For example, Au–ZrO_2 yolk–shell nanoparticles can also be synthesized by etching SiO_2 layer of Au–SiO_2–ZrO_2 core–shell nanoparticles (Figure 18A) [74]. It is reported that such Au–ZrO_2 yolk–shell nanostructure is stable at high temperature and can be used as catalyst for the oxidation of CO. Taking SiO_2 as template, SnO_2 hollow nanosphere, and Au–SnO_2 yolk–shell nanoparticles can be obtained (Figure 18B). Lou et al. show their work on preparation of Au–SnO_2 yolk–shell nanoparticles, etching the SiO_2 layer with HF and they found that controlling the size of SiO_2 template cage-like and double layer shell of Au–SnO_2 can be obtained [75]. The method can also be used for synthesis of yolk–shell with other hybrid material shell [78].

3. The photocatalytic hydrogen evolution applications of metal/ semiconductor hybrid nanocrystals

3.1. Mechanism of photocatalysis on semiconductors and key problems

Since Fujishima and Honda [79] discovered photoelectrochemical formation of H_2 over TiO_2 electrode, photocatalytic water splitting and H_2 evolution are an attractive solution of global

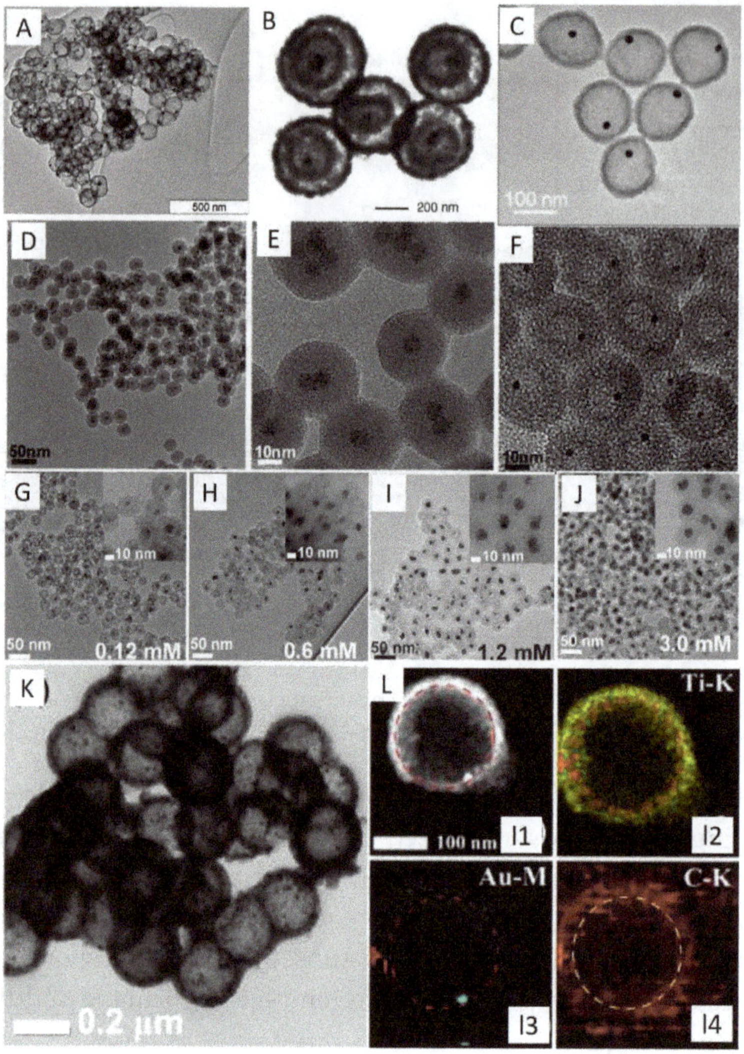

Figure 18. TEM images of (A) Au–ZrO_2 yolk–shell nanoparticles [74]. Copyright: Wiley-VCH, 2006. (B) Au–SnO_2, yolk–shell nanoparticles [75]. Copyright: Wiley-VCH, 2007. (C) Au–Zr0.5Si0.5O_2 yolk–shell nanoparticles [76]. Copyright: Springer, 2014. (D, E) Fe3O_4/Au–SiO_2 yolk–shell nanoparticles. (F) Au–SiO_2 yolk–shell nanoparticles. (G–J) Ag–SiO_2 yolk–shell nanoparticles [77]. Copyright: Royal Society of Chemistry, 2010. (K, L) TEM and HAADF-STEM images and EDS maps of Au–GO/TiO_2 yolk–shell nanoparticles [78]. Copyright: America Chemistry Society.

energy supply and related environmental issues. Numerous researchers had extensively studied water splitting using semiconductor photocatalysts since the finding [80–87]. The fundamental principle of solar water splitting for semiconductor photocatalysts is shown in Figure 19. Semiconductor materials have a band structure in which the conduction band (CB) is separated from the valence band (VB) by a band gap with a suitable width. When the energy of incident light is larger than that of a band gap, the electrons in the VB of the semiconductor photocatalyst are excited to the CB, while the holes are left in the VB. Therefore, it creates the negative-electron (e-) and the positive-hole (h+) pairs. These photogenerated electron–hole pairs may further be involved in the following possible processes: (1) successfully migrate to the surface of semiconductor, (2) be captured by the defect sites in bulk and/or on the surface region of semiconductor, and (3) recombine to release the energy in the form of heat or photon.

If the photoexcited carriers separate and migrate to the surface without recombination, adsorbed species are reduced and oxidized by the photogenerated electrons and holes to produce H_2 and O_2, respectively (Figure 19).

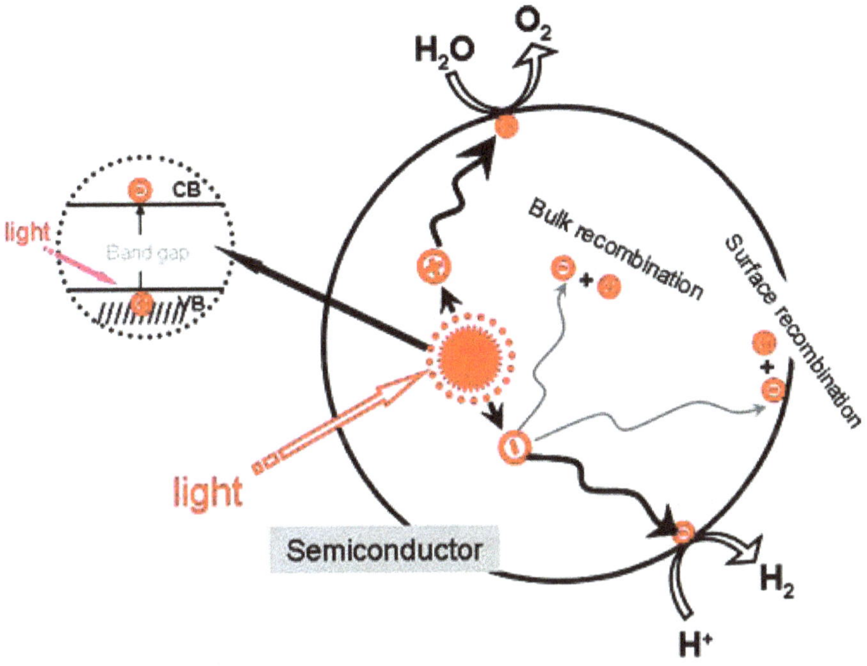

Figure 19. Process in photocatalytic water splitting [80]. Copyright: Royal Society of Chemistry, 2009.

For effective water splitting, important points in the semiconductor photocatalyst materials are their band structure, including the band gap and positions of VB and CB. The bottom level of the CB has to be more negative than the redox potential of H^+/H_2 (0 V vs. NHE), while the top level of the VB be more positive than the redox potential of O_2/H_2O (1.23 V) [80]. Therefore, a minimum band gap of 1.23 eV is required, while a much larger band gap (usually >2.0 eV) is often needed for appreciable water splitting reaction due to additional overpotential associated with each electron transfer and gas evolution step [84]. The band edge positions of some semiconductor photocatalysts are shown in Figure 20. The band structure of a semiconductor is not only determined by its own crystal phase and vacancies but can also be modified by the introduction of foreign elements into the bulk or surface of the semiconductor. Take the most studied photocatalyst TiO_2 as an example, the larger band gap (~3.2 eV) of TiO_2 limits its utilization of the solar spectrum to only the ultraviolet (UV) region (wavelength $\lambda < 400$ nm) [85]. The solar spectrum has a very small fraction of UV light (ca. 5%) in comparison with those of visible light ($400 < \lambda < 800$ nm, ca. 43%) and near-infrared (NIR) light ($800 < \lambda < 2500$ nm, ca. 52%) [83]; therefore, studies on visible-light-driven photocatalysts are more important for practical applications. There are two efficient strategies to make TiO_2 as visible-light-driven photocatalysts as shown in Figure 2B [84]: (1) narrow the band gap of TiO_2 to make it absorb visible light by introducing other elements into TiO_2 and (2) modify the TiO_2 surface with other visible light active materials (dye or quantum dot) as a light harvester to sensitize TiO_2.

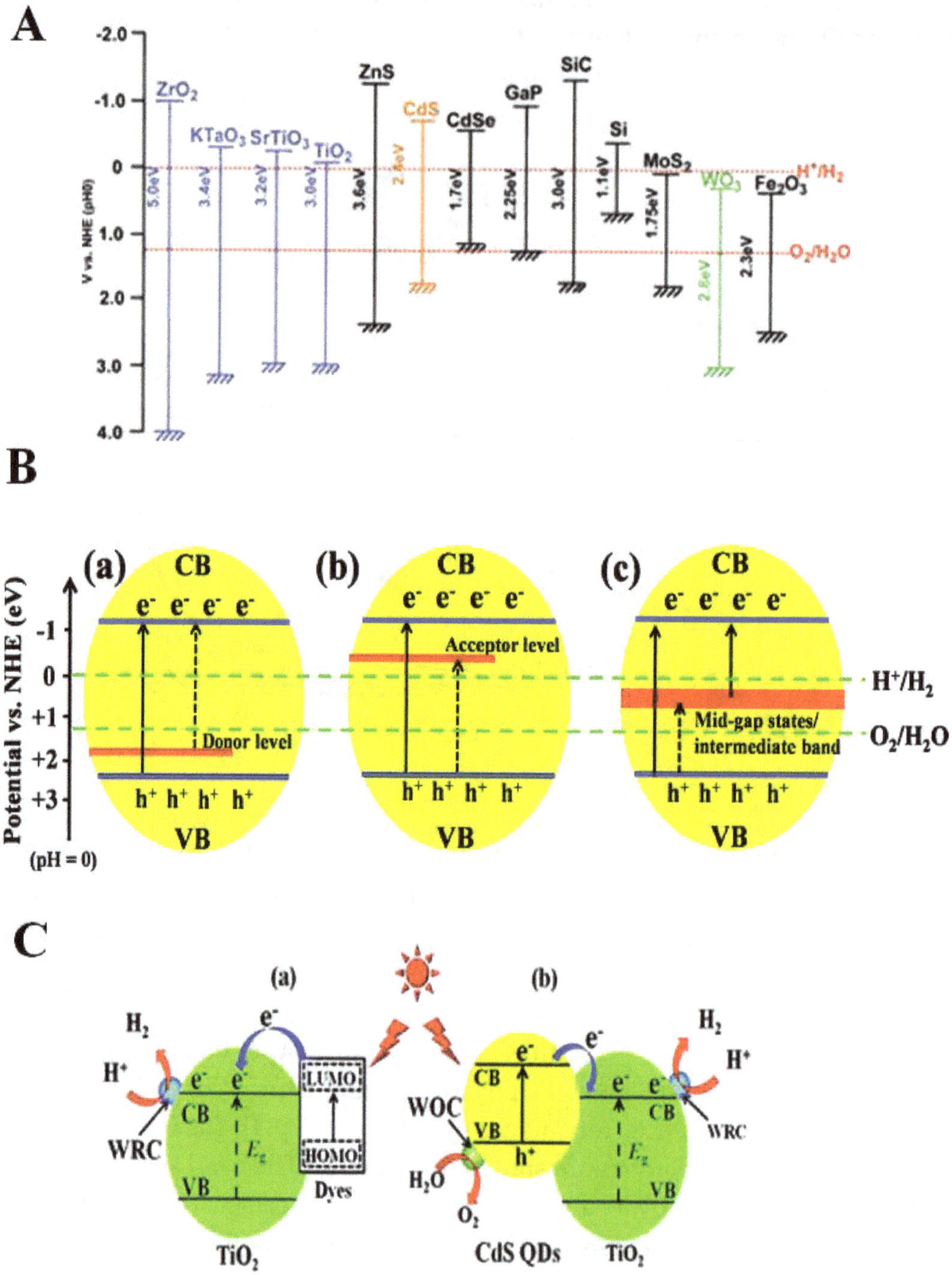

Figure 20. (A) Relationship between band structure of semiconductor and redox potential of water splitting [80]. Copyright: Royal Society of Chemistry, 2009. (B) Donor level (a), acceptor level (b), and mid-gap states (c) formed by metal ion doping. (C) The mechanism of H2 production for (a) dye-sensitized and (b) QDs-sensitized semiconductor [82]. Copyright: Royal Society of Chemistry, 2015.

After excited charges are created, efficient charge separation is next crucial factor determining the light to fuels conversion efficiency. Crystal structure, crystallinity, and particle size strongly affect the step [86]. The defects operate as trapping and recombination centers between photogenerated electrons and holes, resulting poor photocatalytic activity. Therefore, improving the crystalline quality will decrease the amount of defects and improve photocatalytic activity. If the particle size becomes small, the distance that photogenerated electrons and holes have to migrate to reaction sites on the surface becomes short, and this results in a decrease in the recombination probability. Also, the fabrication of junction structure has been recognized as an effective strategy to avoid charge recombination in semiconductors.

The surface catalytic reaction is a successive step of charge separation. The important points for this step are surface character (active sites) and quantity (surface area). Even if the photogenerated electrons and holes possess thermodynamically sufficient potentials for water splitting, they will have to recombine with each other if the active sites for redox reactions do not exist on the surface. Cocatalysts such as Pt, NiO, and RuO_2 are usually loaded to introduce active sites for H2 evolution [87], which will be discussed later.

The processes of the photocatalytic reaction on a semiconductor photocatalyst involve light absorption, charge separation, carrier migration, and surface catalytic reactions. Therefore, developing band gap engineering to narrow down the band gap of semiconductor materials for absorbing broader spectrum of solar energy and materials engineering to tune the physical properties (crystal structure, crystallinity, and particle size) for gaining efficient charge separation and migration and creating enough active sites are the key problems of improving photocatalytic hydrogen evolution efficiencies.

3.2. Metal/semiconductor heterodimer nanocrystals: the role of metal

Integration with metal is a commonly used configuration to improve the photocatalytic hydrogen evolution performance of a semiconductor. The metal may play a variety of roles in the enhancement of photocatalytic performance. In the following sections, we will focus on the role of metal as cocatalyst and the plasmonic effect of noble metals.

3.2.1. The cocatalyst role

Since the work by Kraeutler and Bard in 1978 loading Pt on the surface of TiO_2 [88], the loading of metal nanoparticles onto different semiconductor photocatalysts has been regarded as a popular strategy to improve the photocatalytic performance in photocatalytic water splitting. Besides Pt, other metal cocatalysts, including Pd, Rh, Ru, Ir, Ag, Ni, Co, etc., have been recognized as efficient cocatalysts for photocatalytic hydrogen evolution [89–96]. The processes of charge transfer between metal cocatalyst and host semiconductor photocatalyst are described in Figure 21A.

The metal cocatalysts mainly play two roles in the improvement of photocatalytic performance. One is to assist in electron–hole separation through the formation of Schottky barrier between the metal cocatalyst and the light-harvesting semiconductor. A Schottky barrier is a type of junction resulting from the intimate contact of a metal with a semiconductor (Figure 21B). The metal cocatalyst with larger work function, that is, lower Fermi energy level, should more readily trap the photogenerated electrons migrated to the surface of the host semiconductor photocatalyst. Meanwhile, the photogenerated holes stay at the host photocatalyst and

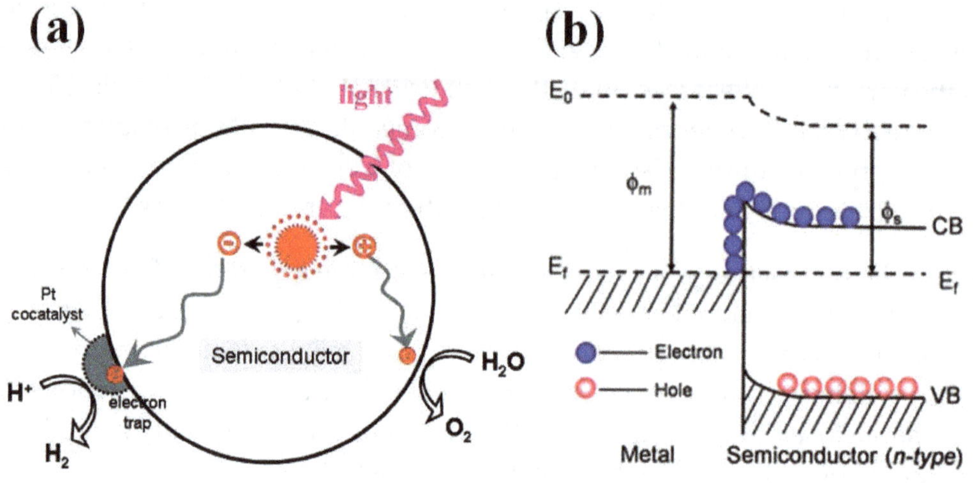

(c)

$$2e^- + 2H^+ \xrightarrow{①} 2[H_{ad}] \xrightarrow{②} H_2$$

Figure 21. (A) Processes of charge transfer between host photocatalyst and cocatalyst, taking Pt as the example of cocatalyst [86]. Copyright: America Chemistry Society, 2010. (B) Schematic of Schottky barrier [90]. Copyright: Royal Society of Chemistry, 2010. (C) Two general steps for proton reduction reaction [89]. Copyright: America Chemistry Society, 2013.

migrate to its surface. This retards the possibility of electron–hole recombination and enhances efficient separation of the photogenerated electrons and holes. It improved the overall photocatalytic activity of the water splitting because it helps to promote charge separation, which in return reduces both bulk and surface electron/hole recombination.

The other role of metal cocatalysts is to serve as the reaction sites to catalyze the proton reduction to H_2. Proton reduction course on cocatalysts goes through at least two steps: a discharge step and catalytic step (Figure 21C) [89]. Most semiconductor photocatalyst, particularly oxides, lack surface H_2 evolution sites rather than O_2 evolution sites since their CB levels are not sufficiently negative for electrons to reduce water to produce H_2 without catalytic assistance whereas their VB levels are usually positive enough for holes to oxidize water and form O_2 even in the absence of cocatalyst [92]. Metal cocatalysts could lower the activation energy or overpotential for H_2 evolution reaction on the surface of semiconductors. It also accelerates the surface chemical reaction by inhibiting the backward reaction.

Overall, the role played by the metal cocatalysts dispersed on the surface of the semiconductor photocatalysts is extremely important. Many factors can affect the capability of H_2 evolution cocatalysts in the semiconductor-based photocatalytic water splitting, such as cocatalyst loading amount and its particle size and structure [89, 94]. For example, there is a volcano-type trend between the loading amount of a given cocatalyst and the photocatalytic activity (regardless of the synthesis method, photocatalyst type, and loaded cocatalyst) [94, 95]. At the same loading amount, metal cocatalysts with smaller size and high dispersion display higher catalytic activity due to their larger specific surface area and more active sites [96].

3.2.2. *The plasmonic metal case: plasmonic-metal/semiconductor heterodimer nanocrystals*

The harvesting and conversion of solar energy become a renewed in recent years. Among various technologies, the direct conversion of solar to chemical energy using photocatalysts has received significant attention. Although heterogeneous photocatalysts are almost exclusively semiconductors, it has been demonstrated recently that plasmonic nanostructures of noble metals (mainly silver and gold) also show significant promise [1]. Herein, the recent progress in using plasmonic metallic nanostructures in the field of photocatalysis is reviewed. We focus on plasmon-enhanced water splitting on composite photocatalysts containing semiconductor and plasmonic-metal building blocks and recently reported plasmon-mediated photocatalytic reactions on plasmonic nanostructures of noble metals.

Plasmonic metallic nanostructures are characterized by their strong interaction with resonant photons through an excitation of surface plasmon resonance (SPR). SPR can be described as the resonant photon-induced collective oscillation of valence electrons, established when the frequency of photons matches the natural frequency of surface electrons oscillating against the restoring force of positive nuclei (Figure 22A). The resonant photon wavelength is different for different metals. For example, gold, silver, and copper nanostructures exhibit resonant behavior when interacting with ultraviolet (UV) and visible (Vis) photons. The utilization of the localized SPR (LSPR) effect of nanostructured Au, Ag, and Cu was examined for the potential photocatalysis [97–99], using Au or Ag nanoparticles (NPs) supported on ZrO_2, AgCl, or TiO_2 for the unselective degradation of organic species under visible light irradiation. As shown in Figure 22 A, LSPR is the resonant photon-induced coherent oscillation of charge at the metal-dielectric interface, established when the photon frequency matches the natural frequency of metal surface electrons oscillating against the restoring force of their positive nuclei [100]. The frequency of the surface plasmon absorption is highly dependent on the type of metal, size, shape, surrounding dielectric medium, distance between neighboring objects, and configuration their ensemble [101]. A wide range of metal/semiconductor heterostructures, including Au/TiO_2, Ag/TiO_2, Au/CdS, and Au/Fe_2O_3, have been explored to achieve enhanced photocatalytic activity [102–110].

Although the exact nature of LSPR effect on enhanced photocatalytic activity is not entirely understood, three possible enhancement mechanisms have been proposed: (1) near-field enhancement, (2) SPR-induced electron transfer from metal to semiconductor, and (3) scattering [103, 104] (Figure 22A–C). The strong SPR-induced electric field of plasmonic metal NPs can interact with the adjacent semiconductor (Figure 22B-a), this interaction may increase the rate of exciton formation and the concentration of the charge carriers generated in this part of the semiconductor [104]. If only the metal excited, the metallic plasmonic NPs can absorb resonant photons and transfer energetic photogenerated charge carrier to the semiconductor during the decay of the LSPR (Figure 22B-b). As a result of the plasmonic sensitization process, a wide band gap semiconductor could perform catalytic reduction reactions under visible light. Plasmonic structures of size larger than 50 nm are efficient in scattering the resonant photons, which increases the path length of photons in semiconductor/plasmonic metal nanostructures. Therefore, the resonant photons that are not absorbed by semiconductor photocatalysts could be scattered by the bigger plasmonic metal particles, ultimately increasing the number of electron/hole pairs. The above-mentioned three mechanisms are governed by the metal/semiconductor configurations and their arrangements in the hybrid system.

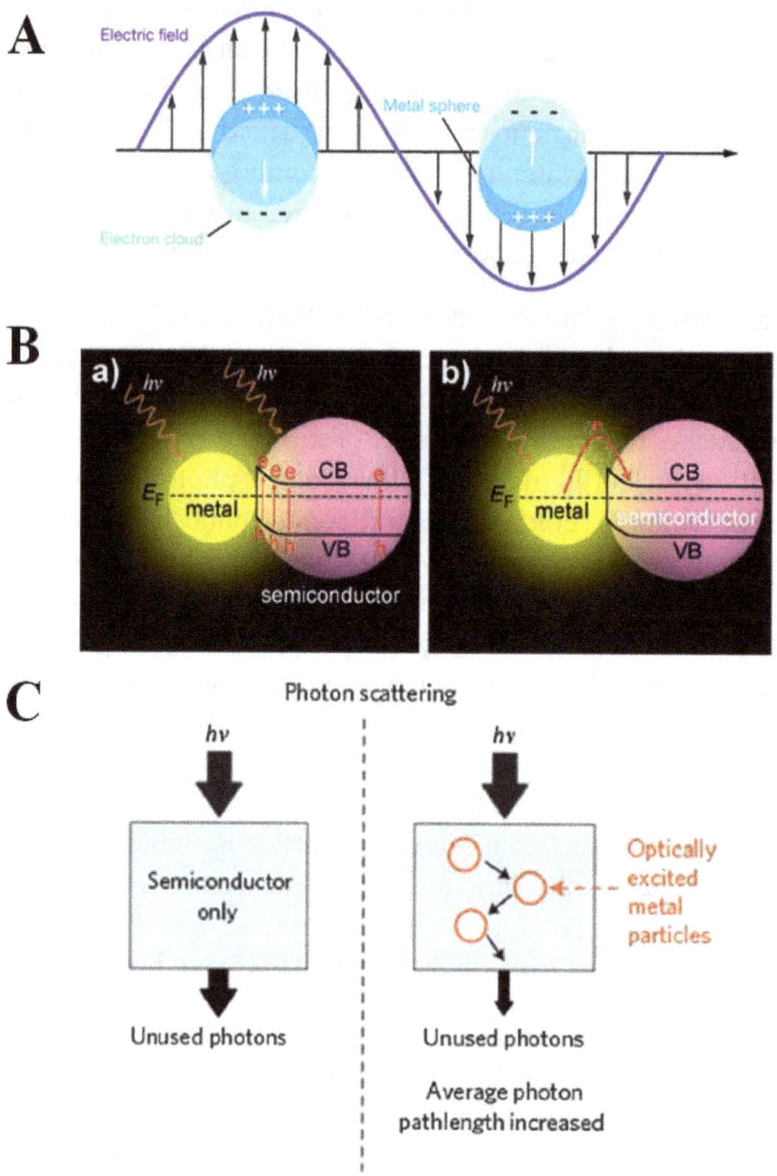

Figure 22. Schematic illustrating (A) localized surface plasmon resonance [102]. Copyright: America Chemistry Society, 2011. (B) The mechanisms for plasmon-enhanced chemical reactions with metal/semiconductor hybrid nanostructures: (a) plasmonic enhancement of light absorption; (b) hot-electron effect [103], with permission. Copyright Wiley-VCH Verlag GmbH & Co. KGaA. (C) The scattering mechanism [104]. Copyright: Royal Society of Chemistry, 2013.

In the case of SPR-mediated charge injection from metal to semiconductor, charge carriers are directly injected from excited plasmonic-metal nanostructures into the semiconductor surface. The metallic plasmonic nanoparticles essentially act as a sensitizer, absorbing resonant photons and transferring the energetic electron, formed in the process of the SPR excitation, to the nearby semiconductor (Figure 23) [1, 111]. Furthermore, the ability to tune the SPR resonance wavelength and intensity by changing the size or shape of Au or Ag nanostructures suggests that plasmon enhancement could be more dramatic when the interface between Au (Ag) and semiconductor, such as TiO_2, CdS, etc. The charge injection mechanism was functional in composite photocatalysts where the plasmonic nanoparticles and semiconductor are in direct contact with each other, allowing a rapid transfer of charge carriers. These composite systems are geometrically similar to the conventional cocatalyst/semiconductor photocatalysts that are often synthesized by an incipient wetness deposition of metal precursors and their subsequent thermal treatment on a semiconductor surface. Therefore, the defect in the heterointerface should be decreased to get more efficient charge separation and injection to be "chemically useful" energetic charge carriers for photoreduction. Otherwise, too many defects on the interface will trap the photoinduced charge.

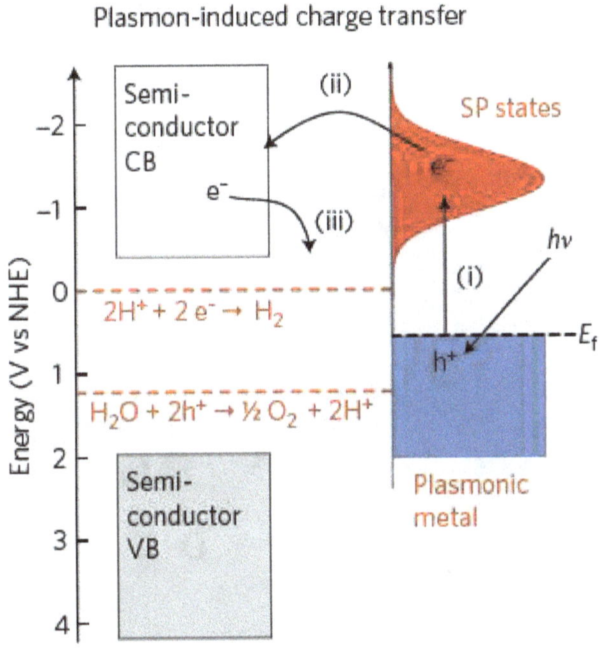

Figure 23. Mechanism of SPR-induced charge transfer with approximate energy levels on the NHE scale. Dashed red lines refer to the water-splitting redox potentials. (i) Electrons near the metal Fermi level, E_f, are excited to surface plasmon (SP) states; (ii) the electrons transfer to a nearby semiconductor particle; (iii) this activates electron-driven processes such as the hydrogen evolution half-reaction [1]. Copyright of Nature group 2011, Macmillan Publishers Limited.

3.2.3. The cocatalystic metal on semiconductor nanocrystals with well-defined crystal face exposure

Based on the photoreduction reaction of Pt^{4+} ($Pt^{4+} + e- \rightarrow Pt$) and photooxidation reaction of Pb^{2+} ($Pb^{2+} + H_2O + h^+ \rightarrow PbO_2$), it is clearly indicated that rutile {110} and {011} facets provide reduction and oxidation sites, respectively (Figure 24Aa, c) [112–113]. The obvious separation of reduction and oxidation sites on faceted rutile crystals is attributed to photoexcited electron and hole transfer between {011} and {110} facets, which is driven by the higher electronic energy levels of {011}. Although the selective distribution of photodeposited Pt and PbO_2 particles on anatase {101} and {001} facets (Figure 24 Ab, d) is not as obvious as that on different rutile facets, anatase {101} and {001} facets can still be considered to be more reductive and more oxidative, respectively [113].

The spatial separation of reduction and the oxidation sites in TiO_2 crystals with appropriate different facets are well known now [114]. Consequently, the corresponding spatial separation of photoexcited charge carriers on different facets improves the photocatalytic activities of photocatalysts, as illustrated in Figure 24B and C. Liu et al. reported that anatase crystals with {101} and {001} facets, with selective deposition of Pt particles by a photoreduction process only on the reductive {101} facets show a much higher photocatalytic hydrogen evolution (reduction reaction) from a mixture of H_2O and methanol than do the same crystals with Pt particles on both {101} and {001} facets (Figure 24D). It is now clear that the difference in spatial distribution of cocatalysts on photocatalysts is an important but unfortunately neglected factor. The spatial separation of photoexcited electrons and holes on different facets was also seen in brookite nanosheets surrounded with four {210} and two {101} facts as reduction sites, and two {201} facets as oxidation sites. Consequently, such spatial separation produces an excellent photocatalytic activity of brookite, which is conventionally inactive [113–116].

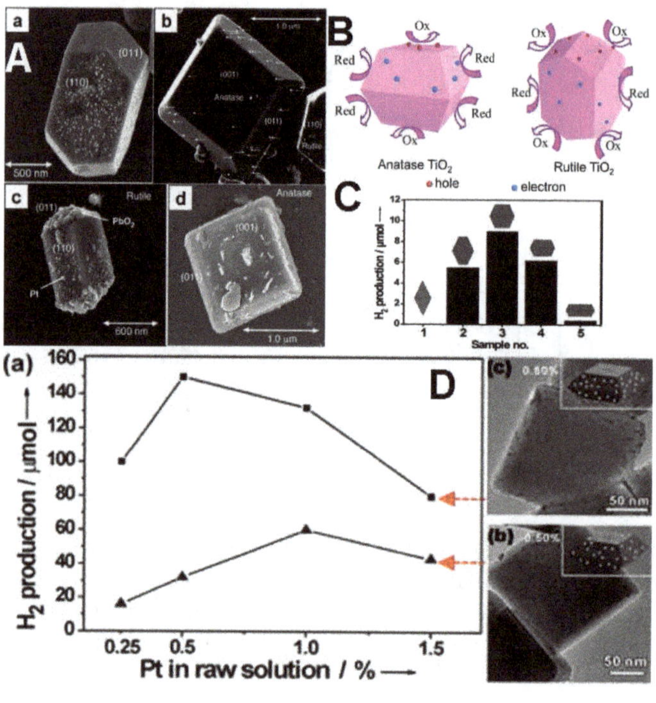

Figure 24. .(A) SEM images of (a) a rutile particle and (b) an anatase particle on which Pt fine particles were deposited by UV-irradiation in a solution of 1.0 mM H_2PtCl_6; SEM images of (c) a rutile particle and (d) an anatase particle showing PbO_2 deposits, which were loaded on the particles by UV irradiation of the Pt-deposited TiO_2 powder in a solution of 0.1 M $Pb(NO_3)_2$. Reprinted with permission from Ohno et al. [113]. Copyright 2002 Royal Society of Chemistry. (B) Schematic of the spatial separation of redox sites on anatase crystals with {101} and {001} facets and rutile TiO_2 particle with {110} and {011} facets. Reprinted with permission from Murakami et al. [115]. Copyright 2009 American Chemical Society. (C) Hydrogen production after 6 h irradiation time using TiO_2 nanocrystals (NCs) with different exposed {001} facets as photocatalysts. Reprinted with permission from Liu et al. [116]. Copyright 2013 Wiley-VCH. (D) (a) Hydrogen production amounts of Pt deposited TiO_2 NCs by means of photochemical-reduction (■) and chemical reduction routes (▲). SEM images of TiO_2 NCs with 0.5% Pt loading amount prepared by (b) chemical reduction and (c) photochemical reduction, respectively. Reprinted with permission from Liu et al. [116]. Copyright 2013 Wiley-VCH.

3.3. Metal–semiconductor core/shell nanocrystals and photocatalysis application

3.3.1. Metal core–semiconductor shell: large lattice mismatch-induced interface control and photocatalytic applications

In metal-semiconductor hetero structures, the contact interface between metal and semiconductor plays a critical role in transfer and separation of charge carriers [117–118]. Our precise symmetry control of Au–CdX can enable different contact interface between metal and semiconductor components at the nanoscale, thus offering a new dimension to control optical and electronic properties of hybrid nanostructures in a highly control manner. Figure 25A highlights evolution of plasmon–exciton coupling in Au–CdS enabled by their symmetry control. As compared with the SPR feature of a pure metal nanoparticles, optical absorption of Au–CdS is significantly broadened in the visible regime (500–700 nm), which can be attributed to the existence of plasmon–exciton coupling [119, 3]. They further performed finite-different time-domain (FDTD) simulation to evaluate optical response in Au–CdS with different structural symmetry and sümmarize the results in Figure 25B. The FDTD simulation reproduces symmetry dependent optical property very well. The strong SPR-induced visible light absorption of Au/CdS colloids can overlap most of the solar spectrum, and particularly important, its tunability makes it feasible to optimize photorelated processes, including solar water splitting.

It is well known that CdS or CdSe possesses conduction and valence bands at potentials appropriate for water reduction, namely, the bottoms of conduction bands locate at a more negative potential than the reduction potential of H^+ to H_2. In order to get high efficiency of H_2 evolution by sunlight irradiation, more visible light should be harvested, and then photo-excited electron/hole should be separated and migrated to the surface without recombination [120, 121]. By integrating with plasmonic Au nanoparticle in a hybrid nanostructure, its tunable optical response achieved in Figure 25A and B should make it feasible to mediate the visible light-induced electron/hole separation in CdX and electron transfer from CdX to Au surface due to the plasmon–exciton coupling [3,122]. As shown in Figure 25C and D, for comparison, photocatalysis with pure 4.5 nm CdS QDs and the CdS/Au heterodimers prepared by reported in situ deposition are also presented [123–124]. The rate of CdS QDs alone is 0.1 mmol h^{-1} g $^{-1}$.The concentric core/shell Au–CdS has the least H_2 evolution activity (0.009 mmol h^{-1} g^{-1}) and the heterodimer structured Au/CdS has better H_2 evolution activity (7.3 mmol h^{-1} g^{-1}) that manifest a dramatic photoactivity enhancement of 730 times. However, the CdS/Au heterodimers has less H_2 evolution activity (0.359 mmol h^{-1} g^{-1}). Figure 25C and D can be understood by the mechanism of SPR enhanced electron–hole separation and collection, as illustrated in the Figure 25E. For an asymmetric Au–CdS heterodimer, under visible light illumination, the energy band alignment between CdS and Au suggests a rapid electron transfer from conduction band of CdS to Au nanoparticle. In the meantime, concentrated electric field due to SPR of metal constituent can significantly enhance light absorption of CdS semiconductor and promote charge separation near the Au–CdS interface. Such synergetic effect enriches the electron (e-) on the Au tip for the efficient water reduction and leads to high efficient H_2 evolution activity. The bad H_2 evolution activity of the CdS/Au heterodimers mainly attributes to the weak SPR and consequent visible light harvest of smaller sized Au (2–3 nm). On the other hand, in a concentric Au–CdS core–shell nanostructure with high structure symmetry, the electrons are retained in the Au core after photo excitation and charge separation and cannot participate in photocatalytic reaction; thus, the photon absorption of CdS is suppressed. As a result, almost no H_2 evolution is observed with concentric Au–CdS.

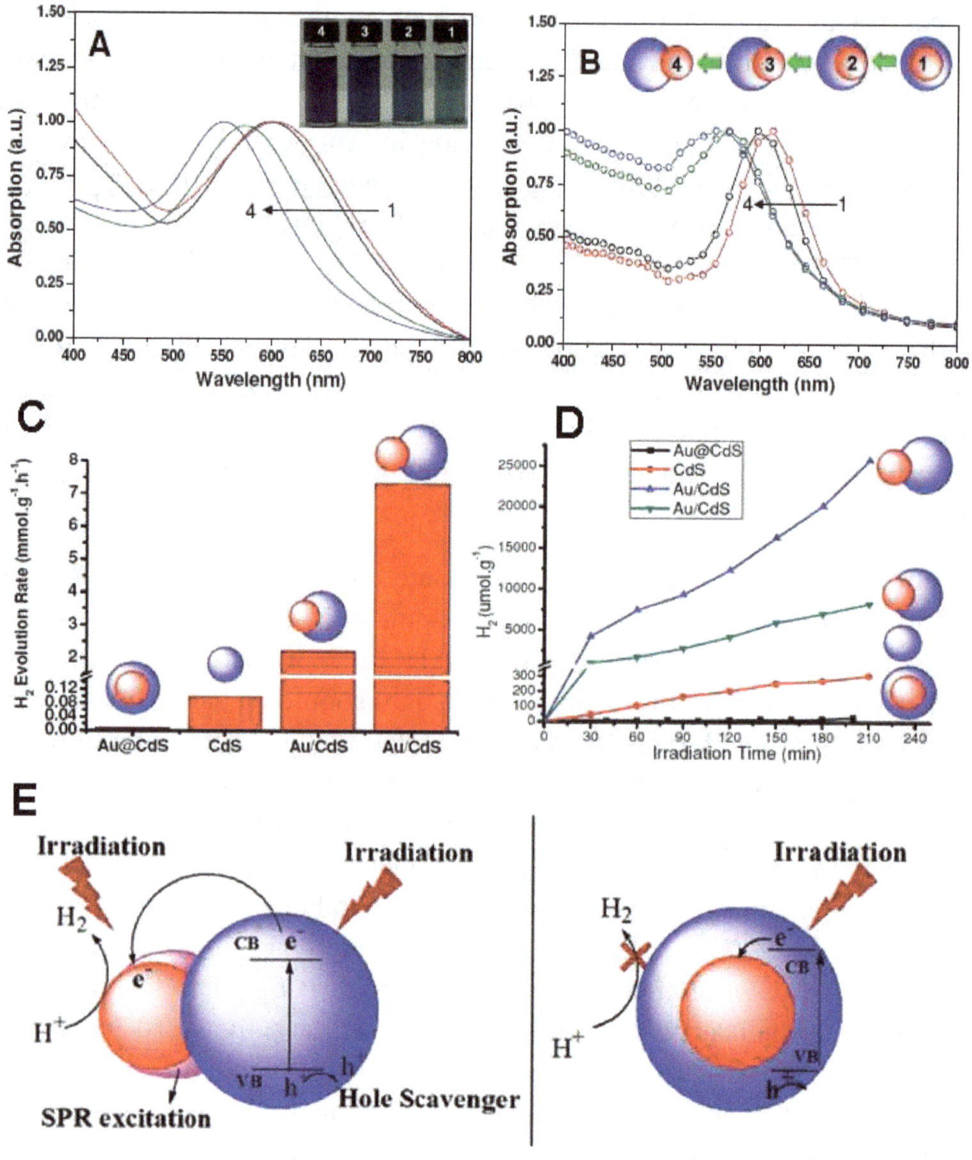

Figure 25. (A) Evolution of experimental UV-Vis absorption spectra with symmetry of Au/CdS colloid. (B) The normalized FDTD simulation of the SPR in these Au/CdS heteronanostructures. (C) Comparison of H_2 evolution activities of different Au–CdS photocatalysts. (D) Time evolution of photocatalytic generation of H_2 evolution amount versus irradiation time for them. (E) Schematic illustrations of visible light-induced electron/hole separation and transfer difference in heterodimer and core/shell Au/CdS when dispersed in aqueous solution containing sacrificial reagents. From Zhao et al. [32] with permission; copyright Wiley-VCH Verlag GmbH & Co. KGaA.

In conclusion, based on the series of metal-semiconductor hybrid nanostructures from concentric core–shell to nonconcentric heterodimer with precisely controlled separation between metal and semiconductor constituents, the gradual symmetry evolution has led to novel control of optical response (plasmon–exciton coupling) and photocatalytic activity in hybrid nanostructures, which highlight the importance of nanoscale interface control for both fundamental understanding and technology applications. Thus, it can allow us to evaluate a possible upper limit of a photocatalytic reaction and guide the design toward high efficient hybrid nanostructures. Due to the plasmon intensity difference from different sized Au NPs, the larger sized Au, such as tens of nanometers to hundreds of nanometers size, the plasmon enhancement for photocatalysis may become strong.

4. The photoreduction of CO_2 applications of metal/semiconductor hybrid nanocrystals

Photoreduction of CO_2 is another important application of metal/semiconductor photocatalyst, and it directly converses CO_2 into organic compound such methane, methanol, formalin, formic acid, and so on [125–130]. Being similar with photocatalysis for hydrogen evolution, metal/semiconductor photocatalysts for photoreduction of CO_2 need some potential to meet the photocatalysis reaction (Schemes 1–5) [125],

$$H_2O \rightarrow H_2 + 1/2\ O_2, \Delta E = 1.23V \quad (1)$$

$$CO_2 + H_2O \rightarrow HCOOH + 1/2\ O_2, \Delta E = 1.40\ V \quad (2)$$

$$CO_2 + H_2O \rightarrow HCOH + O_2, \Delta E = 1.34\ V \quad (3)$$

$$CO_2 + H_2O \rightarrow CH_3OH + 3/2\ O_2, \Delta E = 1.21\ V \quad (4)$$

$$CO_2 + H_2O \rightarrow CH4 + 2\ O_2, \Delta E = 1.06\ V \quad (5)$$

and two pathways of photoreduction of CO_2: formalin pathway and carbine pathway are proposed to explain the mechanism of photoreduction of CO_2. According to these rules, various photocatalysts have been synthesized for conversing CO_2 into fuel and modified for higher conversion efficiency. What is needed to point out here is that the reduction potential of CO_2 depends on the reaction conditions, such as pH, and as the reduction is carried out, a series of reactions occur simultaneously. Therefore, as the reaction conditions vary, the main product of photoreduction of CO_2 is different.

Nowadays, several series of semiconductor have been used for photoreduction of CO_2, such as oxides [126, 131–133], sulfides [134], phosphide [135–136], and so on. In most case, metal will be used as cocatalysts to enhance photoreduction efficiency. Habisreutinger and his coworker have concluded photoreduction of CO_2 using semiconductors (Table 1) and in most of the cases, metal would be used as cocatalyst [132]. However, as using different metal as cocatalysts, the main products of photoreduction of CO_2 are different. It hints that metal nanocrystals play an important role in the photocatalysis system. It provides impossibility for tuning the photoreduction of CO_2 by doping metal cocatalysts.

As a kind of photocatalysts, TiO_2 is one of the most popular semiconductors for photoreduction of CO_2 [130]. For example, Li and her coworkers dope Cu onto TiO_2 nanocrystals supported with mesoporous silica and enhance the photoreduction of CO_2 (Figure 26A–D) [137]. More complex structure of photocatalysts has also been synthesized for photoreduction of CO_2. Wang and his coworkers prepared the CdSe/Pt/TiO_2 and monitor decrement of CO_2 under irradiation of visible light (wavelength > 420 nm). They found that TiO_2/Pt combining with CdSe platform a higher activity for photoreduction of CO_2 (Figure 26E and F) [133].

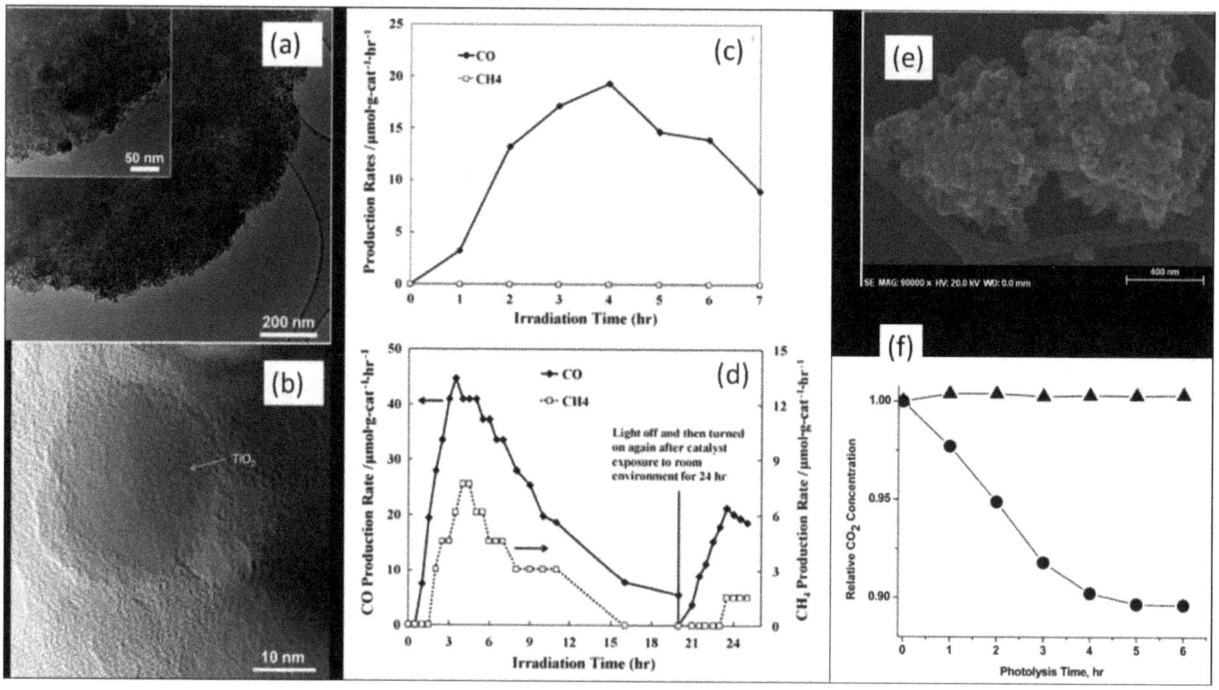

Figure 26. Cu/TiO_2 nanocrystals TEM (A) and HRTEM (B) images and its catalysis properties of photoreduction of CO_2 (C and D). Copyright: Elsevier, 2010. CdSe/Pt/TiO_2 nanocrystals and its photoreduction of CO_2 (E and F). Copyright: America Chemistry Society, 2010.

Beside TiO_2 nanocrystals photocatalyst, sulfides such as CdS nanocrystals and ZnS nanocrystals also show excellent properties for the photoreduction of CO_2. It is reported that using ZnS nanocrystals as photocatalyst, the rate of formic acid production is up to 7000 $\mu mol\ h^{-1}\ g^{-1}$ [132, 134]. What's more, the shape of nanocrystals and other factors on photoreduction of CO_2 attract lots of attention and photoreduction of CO_2 (Figure 27) [138].

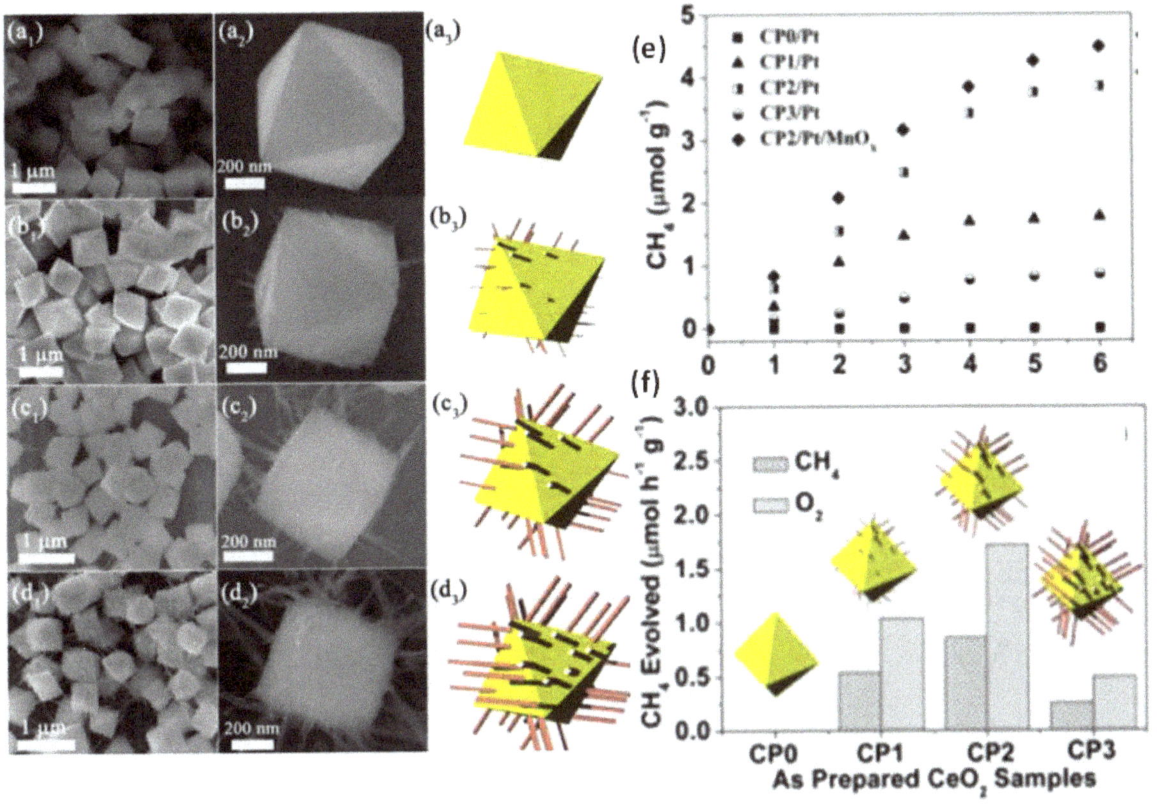

Figure 27. SEM images and schematic of CeO_2 modified with HPO_4 ions (A–D) and CH_4 evolution from photoreduction of CO_2 (E and F). Copyright: America Chemistry Society, 2015.

5. Some other recent photosynthesis applications of metal/semiconductor hybrid nanocrystals

Beside hydrogen evolution and photoreduction of CO_2 form photocatalysis, metal/semiconductor hybrid nanocrystals recently are used for photosynthesis applications. Metal or noble metal is usually employed as catalysts in organic synthesis [139]. Moreover, noble metal has plasmonic effect and induces the organic reaction under visible light and enhances the conversion efficiency [140]. Combining with semiconductor nanocrystals, metal/semiconductor hybrid nanocrystals have potential advantages in organic photosynthesis: higher conversion efficiency, higher selectivity, and lower by-products [141–143].

For instance, Au/CeO_2 is employed for oxidation of Alcohol forming aldehyde or ketone under irradiation of visible light [144]. The reaction rate depends on the surface area of Au nanoparticles, the power of irradiation and the function group of organic compound (Figure 28A–C), and a high selectivity of oxidation is also showed in such hybrid nanocrystals (Figure 28D–F). Beside, metal/semiconductor hybrid nanocrystals, such as Au/ZrO_2, can be used for other organic reaction: reduction of nitroaromatic compound, Suzuki reaction, ester reaction. and so on [140, 142–143].

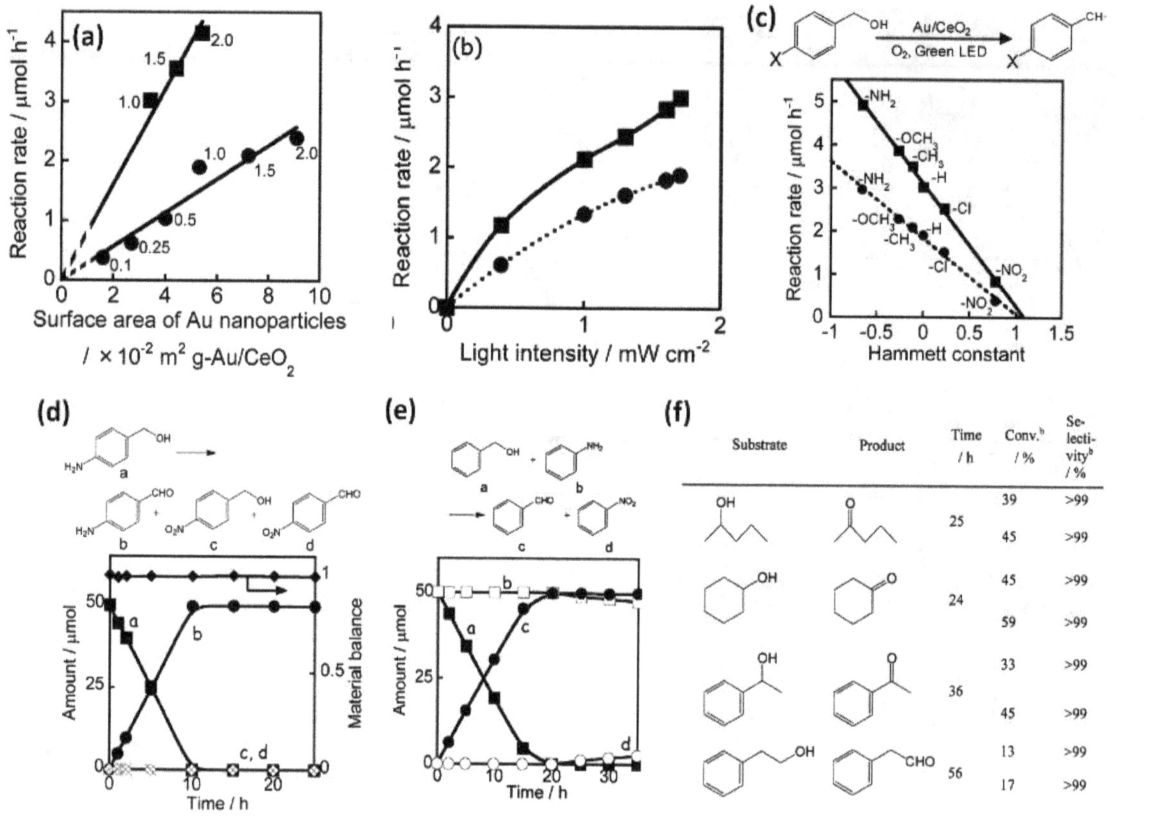

Figure 28. Reaction rate depending on surface area of Au nanoparticles (A), light intensity (B), and substituted functional groups. Copyright: America Chemistry Society, 2012.

In summary, metal/semiconductor photocatalysts have shown their potential application on hydrogen evolution, photoreduction for conversing CO_2 into fuel and organic photosynthesis. Although there is a long way before the metal/semiconductor hybrid nanocrystals being implemented in real-life application, more photocatalysts with different components, special architectures, or structures would be designed for the real-life application.

6. Summary and outlook

Metal/semiconductor hybrid NCs did have advantages on integration of functionalities of noble metal and semiconductors based on well-controlled morphologies and heterointerface control. Then they could have potential applications to improve the quantum yield of photoinduced electrons or holes for photoreduction and photooxidation. This review has demonstrated the recent research efforts to synthesize metal/semiconductor hybrid NCs and to understand and control the photocatalytic applications, such as photocatalytic water splitting, CO_2 photoreduction, and photoinduced organic synthesis. First, on the long term, it can be expected that innovative design and improved synthetic capabilities in the development of elaborate metal/semiconductor NCs with defined topologies and interface control will deliver exciting opportunities in both fundamental understanding and practical exploitation of unconventional properties and functionalities stemming from properly engineered heterostructures with tailored interfaces and structural features. It can also be expected that some recent new synthesis strategy, such as the reaction between coordinated cation ions and semiconductor nanostructures (core/shell, heterodimers, doped NCs), nonepitaxial growth of metal/semiconductor core/shell, and heterodimers will speed up the precise photocatalysts synthesis and design. The new mechanism of photocatalysis, such as plasmon-enhanced photocatalysis, the synergistic cooperation of plasmonic metals, and cocatalysts based on well-defined interface will also boost each of them to an unprecedented level of catalytic performance.

Author details

Jiatao Zhang*, Muwei Ji, Jiajia Liu and Meng Xu

*Address all correspondence to: zhangjt@bit.edu.cn

Beijing Key lab of Construction Tailorable Advanced Functional Materials and Green Applications, School of materials and Engineering, Beijing Institute of Technology, Beijing, China

References

[1] Linic S., Christopher P., Ingram D. B.. Nat. Mater.. 2011; 10:911.

[2] Clavero C.. Nat. Photonics. 2014; 8:95.

[3] Zhang J. T., Tang Y., Lee K., Ouyang M.. Nature. 2010; 466:91.

[4] Jiang R. B., Li B. X., Fang C. H., Wang J. F.. Adv. Mater.. 2014; 26:5274.

[5] Yang J., Wang D., Han H., Li C.. Acc. Chem. Res.. 2013;46:1900.

[6] Xiao M., Jiang R., Wang F., Fang C., Wang J. F., Yu J. C.. J. Mater. Chem. A. 2013;1:5790.

[7] Warren S. C., Thimsen E.. Energy Environ. Sci.. 2012;5:5133.

[8] Ayers J. E. Heteroepitaxy of Semiconductors: Theory, Growth and Characterization. New York: CRC Press; 2007.

[9] Carbone L., Cozzoli P. D.. Nano Today. 2010;5:449.

[10] Casavola M., Buonsanti R., Caputo G., Cozzoli P. D.. Eur. J. Inorg. Chem.. 2008;6:837.

[11] Lee J., Hernandez P., Lee J., Govorov A. O., Kotov N. A.. Nat. Mater.. 2007;4:291.

[12] Gao J. H., Gu H. W., Xu B. Chem. Res.. 2009;42:1097.

[13] Costi R., Saunders A. E., Banin U. Angew. Chem. Int. Ed.. 2010;49:4878.

[14] Wang C., Xu C., Zeng H., Sun S. Adv. Mater.. 2009;21:3045.

[15] Jin Y. D., Gao X. H. Nat. Nanotechnol.. 2004;4:571.

[16] Yang J., Sargent E. H., Kelley S. O., Ying J. Y. Nat. Mater.. 2009;8:683.

[17] Yang J., Ying J. Y. J. Am. Chem. Soc.. 2010;132:2114.

[18] Buck M. R., Bondi J. F., Schaak R. E. Nat. Chem.. 2012;4:37.

[19] Atwater H. A., Polman A. Nat. Mater.. 2010;9:205.

[20] Gu H., Zheng R., Zhang X., Xu B. J. Am. Chem. Soc.. 2004;126:5664.

[21] Wang C., Yin H., Dai S., Sun S.H. Chem. Mater.. 2010;22:3277.

[22] Wu H., Chen O., Zhuang J., Lynch J., Montagne D. L., Nagaoka Y., Cao Y. C. J. Am. Chem. Soc.. 2011;133:14327.

[23] Figuerola A., Fiore A., Corato R. D., Falqui A., Giannini C., Micotti E., Lascialfari A., Corti M., Cingolani R., Pellegrino T., Cozzoli P. D., Manna L. J. Am. Chem. Soc.. 2008;130:1477.

[24] Cozzoli P. D., Pellegrino T., Manna L. Chem. Soc. Rev.. 2006;35:1195.

[25] Mokari T., Rothenberg E., Popov I., Costi R., Banin U. Science. 2004;304:1787.

[26] Costi R., Saunders A. E., Elmalem E., Salant A., Banin U. Nano Lett.. 2008;8:637.

[27] Du Y., Chen B., Yin Z., Liu Z., Zhang H. Small. 2014;10:4727.

[28] Zhang J. T., Tang Y., Lee K., Ouyang M. Science. 2010;327:1634.

[29] Zhang J., Tang Y., Weng L., Ouyang M. Nano lett.. 2009;9:4061.

[30] Anson C. E., Eichhöfer A., Issac I., Fenske D., Fuhr O., Sevillano P., Persau C., Stalke D., Zhang J. Angew. Chem. Int. Ed.. 2008;47:1326.

[31] Gui J., Ji M., Liu J., Xu M., Zhang J. T., Zhu H. S. Angew. Chem. Int.. 2015;54:1.

[32] Zhao Q., Ji M., Qian H., Dai B., Weng L., Gui J., Zhang J. T., Ouyang M., Zhu H. S. Adv. Mater.. 2014;26:1387.

[33] Qian H., Zhao Q., Dai B., Guo L., Zhang J., Liu J., Zhang J. T., Zhu H. S. NPG Asia Mater.. 2015;7:e152.

[34] Son D. H., Hughes S. M., Yin Y., Alivisatos A. P., Science. 2004; 306:1009.

[35] Jeong U. Y., Xia Y. N., Yin Y. D.. Chem. Phys. Lett.. 2005;416:246.

[36] Ryzhii M., Ryzhii V. Physics and Modeling of Tera- and Nano-Devices. In: World Scientific; 2008; Singapore.

[37] McBride J., Treadway J., Feldman L. C., Pennycook S. J., Rosenthal S. J. Nano Lett.. 2006;6:1496.

[38] Battaglia D., Li J. J., Wang Y., Peng X. Angew. Chem. Int. Ed.. 2003;42:5035.

[39] Kim S., Fisher B., Eisler H. J., Bawendi M. J. Am. Chem. Soc.. 2003;125:11466.

[40] Smith A. M., Nie S. Acc. Chem. Res.. 2010;43:190.

[41] Klimov V. I., Ivanov S. A., Nanda J., Achermann M., Bezel I., McGuire J. A., Piryatinski A.. Nature. 2007;441:447.

[42] Caruge J. M., Halpert J. E., Wood V., Bulovic V., Bawendi M. G. Nat. Photonics. 2008;2:247.

[43] Sargent E. H. Nat. Photonics. 2009;3:325.

[44] Chen X. B., Lou Y., Samia A. C., Burda C. Nano Lett.. 2003;3:799.

[45] Peng X. G., Schlamp M. C., Kadavanich A. V., Alivisatos A. P.. J. Am. Chem. Soc.. 1997;119:7019.

[46] Wang Z. L.. J. Phys. Chem. B.. 2000;104:1152.

[47] Palmstrøm C. J.. Annu. Rev. Mater. Sci.. 1995;25:389.

[48] Lee J. S., Shevchenko E. V., Talapin D. V.. J. Am. Chem. Soc.. 2008;130:9673.

[49] Kim H., Achermann M., Balet L. P., Hollingsworth J. A., Klimov V. I.. J. Am. Chem. Soc.. 2005;127:544.

[50] Mokari T., Sztrum C. G., Salant A., Rabani E., Banin U.. Nat. Mater.. 2005;4:855.

[51] Mokari T., Rothenberg E., Popov I., Costi R., Banin U.. Science. 2004;304:1787.

[52] Wang C., Xu C., Zeng H., Sun S.. Adv. Mater.. 2009;21:3045.

[53] Fang C., Jia H., Chang S., Ruan Q., Wang P., Chen T., Wang J.. Energy Environ. Sci.. 2014;7:3431.

[54] Sun Z., Yang Z., Zhou J., Yeung M., Ni W., Wu H., Wang J.. Angew. Chem. Int. Ed.. 2009;48:2881.

[55] Lambright S., Butaeva E., Razgoniaeva N., Hopkins T., Smith B., Perera D., Corbin J., Khon E., Thomas R., Moroz P., Mereshchenko A., Tarnovsky A., Zamkov M. ACS Nano. 2014;8:352.

[56] Priebe M., Fromm K. M.. Chemistry—A European Journal. 2015;21:3854.

[57] Güttel R., Paul M.,Schüth F.. Catal. Sci. Technol.. 2011;1:65.

[58] Li A., Zhang P., Chang X., Cai W., Wang T., Gong J.. Small. 2015;11:1892.

[59] Li G., Tang Z.. Nanoscale. 2014;6:3995.

[60] Fang Q., Xuan S., Jiang W., Gong X.. Adv. Funct. Mater.. 2011;21:1902.

[61] Jun Z., Ling X., Jiali S., Diping H., Huan. J. Surf. Coat. Technol.. 2015;271:119.

[62] Zhang Q., Lee I., Joo J. B., Zaera F., Yin Y.. Acc. Chem. Res.. 2013;46:1816.

[63] Zhao L., Peng J., Huang Q., Li C., Chen M., Sun Y., Lin Q., Zhu L., Li F.. Adv. Funct. Mater.. 2014;24:363.

[64] Zhang N., Fu X., Xu Y.J.. J. Mater. Chem.. 2011;21:8152.

[65] Zhang L., Blom D. A., Wang H.. Chem. Mater.. 2011;23:4587.

[66] Park J. C., Lee H. J., Jung H. S., Kim M., Kim H. J., Park K. H., Song H.. ChemCatChem. 2011;3:755.

[67] Hah H. J., Um J. I., Han S. H., Koo S. M.. Chem. Commun.. 2004;:1012.

[68] Tan L., Chen D., Liu H., Tang F.. Adv. Mater.. 2010;22:4885.

[69] Fan C. M., Zhang L. F., Wang S. S., Wang D. H., Lu L. Q., Xu A. W.. Nanoscale. 2012;4:6835.

[70] Gröger H., Gyger F., Leidinger P., Zurmühl C., Feldmann C.. Adv. Mater.. 2009;21:1586.

[71] Lin Y. S., Wu S. H., Tseng C. T., Hung Y., Chang C., Mou C. Y.. Chem. Commun.. 2009;:3542.

[72] Priebe M., Fromm K. M.. Part. Part. Syst. Charact.. 2014;31:645.

[73] Park J. C., Heo E., Kim A., Kim M., Park K. H., Song H.. J. Phys. Chem. C. 2011;115:15772.

[74] Arnal P. M., Comotti M., Schuth F.. Angew. Chem. Int. Ed.. 2006;45:8224.

[75] Lou X. W., Yuan C., Archer L. A.. Small. 2007;3:261.

[76] Guan B., Wang T., Zeng S., Wang X., An D., Wang D., Cao Y., Ma D., Liu Y., Huo Q.. Nano Res.. 2014;7:246.

[77] Yeo K. M., Shin J., Lee I. S.. Chem. Commun.. 2010;46:64.

[78] Wang M., Han J., Xiong H., Guo R.. Langmuir. 2015;31:6220.

[79] Fujishima A., Honda K.. Nature. 1972;238:37.

[80] Kudo A., Miseki Y.. Chem. Soc. Rev.. 2009;38:253.

[81] Linsebigler A. L., Lu G., Yates J. T.. Chem. Rev.. 1995;95:735.

[82] Li X., Yu J., Low J., Fang Y., Xiao J., Chen X.. J. Mater. Chem. A. 2015;3:2485.

[83] Osterloh F. E.. Chem. Soc. Rev.. 2013;42:2294.

[84] Maeda K.. J. Photoch. Photobio. C. 2011;12:237.

[85] Qu Y., Duan X. F.. Chem. Soc. Rev.. 2013;42:2568.

[86] Chen X., Shen S., Guo L., Mao S. S.. Chem. Rev.. 2010;110:6503.

[87] Ma Y., Wang X., Jia Y., Chen X., Han H., Li C.. Chem. Rev.. 2014;114:9987.

[88] Kraeutler B., Bard A. J.. J. Am. Chem. Soc.. 1978;100:4317.

[89] Yang J., Wang D., Han H., Li C.. Acc. Chem. Res.. 2013;46:1900.

[90] Wang H., Zhang L., Chen Z.G., Hu J., Li S., Wang Z., Liu J., Wang X. Chem. Soc. Rev.. 2014;43:5234.

[91] Bai S., Jiang J., Zhang Q., Xiong Y. J. Chem. Soc. Rev.. 2015;44:2893.

[92] Jing L., Zhou W., Tian G., Fu H. G. Chem. Soc. Rev.. 2013;42:9509.

[93] Kochuveedu S. T., Jang Y. H., Kim D. H. Chem. Soc. Rev.. 2013;42:8467.

[94] Ran J., Zhang J., Yu J., Jaroniec M., Qiao S. Z. Chem. Soc. Rev.. 2014;43:7787.

[95] Thanminimulla C. T. K., Takata T., Hara M., Kondo J. N., Domen K. J. Catal.. 2000;196:362.

[96] Yamada Y., Miyahigashi T., Kotani H., Ohkubo K., Fukuzumi S. Energy Environ. Sci.. 2012;5:6111.

[97] Awazu K., Fujimaki M., Rockstuhl C., Tomingaga J., Murakami H., Ohiki Y., Yoshida N., Watanabe T. J. Am. Chem. Soc.. 2008;130:1676.

[98] Chen X., Zhu H., Zhao J., Zheng Z., Gao X. Angew. Chem. Int. Ed.. 2008;47:5353.

[99] Wang P., Huang B., Qin X., Zhang X., Dai Y., Wei J., Wang M. H. Angew. Chem. Int. Ed.. 2008;47:7931.

[100] SWalclawik., E. R., Zhu H. Green Chem.. 2013;15:1814.

[101] Lang X., Chen X., Zhao J. Chem. Soc. Rev.. 2014;43:473.

[102] Jones M. R., Osberg K. D., Macfarlane R. J., Langille M.R., Mirkin C. A. Chem. Rev.. 2011;111:3736.

[103] Jiang R., Li B., Fang C., Wang J. F. Adv. Mater.. 2014;26:5274.

[104] Kochuveedu S. T., Jang Y. H., Kim D. H. Chem. Soc. Rev.. 2013;42:8467.

[105] Ingram D. B., Linic S. J. Am. Chem. Soc.. 2011;133:5202.

[106] Pu Y., Wang G., Chang K., Ling Y., Lin Y., Fitzmorris B. C., Liu C., Lu X., Tong Y., Zhang J. Z., Hsu Y., Li Y. Nano Lett.. 2013;13:3817.

[107] Seh Z. W., Liu S., Low M., Zhang S., Liu Z., Mlayah A., Han M. Y. Adv. Mater.. 2012;24:2310.

[108] Pradhan S., Ghosh D., Chen S. ACS Appl. Mater. Interfaces. 2009;1:2060.

[109] Zhang Z., Zhang L., Hedhili M. N., Zhang H., Wang P. Nano Lett.. 2013;13:14.

[110] Qian K., Sweeny B. C., Johnston-Peck A. C., Niu W., Graham J. O., Duchenen J. S., Qiu J., Wang Y., Engelhard M. H., Su D., Stach E. A., Wei W. D. J. Am. Chem. Soc.. 2014;136:9842.

[111] Long R., Mao K. K., Gong M., Zhou S., Hu J. H., Zhi M., You Y., Bai S., Jiang J., Zhang Q., Wu X. J., Xiong Y. J.. Angew. Chem. Int. Ed.. 2014;53:3205.

[112] Liu G., Yang H. G., Pan J., Yang Y. Q., Lu G. Q., Cheng H. M. Chem. Rev.. 2014;114:9559.

[113] Ohno T., Sarukawa K., Matsumura M.. New J. Chem.. 2002;26:1167.

[114] Yang H. G., Sun C. H., Qiao S. Z., Zou J., Liu G., Smith S.C., Cheng H. M., Lu G. Q.. Nature. 2008;453:638.

[115] Murakami N., Kurihara Y., Tsubota T., Ohno T.. J. Phys. Chem. C. 2009;113:3062.

[116] Liu C., Han, X.; Xie, S.; Kuang, Q.; Wang, X.; Jin, M.; Xie, Z.; Zheng, L.. Chem. Asian J.. 2013;8:282.

[117] Tian Y., Tasuma T. J. Am. Chem. Soc.. 2005;127:7632.

[118] Silva C. G., Juárez R., Marino T., Molinari R., García H. J. Am. Chem. Soc.. 2011;133:595.

[119] Kelly K. L., Coronado E., Zhao L. L., Schatz G. C. J. Phys. Chem. B.. 2003;107:668.

[120] Maeda K., Domen K. J. Phys. Chem. C. 2007;111:7851.

[121] Hu P., Pramana S. S., Cao S., Ngaw C. K., Lin J. D., Loo S. C. J., Tan T. T. Y. Adv. Mater.. 2013;25:2567.

[122] Cao S. W., Fang J., Shahjamali M. M., Boey F. Y. C., Barber J., Loo S. C. J., Xue C. RSC Adv.. 2012;2:5513.

[123] Yan H., Yang J. H., Ma G. J., Wu G. P., Zong X., Lei Z. B., Shi J., Li C. J. Catal.. 2009;266:165.

[124] Yang J., Yan H., Wang X., Wen F., Wang Z., Fan D., Shi J., Li C. J. Catal.. 2012;290:151.

[125] Liu C., Dasgupta N. P., Yang, P.. Chem. Mater.. 2014;26:415.

[126] Schneider J., Matsuoka M., Takeuchi M., Zhang J., Horiuchi Y., Anpo M., Bahnemann D. W.. Chem. Rev.. 2014;114:9919.

[127] Ma Y., Wang X., Jia Y., Chen X., Han H., Li C.. Chem. Rev.. 2014;114:9987.

[128] Kubacka A., Fernández-García M., Colón, G.. Chem. Rev.. 2012;112:1555.

[129] Wen F., Li, C.. Acc. Chem.Res.. 2013;46:2355.

[130] Cokoja M., Bruckmeier C., Rieger B., Herrmann W. A., Kühn F. E.,. Angew. Chem. Int. Ed.. 2011;50:8510.

[131] Kang Q., Wang T., Li P.; Liu L.; Chang K.; Li M.; Ye J.. Angew. Chem. Int. Ed.. 2015;54:841.

[132] Habisreutinger S. N.; Schmidt-Mende L.; Stolarczyk J. K.. Angew. Chem. Int. Ed.. 2013;52:7372.

[133] Wang C.; Thompson R. L.; Baltrus J.; Matranga C.. J. Phy. Chem. Lett.. 2010;1:48.

[134] Johne P., Kisch H.. J. Photochem. Photobiol. A. 1997;111:223.

[135] Inoue T., Fujishima A., Konishi S., Honda, K.. Nature. 1979;277:637.

[136] Sato S., Arai T., Morikawa T., Uemura K., Suzuki T. M., Tanaka H., Kajino, T.. J. Am. Chem. Soc.. 2011;133:15240.

[137] Li Y., Wang W.N., Zhan Z., Woo M.H., Wu C.Y., Biswas P.. Appl. Catal. B. Environ.. 2010;100:386.

[138] Li P., Zhou Y., Zhao Z., Xu Q., Wang X., Xiao M., Zou Z.. J. Am. Chem. Soc.. 2015;137:9547.

[139] Wu Y., Wang D., Zhao P., Niu Z., Peng Q., Li Y.. Inorg. Chem.. 2011;50:2046.

[140] Wang F., Li C., Chen H., Jiang R., Sun L. D., Li Q., Wang J., Yu J. C., Yan C. H.. J. Am. Chem. Soc.. 2013;135:5588.

[141] Kisch H.. Angew. Chem. Int. Ed.. 2013;52:812.

[142] Xiao Q., Liu Z., Bo A., Zavahir S. Sarina S., Bottle S., Riches J. D., Zhu H.. J. Am. Chem. Soc.. 2015;137:1956.

[143] Zhu H., Ke X., Yang X., Sarina S., Li, H.. Angew. Chem. Int. Ed.. 2010;49:9657.

[144] Tanak, A., Hashimot, K., Kominam, H.. J. Am. Chem. Soc.. 2012;134:14526.

Permissions

All chapters in this book were first published in ACMPOCT, by InTech Open; hereby published with permission under the Creative Commons Attribution License or equivalent. Every chapter published in this book has been scrutinized by our experts. Their significance has been extensively debated. The topics covered herein carry significant findings which will fuel the growth of the discipline. They may even be implemented as practical applications or may be referred to as a beginning point for another development.

The contributors of this book come from diverse backgrounds, making this book a truly international effort. This book will bring forth new frontiers with its revolutionizing research information and detailed analysis of the nascent developments around the world.

We would like to thank all the contributing authors for lending their expertise to make the book truly unique. They have played a crucial role in the development of this book. Without their invaluable contributions this book wouldn't have been possible. They have made vital efforts to compile up to date information on the varied aspects of this subject to make this book a valuable addition to the collection of many professionals and students.

This book was conceptualized with the vision of imparting up-to-date information and advanced data in this field. To ensure the same, a matchless editorial board was set up. Every individual on the board went through rigorous rounds of assessment to prove their worth. After which they invested a large part of their time researching and compiling the most relevant data for our readers.

The editorial board has been involved in producing this book since its inception. They have spent rigorous hours researching and exploring the diverse topics which have resulted in the successful publishing of this book. They have passed on their knowledge of decades through this book. To expedite this challenging task, the publisher supported the team at every step. A small team of assistant editors was also appointed to further simplify the editing procedure and attain best results for the readers.

Apart from the editorial board, the designing team has also invested a significant amount of their time in understanding the subject and creating the most relevant covers. They scrutinized every image to scout for the most suitable representation of the subject and create an appropriate cover for the book.

The publishing team has been an ardent support to the editorial, designing and production team. Their endless efforts to recruit the best for this project, has resulted in the accomplishment of this book. They are a veteran in the field of academics and their pool of knowledge is as vast as their experience in printing. Their expertise and guidance has proved useful at every step. Their uncompromising quality standards have made this book an exceptional effort. Their encouragement from time to time has been an inspiration for everyone.

The publisher and the editorial board hope that this book will prove to be a valuable piece of knowledge for researchers, students, practitioners and scholars across the globe.

List of Contributors

Hiroaki Tada
Department of Applied Chemistry, School of Science and Engineering, Kinki University, Kowakae, Higashi-Osaka, Osaka, Japan

Qiliang Jin
R&D Placement Business Division, WORLD INTEC CO., Sakae, Naka-ku, Nagoya, Aichi, Japan

Arvin Sangalang, Seunghwan Seok and Do Hyun Kim
Department of Chemical and Biomolecular Engineering, KAIST, Daejeon, Republic of Korea

Adnan Qaseem, Fuyi Chen and Nan Zhang
State Key Laboratory of Solidification Processing, Northwestern Polytechnical University, Xian, China

Boris I. Kharisov, Oxana V. Kharissova, Ubaldo Ortiz Méndez, Idalia Gómez de la Fuente, Yolanda Peña and Alejandro Vázquez Dimas
Universidad Autónoma de Nuevo León, Monterrey, México

H.V. Rasika Dias
The University of Texas at Arlington, Arlington, TX, USA

Jingxin Zhou, Qingrui Zhang and Jie Hu
Hebei Key Laboratory of Applied Chemistry, School of Environmental and Chemical Engineering, Yanshan University, Qinhuangdao, P. R. China

Tifeng Jiao
Hebei Key Laboratory of Applied Chemistry, School of Environmental and Chemical Engineering, Yanshan University, Qinhuangdao, P. R. China
National Key Laboratory of Biochemical Engineering, Institute of Process Engineering, Chinese Academy of Sciences, Beijing, P. R. China

Le Li and Minqiang Wang
Electronic Materials Research Laboratory (EMRL), Key Laboratory of Education Ministry, International Center for Dielectric Research, Xi'an Jiaotong University, Xi'an, China

Jiatao Zhang, Muwei Ji, Jiajia Liu and Meng Xu
Beijing Key lab of Construction Tailorable Advanced Functional Materials and Green Applications, School of materials and Engineering, Beijing Institute of Technology, Beijing, China

Index

www.ingramcontent.com/pod-product-compliance
Lightning Source LLC
LaVergne TN
LVHW080723210926
848485LV00003B/510